Engineering Economics

Engineering Economics

J. K. Yates

CRC Press
Taylor & Francis Group
Boca Raton London New York

CRC Press is an imprint of the
Taylor & Francis Group, an **informa** business

CRC Press
Taylor & Francis Group
6000 Broken Sound Parkway NW, Suite 300
Boca Raton, FL 33487-2742

Printed on acid-free paper
Version Date: 20161010

International Standard Book Number-13: 978-1-4987-5085-1 (Hardback)

Library of Congress Cataloging-in-Publication Data

Names: Yates, J. K., 1955- author.
Title: Engineering economics / J.K. Yates.
Description: Boca Raton : Taylor & Francis, a CRC title, part of the Taylor & Francis imprint, a member of the Taylor & Francis Group, the academic division of T&F Informa, plc, [2017]
Identifiers: LCCN 2016027303| ISBN 9781498750851 (hardback : alk. paper) | ISBN 9781498750882 (ebook) | ISBN 9781498750875 (ebook) | ISBN 9781498750868 (ebook)
Subjects: LCSH: Engineering economy.
Classification: LCC TA177.4 .Y38 2017 | DDC 330--dc23
LC record available at https://lccn.loc.gov/2016027303

Visit the Taylor & Francis Web site at
http://www.taylorandfrancis.com

and the CRC Press Web site at
http://www.crcpress.com

Printed and bound in the United States of America by
Edwards Brothers Malloy on sustainably sourced paper

This book is dedicated to all of the students who struggled while learning engineering economics.

In memory of Dr. Thomas J. Kiblen

Contents

Preface

The purpose of this book is to introduce the formulas, methods, and processes required to perform engineering economic analysis to engineers in all of the engineering disciplines. Every design an engineer produces should have a corresponding engineering economic analysis to determine whether the design is economically feasible. In addition, if there are multiple alternatives for the design, the alternatives should be compared to highlight which one is the most cost effective, economical, and profitable. In order to provide analyses and comparisons to owners, engineers should be knowledgeable about performing an engineering economic analysis for their designs.

No matter whether it is a public or a private project, it is important for engineers to be able to design projects that will be the most profitable or bestow the most benefits on the public. Society relies on engineers to not only design structurally sound and safe structures and products but projects that meet the needs of, and generate profits for, clients and the public. Engineering economics is the area that assists engineers in providing economically feasible projects to clients.

This book provides a simplified, easy-to-understand, straightforward approach to explaining engineering economics that is appropriate for members of all of the major engineering disciplines including aerospace, agricultural, biomedical, chemical, civil, electrical, industrial, mechanical, nuclear, petroleum, process, and systems. It is written from the perspective of the main engineering disciplines and includes engineering economic analysis examples and case studies.

This book provides the basic knowledge required for engineers to be able to perform engineering economic analysis for potential equipment, products, services, and projects in both the public and private sectors. It focuses on assisting engineers in mastering the basic engineering economics formulas and using them to analyze engineering and construction projects and products.

This book includes numerous example problems and case studies that illustrate how the formulas introduced in each chapter are applied when analyzing the economics of engineering alternatives. There are several different methods presented for analyzing the economics of alternative equipment, materials, services, and projects including equivalent present worth, future worth, and uniform annual costs; capitalized costs; rate of return and incremental rate of return; break-even comparisons; replacement analysis; and benefit/cost ratios.

This book includes 14 chapters, each of which covers one specific area of the fundamentals of engineering economics. Each chapter contains definitions, explanations, formulas, example problems, case studies, and end-of-chapter problems. The chapters in this book build on the material presented in each of the previous chapters; therefore, they should be read and studied consecutively for increased understanding.

Chapter 1 introduces engineering economics and explains its importance to engineers when they are evaluating the economic viability of their designs. It covers the topic of money as a means of commerce to inform the reader about important aspects of monetary systems and how they are used in commerce. The second part of Chapter 1 explains different methods for improving the economics of projects in the design, fabrication, and construction stages. The last part of this chapter briefly introduces some of the techniques for managing and tracking costs on projects.

Chapter 2 begins by defining some of the most frequently used engineering economic terms related to interest and the time value of money. Along with providing definitions for terms, this chapter explains the process for drawing cash flow diagrams—one of the most important aspects when developing and analyzing engineering economic problems. Being able to properly draw cash flow diagrams is a key step in engineering economic analysis. This chapter also provides a foundation for understanding the material in the subsequent chapters.

Chapter 3 defines the concept of equivalence in relation to analyzing engineering economic alternatives. This chapter explains time value of money and its incorporation into the calculations for future worth of a payment or disbursement and the present worth of future sums. This chapter also explains interpolation, the process for solving for unknown interest rates in time value of money equations, and using interest factor tables to solve engineering economic problems.

Chapter 4 covers the process for converting uniform annual series of payments and disbursements (annuities) into future values, along with solving for the uniform annual worth of a future or present value. In addition, this chapter demonstrates the procedures for calculating the present worth of uniform annual series and annualizing present sums.

Chapter 5 describes the methods for calculating the present or future worth of arithmetic gradients, which are increasing or decreasing uniform payments or disbursements. It also provides techniques for calculating the present or future worth of geometric gradients.

Chapter 6 builds on the material presented in the previous five chapters to demonstrate combining the time value of money formulas to calculate the present or future worth of multiple income or disbursement streams.

Chapter 7 explains capitalized cost calculations, which are calculations for determining the present worth of a project assuming that the project will last forever rather than for a defined period of time. In Chapter 8, methods are provided for comparing alternatives based on their equivalent uniform annual cost. In order to make comparisons, all of the present and future worth income and disbursement streams are converted into equivalent uniform annual amounts and then added or subtracted from any annuities to determine the overall equivalent uniform annual amount.

Chapter 9 demonstrates a technique for calculating the rate of return for potential projects based on developing equations where the equivalent present value of all payments, disbursements, and future values are set equal to zero and then solving for the interest rate where the equations are equal to zero. This technique is also used to solve for the incremental rate of return of the increase in the initial costs between alternatives by developing equivalent present value equations for both alternatives and finding the interest rate at which the difference between the two equations is zero.

Chapter 10 focuses on the decisions required for determining when to replace a piece of equipment or a facility. The first part of the chapter defines the different terms used related to replacement analysis and the second part explains comparing existing equipment or facilities to a proposed alternative by determining the point in time where the original equipment or facility will cost more to operate than the operating cost of the proposed alternative. The original alternative may also be replaced when the proposed alternative has a higher rate of return than the existing equipment or facility.

Chapter 11 illustrates using break-even comparisons to determine which proposed alternative should be selected when two alternatives are being considered for a project. There is a point in time where the number of units being produced, or the number of miles (kilometers) a piece of equipment is driven per year, justifies the selection of a proposed alternative over the original process or equipment. This point, known as the break-even point, is determined by converting the initial fixed cost of each alternative to an equivalent yearly cost, adding it to the yearly operating expenses for the alternative, and plotting the cost per year on the y-axis and the units of production or miles (kilometers) on the x-axis. The point of intersection of these two curves is the number of units or number of miles (kilometers) where the project will break even. If production or the number of miles (kilometers) is above the break-even point the new alternative should be implemented and if production or the number of miles is less than the break-even point the original process or equipment should be retained by the firm.

Chapter 12 demonstrates using the equivalent present worth of benefits divided by the equivalent present worth of costs to determine whether a project is worth pursuing. If the benefit/cost ratio (profitability index) is greater than one, an alternative project is financially viable. Once the benefit/cost ratio is calculated for each project under consideration, they are compared with each other and the project with the highest benefit/cost ratio is retained for further consideration. Benefit/cost ratios are also referred to as discounted profitability indices, profit investment analysis ratios, and value investment ratios.

Chapter 13 begins by defining depreciation as it pertains to equipment, vehicles, office furniture, facilities, and other items used by businesses and then it explains the different components considered and the methods for calculating depreciation. The depreciation methods covered in this chapter include production, straight line, declining balance (accelerated cost recovery system), and sum-of-the-years digits.

Chapter 14 ties together the material in previous chapters to explain after-tax cash flows. This chapter addresses both corporate and individual income taxes and explains the process for using depreciation and business expenses to lower tax rates.

Key terms, which appear in italics when first used, are provided at the end of each chapter.

Appendices A through E provide additional information including engineering economic analysis formulas and their respective cash flow diagrams, interest factor tables, information on using spreadsheets for solving engineering economic problems, derivations of the engineering economic formulas, and a listing of the equations introduced in each chapter.

This book helps engineering students and professionals learn to perform engineering economic analysis for their designs and projects and to be able to provide the most cost-effective or profitable alternatives to clients.

Along with this book, instructional materials are available for educators through CRC Press/Taylor & Francis Group that include a solutions manual to the end-of-chapter problems.

Acknowledgments

The genesis of this book occurred as the author witnessed engineering students struggling to learn engineering economics when she was teaching this course at Texas A&M University, San Jose State University, and North Dakota State University. While teaching engineering economics more than 50 times to thousands of students, it was observed that highly intelligent students from all of the engineering disciplines had difficulty learning the basic concepts of engineering economics. Therefore, each time the course was taught, the lectures were modified to present them in an informative and intelligible manner. Methods for simplifying the presentation of the material and communicating the basic fundamentals of engineering economics were discovered, and the results of this effort are summarized in this book.

I thank all of the students who were in my engineering economics courses for their perseverance and patience in learning the subject, which allowed me to determine the most effective manner for presenting the course material.

What made me realize that the students were finally absorbing the concepts was an event that occurred when I was teaching engineering economics at Texas A&M University. I mentioned to the students that the next lecture was really important and it would make a major difference in their future and to not miss it. When I entered the classroom to present the lecture, the raised lecture platform was covered with tape recorders placed there by the students who did not want to miss a word of the lecture. It was a moment I will always remember since it made me realize the power of effective education. I am hoping to continue the valuable aspects of effective education through the publishing of this textbook.

I acknowledge the suggestions and support of David Kalinowski, former executive with Chevron and Noble Energy, who provided materials for consideration for inclusion in this book and for his moral support during the writing of this book.

I thank Joseph Clements, the editor of this book, for recognizing the value in publishing it and for his continued support and guidance. I also thank Mikaela Kursell, editorial assistant, for helping with the production of this book; Jennifer Ahringer, senior project coordinator, for shepherding this book through production; and the project manager, Niranjana Harikrishnan, who contributed her time and effort into the creation of this book.

Author

J. K. Yates, Ph.D., emeritus professor and currently a civil and construction engineering consultant, is the former dean of the College of Engineering Technology at Ferris State University in Big Rapids, Michigan, which is one of the largest colleges of engineering technology in the United States. She was also formerly the department head and the Joe W. Kimmel Distinguished Professor of construction management in the Department of Construction Management at Western Carolina University in Cullowhee, North Carolina. Dr. Yates was previously a professor and department chair in the Department of Construction Management and Engineering at North Dakota State University, in charge of the construction engineering and management focus area in the Civil and Environmental Engineering Department at Ohio University, and program coordinator for the Construction Engineering program in the Civil and Environmental Engineering Department at San Jose State University in California. Dr. Yates was also a professor in the Civil and Environmental Engineering Department at New York University's Polytechnic School of Engineering (formerly Polytechnic University and Brooklyn Polytechnic) and at Iowa State University, along with being a visiting professor at the University of Colorado for one year.

Dr. Yates received a B.Sc degree in civil engineering with a minor in anthropology/archeology from the University of Washington and a Ph.D. degree in civil engineering from Texas A&M University with minors in global finance and management, global political science, business analysis, and construction science.

Dr. Yates has worked for several of the top-ranked domestic and global engineering and construction firms, as a consultant during legal cases, and on international contracts. Dr. Yates is the author of ten books and numerous refereed journal articles. She is a former member of the American Society of Civil Engineers, the Project Management Institute, and the American Association of Cost Engineers International. Dr. Yates received the Distinguished Professor award from the Construction Industry Institute (CII) in 2010 and Polytechnic University in 1994, was the Associated General Contractors of America's Outstanding Construction Professor in 1997, was one of the Engineering News Records *Those Who Made Marks on the Construction Industry* in 1991, and received the Ron Brown award for industry/academic collaborations with the Hewlett Packard Foundation in 2001. Dr. Yates has traveled for work and pleasure to 25 countries and worked in Bontang, East Kalimantan, on the Island of Borneo in Indonesia.

1 Introduction

For it is just this striving forward that brings us to the fruits which are always falling into our hands and which are the unfailing sign that we are on the right road and that we are ever and ever drawing nearer to our journey's end. But that journeys' end will never be reached, because it is always the still far thing that glimmers in the distance and is unattainable. It is not the possession of the truth, but the success which attends the seeking after it, that enriches the seeker and brings happiness to him.

Max Plank
German Physicist (Brown, 2015, p. 211)

1.1 INTRODUCTION TO ENGINEERING ECONOMICS ANALYSIS

This book introduces engineers to the formulas and techniques for performing engineering economic analysis. In addition to creating designs, engineers also evaluate the economic viability of their designs. If there are multiple design alternatives, each alternative is analyzed to determine which alternative is the most cost-effective, economical, and profitable design. For private projects, engineers are required to design the most profitable project, and government projects need to be the most beneficial project to the public. Members of society rely on engineers to design structurally safe projects and products, but they also have to meet the requirements of clients or the public and generate reasonable profits or benefits. Engineering economic analysis techniques help engineers to evaluate each project as part of the process of providing economically feasible designs to their clients and the public.

In order for engineers to develop the knowledge necessary for economically evaluating projects, this book introduces the fundamentals of engineering economics using a straightforward, easy to understand approach appropriate for members of all of the engineering disciplines. Techniques are presented throughout this book for analyzing the economic viability of equipment, materials, products, projects, and services, all of which will be referred to as projects throughout this book. The economic viability of projects is determined by analyzing the costs and tangible and intangible benefits using engineering economic formulas to calculate the rate of return on an investment or determining whether accounting for the time value of money the benefits or profits are greater than the costs.

Time value of money considers that a sum of money could be invested at a particular interest rate and the value of the sum invested would increase over time due to the accumulation of interest. Clients are not willing to invest in a project unless the return on the investment will be greater than what the client could earn if he or she invested the funds in some type of interest-bearing account. Therefore, engineers are responsible for demonstrating to clients that the designs being proposed will result in a client being able to realize a higher rate of return on the investment than the rate of return he or she would be able to earn by investing the funds instead of creating a project.

The techniques and formulas presented in this book provide engineers with the fundamental knowledge required to economically analyze designs, projects, or products and present the results of the analysis to clients. This book also includes examples of how engineering economic analysis techniques are integrated into other aspects of society to determine the economic viability of proposed purchases and investments such as the following:

- Calculating the remaining balance and balloon payments on loans.
- Calculating the amount of future social security income.
- Comparing different alternatives when trying to decide which alternative to select.

- Consequences of financing a university education using student loans.
- Determining the amount of interest paid when borrowing money to finance items such as computers, equipment, furniture, machinery, property, vehicles, and other assets.
- How starting a retirement account when someone is in his or her 20s or 30s could greatly increase the funds available at retirement.
- Purchasing a home using a home mortgage.
- Using engineering economic analysis techniques to help make decisions on whether to invest funds now or wait until some future time to investment the funds.

The engineering economic analysis techniques introduced in this book include:

- Present worth
- Future worth
- Annuities—uniform annual series
- Gradients—arithmetic and geometric
- Capitalized cost evaluations
- Rate of return
- Incremental rate of return
- Replacement analysis
- Breakeven comparisons
- Benefit/cost ratios
- Deprecation
- After-tax rate of return

1.1.1 Organization of This Book

This book includes 14 chapters, each of which covers one specific area of the fundamentals of engineering economics. Each chapter contains definitions, explanations, formulas, example problems, case studies, and end-of-chapter problems. Every chapter builds on the material presented in previous chapters; therefore, each chapter should be read and studied consecutively for increased understanding of the material being presented in each chapter.

The first part of Chapter 1 provides an introduction to the topic of engineering economics and how money is used in commerce. The second part discusses improving the economics of projects and managing and tracking costs. Chapter 2 provides definitions for engineering economic terms such as interest and time value of money. This chapter also explains how to construct cash flow diagrams, which are an integral part of being able to develop and solve engineering economic problems.

Chapter 3 addresses equivalence and its relationship to engineering economic analysis and the time value of money. This chapter introduces formulas for calculating future worth and the present worth of future sums. It also explains the procedures for solving for rates of return and the process for using interpolation in rate of return calculations. Chapter 4 presents formulas for calculating the present and future worth of repetitive payments or disbursements called annuities (uniform series). It also provides formulas for solving for the uniform annual worth of a future or present value.

Chapter 5 introduces arithmetic and geometric gradients and explains the procedures for calculating the present and future worth of both of these types of payment or disbursement streams. In Chapter 6, the formulas introduced in Chapters 1 through 5 are combined to demonstrate the process for calculating the present and future worth of multiple income and disbursement streams.

Chapter 7 explains capitalized cost calculations and introduces equations for calculating the present worth of projects lasting forever (perpetual life) rather than for a defined period of time.

Government agencies and private firms use capitalized cost calculations to determine the amount of funds that need to be deposited into an interest-bearing account in order for a uniform amount to be withdrawn from the account every month or year and used to pay for the upkeep or maintenance and repairs of a project forever. Chapter 8 combines the formulas from previous chapters to calculate the equivalent uniform annual worth of all present and future worth and gradient values.

Chapter 9 integrates the formulas from previous chapters into the development of equations for calculating the rate of return on investments. In order to calculate the rate of return, equations are derived where the equivalent present worth or equivalent uniform annual worth of all income, disbursements, and future values are set equal to zero and the interest rate where the equation sums up to zero is the rate of return.

Chapter 10 covers the processes for calculating when to replace a facility or a piece of equipment. Existing facilities or equipment are compared to proposed alternatives to determine when a facility, or piece of equipment, will cost more to operate than the proposed new equipment. Chapter 11 introduces breakeven comparisons and discusses their use when analyzing different alternative projects, products, or equipment. There is a point where the selection of a new alternative is justified over retaining the existing alternative due to economies of scale or increased maintenance costs and repairs of the old equipment.

Chapter 12 explains the process for calculating benefit/cost ratios, which is another method for determining whether a project or product is worth pursuing. If there are multiple projects or products under consideration, a benefit/cost ratio is calculated for each alternative and then the benefit/cost ratio for each alternative is compared to the other alternatives to determine which project has the highest benefit/cost ratio. If the alternatives are mutually exclusive, then the alternatives are compared to each other using incremental benefit/cost ratios where the incremental increase in the cost of each alternative from the lowest initial cost to the highest initial cost is analyzed to determine if the benefit/cost ratio justifies the increasing cost of each alternative.

Chapter 13 provides formulas for calculating depreciation. The Internal Revenue Service (IRS) allows for a reduction in the value of the fixed assets of a firm on a yearly basis through depreciation and this reduces the business tax burden of a firm. Firms may depreciate any type of capital purchase for a time period determined by the IRS. The last chapter, which is Chapter 14, discusses federal income taxes and the process for calculating after-tax rates of return for corporations and the effective tax rate for businesses and private individuals.

At the end of each chapter, the key terms for each chapter are provided and the key terms are in italics when they are first used in the chapters.

Appendices A through E provide additional information including engineering economic analysis formulas and their respective cash flow diagrams, interest factor tables, information on using spreadsheets for solving engineering economic problems, derivations of the engineering economic formulas introduced in the chapters, and a summary of the equations introduced in each of the chapters.

By the end of this book, engineers, or engineering students, should have mastered the fundamentals of engineering economics and be able to evaluate the economic viability of any type of engineering design, equipment, project, or product.

1.2 MONEY AS A MEANS OF COMMERCE

Before an engineer is able to start analyzing economic alternatives, he or she needs to have a basic understanding of how money is used as a means of exchange for services rendered, and items purchased, in economies. One example of this is when university students pay a university for the knowledge imparted in classes during lectures by professors. Another example is when a client hires an engineer to design a project for him or her and then pays the engineer for his or her services rendered in creating the design. Sections 1.2.1 through 1.2.11 explain how money is used as a means of commerce.

1.2.1 Money as a Means of Storing Value

Money is used as a means to store value, but it only stores value if the country where it is being stored does not have high inflation since inflation causes money to lose value. If money is invested in interest-bearing accounts, it is not only storing value but it is also increasing in value. Before countries had viable currencies, value was stored using gold, silver, and other precious metals. Transactions involving exchanges might also use *bartering* to procure goods or services. Bartering occurs when one type of good or service is exchanged for another type of good or service. Bartering is still a viable means of conducting transactions in cultures where some of the citizens are in a *nonmonetary economy* (they do not earn an income) or governments do not have the funds to pay for required products or services.

1.2.2 Origin of Money

Eventually, the governments of many countries started using paper money backed by gold rather than using precious metals as a means of exchange. In 1968, the United States stopped using the *gold standard*, which means currency in the United States is no longer backed by an equivalent amount of gold stored in a secure location. The U.S. government had approximately $11 billion worth of gold at the end of 2015 (U.S. Department of the Treasury, 2015). The amount of gold in metric tons and tons being held in the countries with the top 10 highest *gold reserves* in 2014 is shown in Table 1.1.

TABLE 1.1
Top 10 Largest Gold Reserves by Country (2014)

Ranking	Country	Metric Tons of Gold Reserve	Tons of Gold Reserve
1	United States	6,133.5	6,761.03
2	Germany	3,384.2	3,730.44
3	International Monetary Fund	2,814.0	3,101.90
4	Italy	2,451.8	2,702.65
5	France	2,435.4	2,684.57
6	Russia	1,094.7	1,206.70
7	China	1,054.1	1,161.95
8	Switzerland	1,040.0	1,146.40
9	Japan	765.2	843.49
10	Netherlands	612.5	675.17

Source: Data from Duran, Y., 20 Largest gold reserves by country, *Futures Magazine*, August 28, 2014, http://www.futuresmag.com/2014/08/28/top-20-largest-gold-reserves-country-2014-edition?page=13, accessed on November 16, 2015.

When currencies are no longer backed by precious metals, governments have the ability to print paper currency at their discretion. If too much paper money is in circulation, it could lead to *hyperinflation* where the value of the paper currency decreases rapidly, as was the case in Germany between WWI and WWII when the government was printing extra money in an effort to try and more rapidly reduce the debt it owed to allied countries for war reparations. Some of the countries in South America have also experienced hyperinflation when their governments have injected too much currency into the economy in times of economic crisis.

1.2.3 Money Used as Units of Accounting

Money is used throughout the world as a means of accounting. Without money, it would be difficult for members of firms or individuals to track income and disbursements. Balance sheets provide a means for tracking money since they provide a record of all income and disbursements. Managers of firms use accounting systems to track company transactions. Accounting systems provide a written record of all of the transactions a firm enters into when they are conducting business.

1.2.4 Money as a Commodity Bought and Sold

Money in the form of different currencies is also a commodity bought and sold throughout the world on exchanges such as:

- Bolsa de Comercio de Santiago (Chile—Santiago Stock Exchange)
- BM&F Bovespa (Brazil—Bolsa de Valores, Mercadorias, and Futuros de São Paulo)
- CAC 40 (France—Cotation Assistée en Continu)
- Canada S&P (Canada—Standard and Poor's)
- DAX (Germany—Deutsche Boerse AG German Stock Index)
- Dow Jones Industrial Average (United States)
- FTSE 100 (United Kingdom—Financial Times Stock Exchange 100)
- Hong Kong Stock Market (Hong Kong—Hang Seng)
- BMV (Mexico—Bolsa Mexicana de Valores)
- NASDAQ (United States—National Association of Securities Dealers Automated Quotations)
- Nikkei Heikin Kabuka (Japan—Tokyo Stock Exchange)
- Shanghai Stock Exchange (China—Shanghai Composite)
- Standard and Poor's 500 (United States)

Governments sell interest in their countries to other countries, firms, or individuals when they need to raise funds to pay for government-related expenses when tax revenues do not generate enough funds to cover government expenses. By doing this, governments acquire *debt* and debt is usually measured as a percentage of the *gross domestic product* (GDP). Table 1.2 shows government debt in 2012 for various countries as a percentage of GDP for each country.

TABLE 1.2
Government Debt in 2012 as a Percentage of Gross Domestic Product

Country	Debt as a Percentage of GDP
Australia	40.5
Austria	78.5
Belgium	89.5
Canada	53.2
Columbia	65.5
Cyprus	119.6
Denmark	47.2
Finland	50.8
France	101.1
Germany	55.2
Greece	163.6
Hungary	84.7
Iceland	112.2
India	50.3
Indonesia	28.4
Ireland	120.5
Italy	127.2
Netherlands	67.8
New Zealand	67.9
Norway	20.5
Portugal	123.7
Russian Federation	9.4
Singapore	109.7
Spain	65.9
Sweden	35.3
Ukraine	33.7
United Arab Emirates	1.8
United Kingdom	97.2
United States	94.3

Source: Data from World Bank, 4.12 World development indicators: Central Government Finances, World Bank, Washington, DC, 2015, http://wdi.worldbank.org/table/4.12, accessed on November 16, 2015.

In November 2015, the United States had a debt of approximately $18 trillion (U.S. Department of the Treasury, January 12, 2016). In 2014, 65% of the U.S. debt was owned by U.S. government agencies and the largest debt holders in the U.S. government are the Social Security Administration—16% and the Federal Reserve System—12%. Foreign governments own 34% of the U.S. debt, countries such as China—7.2% and Japan—7%, along with 32 other countries (Patton, 2014).

When interest rates increase in one country, the governments in other countries or investors purchase the debt of the higher interest-bearing country, since it means they will be paid more in interest on their investment. When interest rates decline in one country, other governments or investors may sell their debt and use the funds they realize from the sale to buy debt in another country with a higher interest rate. It is more beneficial for countries to keep their interest rates low, as it lowers the amount owed in interest on government debt, which in turn is a reduction in the total amount owed to investors. However, keeping interest rates low also makes a country less attractive to investors. When interest rates rise in a country, the government has to pay more interest to investors, and this increases the total amount of government debt.

1.2.5 Current Status of Money

The *European Union* has created a common currency, the *Euro*, for use within many of the countries in the European Union, but the rest of the countries in the world rely on their own currencies, as is illustrated by the sample currencies listed in Table 1.3. As of January 2016, there were 167 official national currencies in the world, although there are 197 independent countries (Currencies of the World, 2015). Some countries use the currencies of other countries such as the U.S. dollar being used in 10 countries, the West African CFA franc in 8, the Central African CFA franc in 6 African states, and the East Caribbean dollar in 6 Caribbean nations (Countries of the World, 2015).

TABLE 1.3
Samples of Currencies throughout the World

Country	Currency
Argentina	Argentine peso
Australia	Australian dollar
Bangladesh	Bangladesh taka
Botswana	Botswana pula
Cambodia	Cambodian riel
Canada	Canadian dollar
Chile	Chilean peso
China	Chinese yuan renminbi
Costa Rica	Costa Rican colon
Denmark	Danish krone
Egypt	Egyptian pound
Guatemala	Guatemalan quetzal
Hong Kong	Hong Kong dollar
Indonesia	Indonesian rupiah
Israel	Israel new sheqel
Japan	Japanese yen
Kenya	Kenya shilling
Kuwait	Kuwaiti dinar
Nigeria	Nigerian naira
Qatar	Qatari riyal
Romania	Romanian leu
Russia	Russian ruble
Serbia	Serbian dinar
South Africa	South African rand
Turkey	Turkish lira

Source: Data from Countries of the World, List of currencies of the world, Countries of the World, 2015, https://www.countries-ofthe-world.com/world-currencies.html, accessed on November 16, 2015.

Government administrative agencies are responsible for printing money. The *Bureau of Engraving and Printing* is responsible for printing currency in the United States. Many smaller countries do not have the ability or the security to print their own currency; therefore, they may outsource the production of their currency to other countries. Germany produces paper currency and Canada coins for more than 60 countries (Countries of the World, 2015).

Banks were created to foster the movement of money within countries and around the world. Banks extend credit and provide loans to firms or individuals. Banks also use their deposits to buy *securities* (*bonds* and *mutual funds*) and other investment instruments to increase the assets of the bank. Banks are required to keep a *cash reserve*, which is a percentage of their deposits, and this percentage is set by the government. The *deposit reserve rate* keeps changing depending on the state of the economy. Typical percentages for reserve requirements in the United States for 2016 are listed in Table 1.4.

TABLE 1.4
U.S. Bank Reserve Requirements as of January 2016

	Requirement	
Liability Type	**% of Liabilities**	**Effective Date**
Net transaction accounts[a]		
\$0–\$15.2 million[b]	0	1-21-16
More than \$15.2–\$110.2 million[c]	3	1-21-16
More than \$110.2 million	10	1-21-16
Nonpersonal time deposits	0	12-27-90
Eurocurrency liabilities	0	12-27-90

Source: Federal Reserve, Reserve Requirements, Board of Governors of the Federal Reserve System, Washington, DC, November 12, 2015, http://www.federalreserve.gov/monetarypolicy/reservereq.htm#table1, accessed on November 16, 2015.

[a] Total transaction accounts consist of demand deposits, automatic transfer service accounts, *NOW accounts* [checking accounts earning interest], share draft accounts, telephone or preauthorized transfer accounts, ineligible bankers acceptances, and obligations issued by affiliates maturing in 7 days or less. Net transaction accounts are total transaction accounts less amounts due from other depository institutions and less cash items in the process of collection. For a more detailed description of these types of deposits, see Form FR 2900 at http://www.federalreserve.gov/apps/reportforms/default.aspx.

[b] The amount of net transaction accounts subject to a reserve requirement ratio of 0% (the *exemption amount*) is adjusted each year by statute. The exemption amount is adjusted upward by 80% of the previous year's (June 30 to June 30) rate of increase in total reservable liabilities at all depository institutions. No adjustment is made in the event of a decrease in such liabilities.

[c] The amount of net transaction accounts subject to a reserve requirement ratio of 3% is the *low-reserve tranche*. By statute, the upper limit of the low-reserve tranche is adjusted each year by 80% of the previous year's (June 30 to June 30) rate of increase or decrease in net transaction accounts held by all depository institutions.

1.2.6 U.S. Federal Reserve System

In the United States in 1913, the federal government implemented the *Federal Reserve System*. There are 12 district reserve banks and the district numbers and cities where they are located are listed in Table 1.5. Figure 1.1 shows the range of each Federal Reserve district bank.

TABLE 1.5
District Numbers and Locations for Federal Reserve District Banks

District Number	Location
1	Boston, Massachusetts
2	New York, New York
3	Philadelphia, Pennsylvania
4	Cleveland, Ohio
5	Richmond, Virginia
6	Atlanta, Georgia
7	Chicago, Illinois
8	St. Louis, Missouri
9	Minneapolis, Minnesota
10	Kansas City, Kansas
11	Dallas, Texas
12	San Francisco, California

Source: Data from Federal Reserve Board, The Twelve Federal Reserve Districts, Federal Reserve Board, Washington, DC, December 13, 2005, http://www.federalreserve.gov/otherfrb.htm, accessed on January 4, 2016.

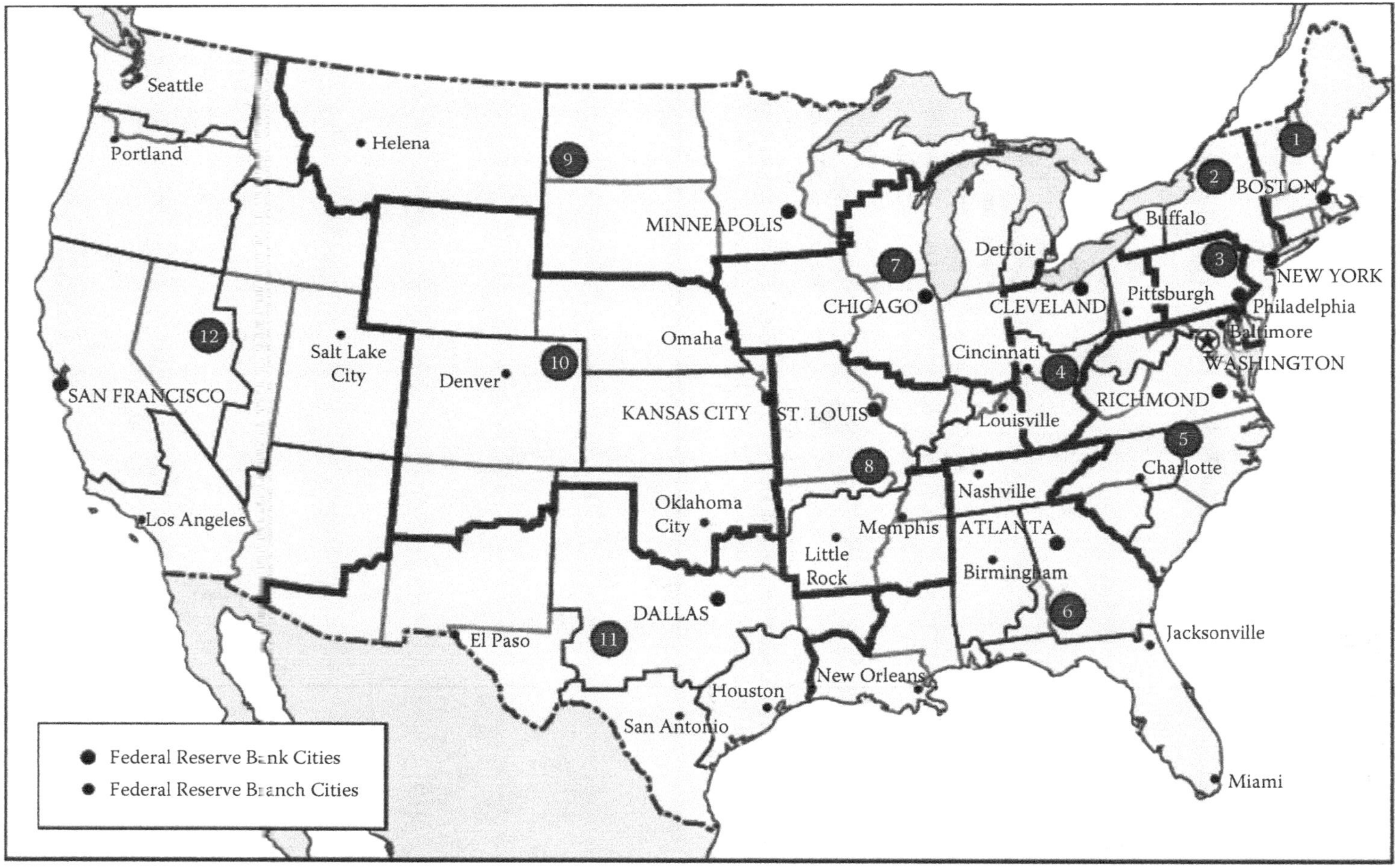

FIGURE 1.1 Map of the Federal Reserve System district banks showing the range of each district bank. Data from Federal Reserve Board, *The Twelve Federal Reserve Districts*, Federal Reserve Board, Washington, DC, December 13, 2005, http://www.federalreserve.gov/otherfrb.htm, accessed on January 4, 2016.

In addition to the Federal Reserve setting monetary policy, it also varies the amount of money in the economy by either injecting additional money or withdrawing money to reduce the amount of money in circulation. The amount of money in the economy is usually increased during recessions to try and stimulate spending. Interest rates are reduced during recessions to allow for more borrowing of money, which also helps stimulate the economy. The amount of money in circulation is decreased and interest rates are raised during prosperous economic times to inhibit spending.

The Federal Reserve is responsible for preventing deflation and high levels of inflation. Deflation occurs when too many goods are available or there is not enough money in circulation to purchase goods. Inflation occurs when the price of goods and services rises. When inflation occurs, the interest rate used in any type of economic calculation is the real interest rate plus the percentage of inflation, and this is called the nominal interest rate. Equation 1.1 is the formula for the nominal interest rate including the effects of inflation:

$$i_n = r + e \tag{1.1}$$

where

i_n is the nominal interest rate
r is the real interest rate
e is the inflation rate

One example of nominal interest rates is the rates paid on U.S. *treasury bills* (*T-bills*) purchased by the public from the government. Treasury bills are sold at a discounted value from their face value (*par amount*). The interest rate paid on T-bills depends on how long it takes for them to reach maturity. Treasury bills are sold in denominations of $100.00 and the maximum amount that may be purchased noncompetitively is $5 million. Treasury bills pay the bill holder the face value of the T-bill in terms that range from a few days to 52 weeks. The difference between what is paid for a T-bill and the amount it is worth when it matures, as a percentage of the T-bill total value, is the interest on the T-bill. Treasury bills may be sold before they mature or held onto until they reach their maturity value. When T-bills are bought and sold in the open market, the interest rates for different time periods vary. One example of how T-bills mature is if someone pays $990.00 for a T-bill and when it matures in 52 weeks it is worth $1,000.00. The interest paid on this T-bill is 1.01%.

Sections 1.2.7 through 1.2.10 explain some of the processes available whereby vendors and suppliers conduct monetary transactions.

1.2.7 Promissory Notes

One type of transaction commonly used for paying for goods and services is *promissory notes*. Promissory notes are a written promise by one person to another person, such as a vendor or a supplier, to pay on demand or at a specified date, a certain sum of money, and then the note is presented by the seller to the bank of the purchaser for collection.

1.2.8 Trade Acceptance

Trade acceptance is a process whereby a supplier submits shipping documents through a local bank, and this is a *draft*. The trade acceptance is an order for the person making the purchase to pay the amount owed to the supplier. The purchaser assigns the debt to the supplier and it is payable by a specific date. Then the bank returns the trade acceptance to the supplier and the supplier sells it to his or her bank and is paid the amount owed by the purchaser.

1.2.9 Trade Credit

Another method for using money as a means of commerce is through the use of *trade credits*. Trade credits are used by firms involved in a substantial amount of business with a particular supplier or vendor. Rather than having to pay cash, or use a charge card, to pay for their purchases, the vendor or supplier establishes an *open account* for the purchaser. Whenever the purchaser needs to buy items, they contact the seller and then send the seller a *purchase order* (commercial document issued by a buyer to a seller, indicating types, quantities, and prices agreed upon for products or services). The seller then sends an *invoice* with the merchandise describing the items shipped and the selling price. Purchase orders are tracked by entering information about the purchase orders into a purchase order log, and information about invoices is entered into an invoice log.

1.2.10 Trade Credit Terms

In addition to the previously mentioned methods for paying for goods and services, suppliers, and vendors may allow purchasers to use other trade credit terms for paying for items purchased, and these techniques are discussed in Sections 1.2.11 through 1.2.15.

1.2.11 Cash before Delivery

If a supplier or vendor is not familiar with a purchaser, or the purchaser does not have good credit, the supplier or vendor may require the purchaser to pay cash before the supplier or vendor delivers or releases the items purchased, and this is *cash before delivery* (CBD). If the purchaser pays by check, the vendor or supplier may wait until the check clears before delivering or releasing the items.

1.2.12 Cash on Delivery

Cash on delivery (COD) occurs when a vendor or supplier requires the purchaser to pay cash when the items are delivered to him or her or to his or her place of business.

1.2.13 Single Draft Bill of Lading

A *single draft bill of lading* is a process whereby a vendor or supplier sends a *sight draft* (commercial paper payable when it is presented) and a *bill of lading* (document issued by a carrier detailing what is in a shipment of merchandise and giving title of the shipment to a specified party) to the bank of the purchaser when the supplier ships the items ordered to the purchaser. The bank then pays the vendor or supplier from the bank account of the purchaser the amount shown on the sight draft and bill of lading.

1.2.14 Net Cash and Cash Terms

The process *net cash* is used by vendors and suppliers to prompt purchasers to pay their bills in a timely manner. One example of how net cash is used is when a bill includes a Net10 statement, which indicates that the bill needs to be paid within 10 days.

Cash terms are when a vendor or supplier requires the bill for the cost of previously delivered items to be paid in full at the time of a new delivery.

1.2.15 Ordinary Terms, Seasonal Dating, and Consignment

Vendors and suppliers may use additional methods to prompt purchasers to pay their bills more quickly. Bills may indicate there is a cash discount if the bill is paid within a certain period of time, such as 10 or 20 days after the date of the invoice. An example would be if a vendor lists 10Net2 on a bill. This would indicate that if the bill were paid within 10 days, there would be a 2% discount on the total amount of the bill.

Vendors or suppliers might also offer a discount to buyers to encourage them to send their orders for seasonal goods before the peak buying period, and this is *seasonal dating*.

Some vendors and suppliers are willing to grant credit to a purchaser for the entire period in which goods are held for sale and accept payment for the goods when the purchaser has sold them and this form of credit is *consignment*.

1.3 IMPROVING THE ECONOMICS OF PROJECTS

The greatest monetary impact to a project occurs during the design stage. If design engineers have an understanding of the economic ramifications of their designs, they are able to create more economically feasible designs. In order to assist design engineers in creating economically feasible designs, many firms incorporate constructability reviews as part of the design process.

1.3.1 Constructability Reviews

Constructability reviews are a design review conducted by construction or production personnel to evaluate the buildability (the extent to which the design facilitates the ease of construction (or production) of a facility or products, subject to overall requirements for the completed structure (or product); CIRIA, 1983, p. 26). It is possible for engineers to create designs that contractors or fabricators are not able to build economically, and examples of some of the types of items contributing to this occurring include the following:

- Beam to column connections not available or they are excessively expensive
- Situations where it is not possible to fabricate the structural formwork inexpensively or the formwork cannot be removed due to its proximity to another structure
- Design requirements stipulating equipment not normally available
- Extremely heavy elements requiring lifting by expensive rigging equipment

In order to improve the economics of projects, economy could be achieved through several avenues including:

- Designing to reduce costs
- When a contract includes a *value-engineering* clause—where the contractor or fabricator finds methods for reducing costs and the contractor or fabricator receives half of the money saved by implementing the new method
- During maintenance and operation—could money be saved later if sustainable practices are incorporated during construction or fabrication
- Using environmentally friendly or sustainable practices such as wind energy, biofuel, and so forth

Some of the following techniques for increasing cost savings could be incorporated during the design stage:

- Designing to minimize labor use
- Eliminating unnecessary production requirements
- Furnishing adequate foundation information
- Simplifying designs
- Using duplicated elements for items such as formwork where a steel form could be used to create numerous elements with the same formwork

- Using inspectors with sufficient experience
- Specifying local materials, if possible
- Using standard specifications
- Writing simplified specifications

Another technique for improving the economy of projects is trying to reduce uncertainty on the part of the contractor or fabricator. Some techniques for reducing uncertainty are the following:

- Engineer should hold pre-bid conferences to help clarify unknown or questionable information.
- Providing pre-bid studies such as
    - Site studies
        - Geology reports
        - Soil samples (have the engineer visit the site and maybe even take his or her own samples)
        - Climate reports
        - Weather reports
- Providing information on
    - Materials
        - Sources
        - Storage
        - Access
    - Labor supply
        - Quality
        - Supervision
    - Equipment
        - Renting or owning equipment
    - Subcontractors
        - Who, what, where, and cost
    - Utilities
        - Who will pay for temporary utilities
        - Access cost
        - Reliability
        - Backup utilities
    - Jobsite
        - Communication network
        - Job staff conferences
        - Realistic safety practices
        - Bonus pay for key personnel

1.4 MANAGING COSTS AND PROFITS

This section presents information on managing costs and profits.

1.4.1 Managing Costs and Profits during Projects

The following are some of the techniques that could be implemented to help manage costs and profits during any type of project:

- Closely monitoring the project
- Controlling project costs—having a viable project controls system
- Improving profitability

- Setting a reasonable *labor burden markup rate* (overhead)
- Setting minimum profit margins
- Tracking overhead for budgets
- Using management by exception reporting—only reporting items that are behind schedule or over the budgeted amount for each stage of the project

1.4.2 Project Accounting Systems

Project accounting systems should be implemented prior to the inception of a project and they are used for:

- Ensuring the project and general overhead are accurately tracked
- Ensuring there is an effective cost accounting system that includes tracking data using spreadsheets and databases (could be an owner required cost accounting system)
- Forecasting and trend analysis
- Preparing all required financial statements

During projects, there are three main ledgers for tracking projects and they are:

1. *General ledger*—provides financial data for each part of a project
2. *Job cost ledger*—includes detailed information for the general ledger and information about specific jobs
3. *Equipment ledger*—tracks use of equipment and vehicles

1.4.3 Balance Sheets

Balance sheets summarize all of the resources owned by a firm including current assets, which are short-term capital assets, and fixed assets such as land and equipment. In addition, balance sheets also track liabilities, which include all financial obligations.

1.4.4 Calculating Net Worth

All organizations determine their net worth to provide a measure of the success or failure of the organization. The formula for calculating net worth is Equation 1.2.

$$\begin{aligned}\text{Net worth} = {} & \text{Equity of the firm (summary of the value of the company and its assests)} \\ & - \text{The liabilities of the firm}\end{aligned} \tag{1.2}$$

1.4.5 Income Statements

Income statements track revenue and expenses. Revenue includes all income from sales or work performed and the interest received during the accounting period. Expenses are summarized on income statements for the same time period used for tracking revenue.

1.4.6 Assets

There are many items included as assets of a firm, and they include, but are not limited to, the following:

- Buildings
- Copyrights

- Equipment
- Goodwill
- Machinery
- Patents
- Property

1.4.7 Equities

Equities are items subtracted from assets to determine the value of a firm, and they include:

- Claims against assets such as mortgages or loans
- Claims of creditors
- Liabilities

1.4.8 Accounting Transactions Tracked and Accounting Categories

During projects, there are many different types of items related to project transactions tracked with accounting systems including items such as:

- Direct costs—labor, material, and equipment
- Employee pay
- Equipment bought or leased
- Indirect costs and overhead
- Money borrowed including how much, when, and how it has to be repaid
- Purchases

The five accounting categories used in accounting systems are the following:

1. Assets
2. Liabilities
3. Net worth
4. Revenue
5. Expenses

1.5 ACCOUNTING METHODS

Managers of firms select the accounting method that will be used within their firm. The major difference in methods is the point in time where revenue or expenses are recognized in the accounting system. Three of the most frequently used accounting methods are cash, accrual, and long-term contract.

1.5.1 Cash Accounting

In the *cash accounting* method, income or expenses are only included when they are received or expended by a firm. Income is calculated as the difference between the cash collected and the cash expended to date.

1.5.2 Accrual Accounting

In *accrual accounting*, expenses are recognized when events occur that establish the firm has incurred a liability and the amount of the liability is known. Income is calculated as the difference between the amounts billed to customers and the expenditures paid and unpaid.

1.5.3 Long-Term Contract and Completed Contract Accounting

The *long-term contract accounting* method is unique to construction and it relies on using *percentage of completion* when reporting financial data. This accounting method recognizes income on a current basis since it results in a more regular cash flow. One disadvantage of this method is it is dependent on cost estimates and their accuracy. This system pays contractors progress payments based on a percentage of completion of the total work required in a bid package.

Before a contractor is paid, he or she provides an estimate of the work completed for each bid package in monthly progress reports submitted to the engineer or construction manager. This system recognizes income based on what is earned rather than on what has been collected in cash or what has been billed thus far. It is common for overbilling or underbilling to occur in this system. A company may not bill in proportion to the percentage of expenditures since the bills are submitted based on the percentage of work completed for each particular time period. Front-end loading could also occur in this system. *Front-end loading* is when a contractor bills a substantially higher amount for front-end items, such as mobilization, so that he or she is receiving funds from the owner early in the process, which prevents him or her from having to use his or her own funds for operating expenses. In order to front-end load a bid, the overall bid has to be *unbalanced* to compensate for the increased early amount being billed to the owner by lowering the estimated cost of an item occurring later in the project.

Another accounting technique is the *completed contract accounting* method. This method is mainly used on projects with a short duration. In this system, a firm performing the work is paid when the work is completed, not at any other time during a project.

1.5.4 Accounting Cycles

Accounting systems are used to record financial transaction in journals or computer software programs such as spreadsheets or databases. Accounting information is summarized and entered into ledgers, and this is *posting*. Journal entries are usually performed every day and posting to ledgers is performed monthly. Journals normally include information on current activities and documents such as invoices (in logs), sales receipts, cash register tapes, and computer-generated receipts. All of these items are dated as they are entered into the journals and a note is included as to what type of transaction each item represents. Ledgers summarize all debts and credits, and separate ledger pages (spreadsheets) are used for each type of account.

During projects, labor is tracked and this is accomplished using timekeeping methods for monitoring the total amount of time each employee allocates to various project work items. There are also systems used during projects to monitor materials and equipment to help control purchasing and receiving and to identify materials and where they are used on a project.

1.6 SOURCES OF FUNDING FOR PROJECTS

Before an owner undertakes any type of project, members of engineering and construction firms are hired to analyze the feasibility of a project and determine whether it will be profitable. But before any type of engineering economic analysis is performed, several steps are followed to produce a capital budget, and once the capital required for a project is estimated, then the owner locates sources of funding for the project.

1.6.1 Capital Budgeting

The first two steps in determining a *capital budget* are to estimate the cash flow for the project and then the value of the asset or project at a specific terminal date, which is referred to as the asset or project salvage value. Salvage values are discussed in detail in Section 2.3.4. The third step in capital budgeting is to estimate the cost of the project by developing an engineer's estimate and a project cash flow.

The fourth step in capital budgeting is to select an appropriate *discount rate* or *minimum attractive rate of return* (MARR), both of which are the cost of capital. The MARR is the lowest acceptable interest rate a project would earn in order for a firm to undertake the project. Many firms set their MARR at a rate higher than what they could earn if they invested the money that would be used to fund the project in a secure investment such as a bank account or *certificate of deposit*, plus an amount representing the risks associated with the project.

The fifth step requires a review of the intangibles associated with the project. The following are some of the types of intangibles that might be analyzed during the capital budgeting process:

- Environmental issues effecting or delaying execution of a project
- Issues either negatively or positively effecting the corporate image of a firm
- Legal constraints negatively effecting a project
- Potential product liability issues that might lead to litigation
- Ways to save time during the execution of a project, which in turn helps save money

After the capital budgeting process is complete, and the intangibles have been explored, then the owner develops a strategy for obtaining funding for a project.

1.6.2 Sources of Funding

There are potential sources of project funding both internal and external to organizations. Internal sources include (1) capital earned from operations, (2) *retained earnings* (reserves), and (3) retained earnings in excess of the *after-tax net earnings* divided by the dividends paid to stockholders.

There are more sources of funding for projects external to organizations than the ones available internally. One of the most prevalent methods is to borrow money through a commercial loan. The major disadvantage of this option is the high interest rates charged on commercial loans for engineering and construction projects. Interest rates fluctuate based on the credit worthiness of a firm and the risks associated with a project, and they could be in the double digits for risky projects.

A second source of project funding is to borrow in the *open market*. This entails drawing up a note to the order of the bearer of the note and *discounting* it through a dealer. Notes are sold to private individuals or companies that are essentially funding the project through their purchase of the notes. The notes are repaid with interest at a future date specified in the note.

A third option is to use *open market paper* or a *banker's acceptance* (bank vouchers), which after being approved by the bank provide funds to a firm as the firm orders items and they are delivered to the firm.

A fourth option is to enter into a *deferred payment contract*. This type of a contract requires borrowers to sign a note for a series of payments over a set period of time. In some instances, the note holder holds title to the equipment owned by the borrower until the debt is repaid.

A fifth option is to use *term loans*, which are regular bank loans with a maturity date and either a set or variable interest rate.

One additional option is long-term loans, which are loans maturing beyond 10 years. The following are several different types of long-term loans:

- *Bonds*—When bonds are sold, they either pay dividends to the bondholders or they are purchased at a discount rate and paid in full at maturity or both.
- *Convertible bonds*—Bonds that could be converted to a specified number of shares of common stock in the issuing company or cash of equal value.

- *Debentures*—Debt instruments not backed by assets or collateral but backed only on the creditworthiness and reputation of the issuer.
- *Mortgages*—These are secured by assets such as real property.
- *Stockholder's equity*—Stockholders receive dividends before the firm reinvests funds.

These long-term loan options are not mutually exclusive (one used to the exclusion of the others) but could be used in combination to secure the funds required for a project.

1.7 SUMMARY

This chapter introduced engineering economics and explained its role in evaluating the economic viability of projects. A synopsis of the organization of this book was provided, including a brief description of what is presented in each chapter. This chapter explained money and how it is used as a means of commerce and as a means of storing value, where it comes from, and how it is used as units of accounting and as a commodity bought and sold. The current status of money was covered along with an introduction to the U.S. Federal Reserve System, promissory notes, trade acceptance, and trade credit.

Information was provided in this chapter on increasing cost savings and improving the economy of projects. Constructability reviews were explained in terms of how they are included during the design stage to try and increase the economic feasibility of designs.

The last part of this chapter covered information on managing costs and profits; project accounting systems; ledgers; balance sheets; net worth; income statements; accounting ratios; assets; equities; and accounting transactions tracked, accounting categories, methods, and cycles. Sources of funding for projects were also introduced along with an explanation of the capital budgeting process.

KEY TERMS

Accrual accounting
After-tax net earnings
Assets
Balance sheets
Banker's acceptance
Bartering
Bill of lading
Bonds
Bureau of Engraving and Printing
Capital budgeting
Cash accounting
Cash before delivery
Cash on delivery
Cash reserve
Cash terms
Certificate of deposit
Completed contract accounting method
Consignment
Constructability review
Convertible bonds
Debentures
Deferred payment contract

Deposit reserve rate
Discount rate
Discounting
Draft
Equipment ledger
Equities
Equity ratio
Euro
European Union
Federal Reserve System
Front-end loading
General ledger
Gold reserves
Gold standard
Gross domestic product (GDP)
Hyperinflation
Income statements
Invoice
Job cost ledger
Labor burden markup rate
Liabilities
Long-term contract accounting
Minimum attractive rate of return
Mortgages
Mutual funds
Net cash
Net worth
Nonmonetary economy
Open account
Open market
Open market paper
Par amount
Percentage of completion
Posting
Promissory notes
Purchase order
Retained earnings
Seasonal dating
Securities
Sight draft
Single draft bill of lading
Stockholder's equity
Treasury bills
Term loans
Trade acceptance
Trade credits
Treasury bills
Unbalanced bid
Value-engineering

PROBLEMS

1.1 Explain the steps managers of firms perform during the capital budgeting process.
1.2 What are sources of funding for projects internal to a firm?
1.3 Discuss what hyperinflation is and why it occurs in economies.
1.4 What are the five commonly used accounting ratios?
1.5 What types of transactions are banks involved in throughout the world?
1.6 Explain how trade credits are used as a means of commerce.
1.7 What are some of the processes that help manage costs and projects during construction?
1.8 In addition to setting monetary policies, what else is the U.S. Federal Reserve System responsible for in the United States?
1.9 Explain the difference between cash and accrual accounting systems.
1.10 What are the three main types of ledgers used during a project?
1.11 How are promissory notes used to cover the cost of goods and services?
1.12 How does the trade acceptance process facilitate monetary transactions?
1.13 What items that help increase cost savings could be incorporated during the design stage of projects?
1.14 Discuss what types of items are tracked with accounting systems.
1.15 What types of project funding are external to firms?
1.16 What happens when one country is paying higher interest rates on the debt it issues than other countries?
1.17 How do single draft bill of ladings facilitate the transfer of merchandise?
1.18 Explain the gold standard and whether the United States uses the gold standard.
1.19 What are constructability reviews and how do they help improve the economics of projects?
1.20 Before countries had viable currencies, what was used to store value?
1.21 Project accounting systems are implemented prior to a project's inception in order to accomplish what types of things?
1.22 Explain what net worth is and how firms calculate it.
1.23 Explain what bartering is and how it is used instead of currencies in economic transactions.
1.24 How do governments raise funds to pay for government-related expenses if they are not able to raise enough funds through taxes?
1.25 What types of items are considered to be assets of a firm?
1.26 Explain what the deposit reserve rate is and how it helps to regulate banks in the United States.
1.27 What are some of the items in designs that prevent contractors from building economically?
1.28 Explain net cash and how it relates to ordinary terms stipulated in bills for merchandise.
1.29 Explain how long-term contract accounting is used in construction.
1.30 Explain the difference between before cash delivery and cash on delivery.

REFERENCES

Brown, R. 2015. *Plank: Driven by Vision, Broken by War.* New York: Oxford University Press.

CIRIA (Construction Industry Research and Information Association). 1983. *Buildability: An Assessment.* London, England: CIRIA.

Countries of the World. 2015. List of currencies of the world. Countries of the World. Countries-ofthe-World.com. Accessed on November 16, 2015. https://www.countries-ofthe-world.com/world-currencies.html.

Duran, Y. August 28, 2014. 20 Largest gold reserves by country. *Futures Magazine.* Accessed on November 16, 2015. http://www.futuresmag.com/2014/08/28/top-20-largest-gold-reserves-country-2014-edition?page=13.

Federal Reserve. November 12, 2015. Reserve requirements. Washington, DC: Board of Governors of the Federal Reserve System. Accessed on November 16, 2015. http://www.federalreserve.gov/monetarypolicy/reservereq.htm#table1.

Federal Reserve Board. December 13, 2005. The Twelve Federal Reserve Districts. Washington, DC: Federal Reserve Board. Accessed on January 4, 2016. http://www.federalreserve.gov/otherfrb.htm.

Patton, M. October 28, 2014. Who owns the most U.S. debt? *Forbes Advisor Network*. Accessed on November 16, 2015. http://www.forbes.com/sites/mikepatton/2014/10/28/who-owns-the-most-u-s-debt/.

U.S. Department of the Treasury. January 12, 2016. U.S. National Debt. Washington, DC. Accessed on January 12, 2016. http://www.usdebtclock.org/index.html.

U.S. Department of the Treasury. September 30, 2015. Status Report of the U.S. Government Gold Reserve. Washington, DC: Bureau of the Fiscal Service. Accessed on November 16, 2015. https://www.fiscal.treasury.gov/fsreports/rpt/goldRpt/current_report.htm.

World Bank. 2015. 4.12 World Development Indicators: Central Government Finances. Washington, DC: World Bank. Accessed on November 16, 2015. http://wdi.worldbank.org/table/4.12.

2 Time Value of Money, Interest, and Cash Flow Diagrams

This chapter introduces the concept of time value of money and explains the responsibility of engineers for incorporating the time value of money into the economic analysis of alternatives they perform for clients. This chapter also provides definitions for different types of interest including simple, compound, nominal, effective, and continuously compounded and for engineering economic terms such as present worth, future worth, annuities, salvage value, and sunk costs. Cash flow diagrams are introduced and the procedures for drawing them are discussed along with an explanation on how they are integrated into engineering economic analysis.

2.1 TIME VALUE OF MONEY

Knowledge of the fundamentals of engineering economics allows engineers to recommend alternatives to clients on the basis of either the:

1. *Least equivalent cost*
2. *Greatest equivalent net worth*

These two choices indicate the difficultly of comparing project alternatives directly against one another since the variables involved in each alternative may occur over different time frames or have different interest rates. Therefore, in order to compare alternatives, the variables associated with each alternative, and the time frame over which each alternative is being evaluated, have to be converted into equivalent monetary terms. The formulas introduced in this book provide methods for converting alternatives into equivalent costs, revenue, and equivalent net worth, so all of the alternatives under consideration may be compared to each other directly when analyzing alternatives.

Time value of money principles apply to all types of engineering economic evaluations. Time value of money accounts for the interest an investment earns and it indicates that an amount of money with a certain value now will increase in value in the future due to the interest the money earns during the intervening time period. Therefore, in most circumstances, investments are worth more in the future than they are in the present. Time value of money is also considered in situations where funds are not invested but are used to fund a project and the resulting project is compared to what the funds would have earned if they had been invested in an interest-bearing investment instrument.

To foster an understanding of the time value of money, and the fundamentals of engineering economic formulas, three types of terms are defined in Sections 2.2 through 2.4:

1. Interest
2. Engineering economic terms
3. Cash flow diagrams

2.2 DEFINITIONS FOR INTEREST

To understand how *interest* is defined in terms of monetary policies, it needs to be viewed as a commodity such as rent on money borrowed or loaned from one individual to another, from one institution to another institution, or from an institution to an individual. The interest or rent charged is normally a percentage of the total amount borrowed during the transaction.

Whenever interest calculations are performed, the interest rate given in percent is divided by 100 to convert it into the proper format for use in engineering economic formulas. The following equation is the formula for calculating the amount of interest owed on a transaction at the end of one year:

$$\text{Interest owed at the end of 1 year} = \frac{\text{Interest rate}}{100} \times \text{Amount borrowed or loaned} \tag{2.1}$$

Using this, the interest owed on a loan of \$1,000.00 at the end of one year if the interest rate is 5% is calculated as follows:

$$\text{Interest owed at the end of 1 year} = \frac{5\%}{100} \times 1{,}000.00 = \$50.00$$

Interest payments are normally due at the end of the interest period unless stated otherwise in loan agreements. Banks and other lending institutions are able to set the terms they offer for their loan; therefore, it is important to always be aware of the terms of a loan including the following:

- Amount being borrowed or loaned
- Type of interest rate
- Interest rate
- Interest period
- Other terms of the loan

These are the main items stipulated in loan agreements and additional terms used in loan transactions are discussed throughout this book.

Sections 2.2.1 through 2.2.5 provide definitions and formulas for simple, compound, nominal, effective, and continuously compounded interest.

2.2.1 Simple Interest

Simple interest pertains to financing but it is also used in modeling, and an example of this would be modeling population growth such as the population increasing by a certain percentage each year. Equation 2.2 is the formula for calculating the total amount owed at the end of one year if the interest rate is simple interest. Simple interest is interest charged on the *principal* (the initial amount borrowed or loaned) for that year:

$$\text{Total amount owed} = \text{Principal} + \left(\frac{\text{Interest rate}}{100} \times \text{Principal}\right) \tag{2.2}$$

Applying this to the previous data results in the following solution:

$$\begin{aligned}\text{Total amount owed} &= \$1{,}000.00 + \left(\frac{5\%}{100} \times \$1{,}000.00\right)\\ &= \$1{,}000.00 + \$50.00\\ &= \$1{,}050.00\end{aligned}$$

In Equation 2.2, if the interest is simple interest, the amount of interest owed each year will be \$50.00 per year regardless of how many years are required to repay the loan. In addition to paying \$50.00 per year in interest each year, the amount borrowed—\$1,000.00—has to be repaid at the end of the loan period.

If a loan is due at the end of the first year, the future amount owed at the end of the first year is calculated using the following equation:

$$F_1 = \text{Principal} + \left(\frac{\text{Interest rate}}{100} \times \text{Principal} \right) \tag{2.3}$$

For the data used in the previous example, the future worth would be the following:

$$F_1 = \$1,000.00 + \left(\frac{5\%}{100} \times \$1,000.00 \right)$$

$$= \$1,050.00$$

2.2.2 Compound Interest

Both simple and compound interest represent the rate of return (ROR) on investments. In the previous example, the bank is the entity investing the money by loaning it to the person borrowing the money. Five percent interest represents the rate of return the bank will receive on its investment. If the ROR is 5% simple interest, the bank will receive the amount borrowed—\$1,000—plus the unpaid interest of \$50.00 per year at the end of the investment period. But if the bank is charging 5% compound interest, rather than just receiving 5% per year on the investment, the bank will receive an amount greater than 5% due to the compounding of the unpaid interest and the amount borrowed or owed each year.

There are situations where loans stipulate that the initial amount (principal) will not be repaid at the end of one year, and in these situations, the amount borrowed plus the amount of interest owed for the first year are carried forward into the second year, and the interest for the second year is calculated by multiplying the interest rate by the principal plus the interest from the first year on the principal. Equation 2.4 is used for calculating the interest owed at the end of the second year:

$$\text{Interest for year 2} = (\text{Principal} + \text{First year's interest}) \times \frac{\text{Interest rate}}{100} \tag{2.4}$$

For the data in the previous calculation, the interest owed at the end of year two would be the following:

$$\text{Interest owed at the end of year 2} = (\$1,000.00 + \$50.00) \times \frac{5\%}{100}$$

$$= \$52.50$$

The total amount owed at the end of year two is calculated using Equation 2.5:

$$F_2 = \text{Principal} + \text{First year's interest} + \left[(\text{Principal} + \text{First year's interest}) \times \frac{\text{Interest rate}}{100} \right] \tag{2.5}$$

For the data in the previous calculation, the future value at the end of year two is the following:

$$F_2 = \$1{,}000.00 + \$50.00 + \left[(\$1{,}000.00 + \$50.00\,) \times \frac{5\%}{100} \right]$$

$$= \$1{,}102.50$$

When interest is not repaid yearly and the amount of interest owed is carried over into subsequent years and used in the calculations to determine subsequent interest, this is *compound interest.* If an interest rate (i) is provided, unless it states otherwise, the interest rate represents compound interest.

2.2.3 Nominal Interest

Nominal interest rates apply when the *compounding period* (interest period) is less than one year. Compounding periods are the point in time when the amount borrowed or loaned has the interest owed calculated for that particular period. If the interest rate expressed over a period of time is less than a year, such as 2% per month or 8% per six months, then the nominal interest rate has to be calculated to determine the amount of interest owed at the end of the interest period. Equation 2.6 is the formula for calculating nominal annual interest:

$$i_n = i \times m \tag{2.6}$$

where
- i_n is the yearly interest rate
- i is the interest rate per interest period
- m is the number of interest periods per year

Using Equation 2.6 to calculate the nominal annual interest rate for an interest rate of 1% per month results in the following:

$$i_n = \frac{1\%}{\text{Period}} \times \frac{12 \text{ periods}}{\text{Year}}$$

$$= 12\% \text{ per year}$$

Using Equation 2.6 to calculate the nominal interest rate for 2% interest compounded quarterly results in the following:

$$i_n = \frac{2\%}{\text{Period}} \times \frac{4 \text{ periods}}{\text{Year}}$$

$$= 8\% \text{ per year}$$

2.2.4 Effective Interest

When the time value of money is considered while calculating the annual interest rate using the period interest rate, this is the *effective interest rate* (i_e). The frequency of payments within one year is the *payment period*. The payment period may or may not coincide with the compounding period. Equation 2.7 is the formula for calculating effective interest:

$$i_e = \left(1 + \frac{i_n}{m}\right)^m - 1 \quad \text{or} \quad i_e = (1+i)^m - 1 \tag{2.7}$$

where

i_e is the effective interest rate
i is the interest rate per compounding period
i_n is the nominal interest rate
m is the number of compounding periods per year

Example 2.1 calculates the effective interest rate per year using Equation 2.7.

Example 2.1

If $i = 2\%$ per quarter, what is the effective interest rate per year?

Solution

$$\begin{aligned} i_e &= (1+i)^m - 1 \\ &= (1+0.02)^4 - 1 \\ &= 0.082432 \\ &= 8.24\% \text{ per year} \end{aligned}$$

Example 2.2 illustrates the difference between the payment period and the compounding period.

Example 2.2

Every month a student deposits part of her paycheck into a savings account at a bank paying a nominal interest rate of 6% per year compounded semiannually. What is the payment period and what is the compounding period?

Solution

The payment period is monthly and the compounding period is every six months.

Example 2.3 demonstrates calculating nominal and effective interest rates.

Example 2.3

A student deposits part of his paycheck into a savings account paying 1.5% interest every three months. Part I: What are the nominal and effective interest rates? Part II: What would be the nominal and effective interest rates if the account pays 4% interest semiannually?

Solution

Part I

Solve for the nominal interest rate using Equation 2.6:

$$\begin{aligned} i_n &= i \times m \\ &= \frac{1.5\%}{\text{Period}} \times \frac{4 \text{ periods}}{\text{Year}} \\ &= 6.0\% \text{ per year} \end{aligned}$$

Solve for the effective interest rate using Equation 2.7:

$$\begin{aligned} i_e &= \left(1 + \frac{i_n}{m}\right)^m - 1 \\ &= \left(1 + \frac{0.06}{4}\right)^4 - 1 \\ &= (1 + 0.015)^4 - 1 \\ &= 0.0614 \\ &= 6.14\% \end{aligned}$$

Part II

$$\begin{aligned} i_n &= i \times m \\ &= \frac{4\%}{\text{Period}} \times \frac{2 \text{ periods}}{\text{Year}} \\ &= 8.0\% \text{ per year} \end{aligned}$$

$$\begin{aligned} i_e &= \left(1 + \frac{i_n}{m}\right)^m - 1 \\ &= \left(1 + \frac{0.08}{2}\right)^2 - 1 \\ &= (1 + 0.0816)^2 - 1 \\ &= 0.0816 \\ &= 8.16\% \end{aligned}$$

2.2.5 Continuous Compounding of Interest

In addition to simple, compound, and nominal interest, there is also *continuously compounded interest* and it is represented by the symbol ($\bar{i}_e$). Continuous compounding of interest means the interest on the amount in an account is calculated all of the time. When funds are deposited into an interest-bearing account, the funds immediately start to draw interest and continually accumulate interest until the funds are withdrawn from the account. The formula for continuously compounded interest is Equation 2.8:

$$\bar{i}_e = e^{i_n} - 1 \tag{2.8}$$

where

$\bar{i}_e$ is the continuously compounded interest rate

i_n is the nominal interest rate

e is 2.178282…[Euler's number $\left(1+\dfrac{1}{n}\right)^n$ as n approaches ∞]

Examples 2.4 and 2.5 demonstrate how to calculate continuously compounded interest.

Example 2.4

If the nominal interest rate is 8%, what is the continuously compounded interest rate?

Solution

Use Equation 2.8 to solve for the continuously compounded interest rate:

$$\begin{aligned}\bar{i}_e &= e^{i_n} - 1\\ &= e^{0.08} - 1\\ &= 0.0833\\ &= 8.33\%\end{aligned}$$

Example 2.5

If the interest rate is 12%, what is the continuously compounded interest rate?

Solution

Use Equation 2.8 to solve for the continuously compounded interest rate:

$$\begin{aligned}\bar{i}_e &= e^{i_n} - 1\\ &= e^{0.12} - 1\\ &= 0.1275\\ &= 12.75\%\end{aligned}$$

2.2.6 Rate of Return

Before an individual or a manager of a firm decides whether to invest his or her or the firm's money, he or she needs to know if the investment will be profitable, which means the money invested will be worth more at the end of the investment period. The amount of money received in addition to the original investment is the *rate of return* on the investment. The formula for the ROR in percent for the interest period is Equation 2.9:

$$\text{Rate of return (ROR) (in percent)} = \frac{\text{Total amount of money received} - \text{Original investment}}{\text{Original investment}} \times 100\% \tag{2.9}$$

Equation 2.10 is an alternative formula for rate of return using profit to represent the total amount of money received minus the original investment:

$$\text{Rate of return (ROR) (in percent)} = \frac{\text{Profit}}{\text{Original investment}} \times 100\% \quad (2.10)$$

The procedure for calculating unknown rates of return is explained in Section 3.5.

2.3 DEFINITIONS FOR ENGINEERING ECONOMIC TERMS

In Sections 2.3.1 through 2.3.5, definitions are provided for five of the commonly used engineering economic terms:

1. Present worth
2. Future worth
3. Annuities—uniform series
4. Salvage value
5. Sunk cost

2.3.1 Present Worth

Present worth (*PW*), *present value* (*PV*), and principal (*P*) all represent the value of money at time zero, which is the beginning of the engineering economic analysis period under investigation. In formulas, the present sum of money may be labeled as (*PW*), (*PV*), (*P*), or (P_0). All four of these symbols represent the same initial time frame, which is *time zero.* In performing engineering economic analysis, a formula for calculating present worth is introduced in Section 3.3 that is the *present worth factor*, and it is used to calculate the current value of future values with the interest discounted back to time zero.

2.3.2 Future Worth

Future worth (*FW*), *future value* (*F*), or (F_n) represent the future sum of money including principal plus interest. Future values occur at any point in time in the future and they are usually designated as the end of the engineering economic analysis period if they are the last activity to occur in the analysis period. The future worth of present values, and payments and disbursement streams, includes interest on the money invested or withdrawn from an account. The formula for future worth is introduced in Section 3.2.

2.3.3 Annuities: Uniform Series

An *annuity* (*A*) is the term used to describe a series of uniform, end-of-period payments or disbursements. Annuities represent a payment or disbursement stream deposited or withdrawn at equal set intervals such as daily, weekly, monthly, or yearly. As each annuity is deposited into an interest-bearing account, it begins to draw interest at the end of each compounding period. The annuities deposited, plus any previous interest earned, are used when calculating the interest on the funds in the account at the end of each period. The formulas for calculating present and future worth of uniform series are introduced in Sections 4.1 and 4.2.

2.3.4 Salvage Value

The *salvage value* is what an asset is worth at the end of its useful life. In engineering economic analysis, the salvage value is represented by a future value occurring at the end of the analysis period. It is not always possible to accurately determine what a future salvage value of an asset will be; therefore, for the purpose of an analysis, a reasonable salvage value is assumed and included in the calculations. Many times, salvage values for similar items from previous projects are incorporated into a new analysis.

2.3.5 Sunk Cost

Sunk cost is a difficult concept to understand when performing engineering economic analysis. Sunk cost represents funds not recoverable because they have already been expended some time in the past. Sunk costs are discussed in more detail in Section 8.2.6.

2.4 CASH FLOW DIAGRAMS

This section discusses an important aspect of engineering economic analysis—*cash flow diagrams.* If cash flow diagrams are drawn properly, then the diagrams are helpful when developing engineering economic solutions. Cash flow diagrams are a visual representation of the flow of funds into and out of investment instruments, maintenance accounts, projects, and any other type of activity involving the movement of money.

2.4.1 Drawing Cash Flow Diagrams

Cash flow diagrams are drawn either from a *borrower's point of view* or from a *lender's point of view.* Once a cash flow diagram is drawn from one of these two perspectives, the other perspective is represented by the inverse cash flow. Figure 2.1 illustrates the borrower's perspective in a cash flow diagram and Figure 2.2 shows the lender's perspective. For both examples, the initial principal amount is $1,000.00, the interest rate is 5%, and there is only one interest period in the year.

Figures 2.1 and 2.2 also demonstrate the concept of *equivalency.* For both figures the principal amount of $1,000.00 at time zero is equivalent to $1,050.00 at the end of one year at an interest rate of 5%.

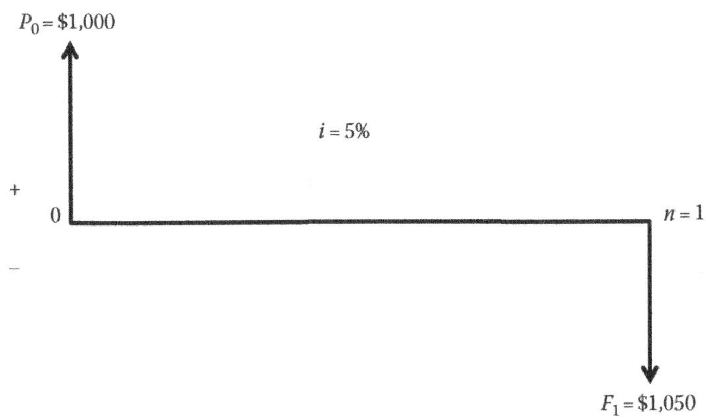

FIGURE 2.1 Cash flow diagram for one year—from the borrower's perspective.

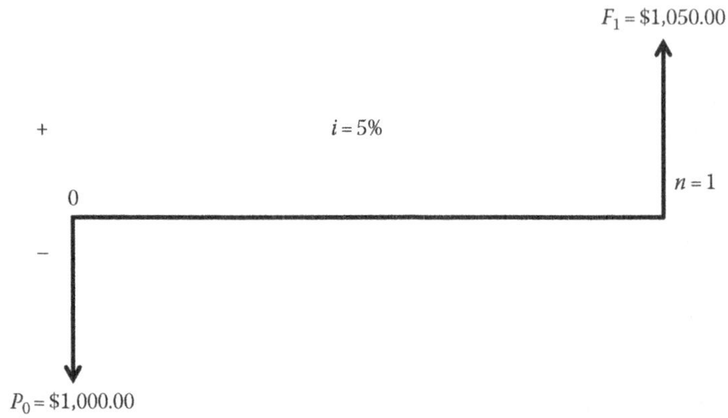

FIGURE 2.2 Cash flow diagram for one year—from the lender's perspective.

If the principal amount at time zero of \$1,000.00 is not repaid at the end of the first year but is repaid at the end of the second year, the cash flow diagram would be the diagram from the borrower's point of view shown in Figure 2.3.

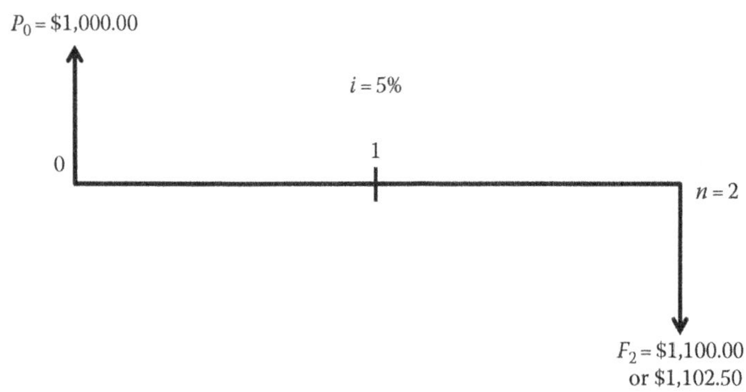

FIGURE 2.3 Cash flow diagram for repaying a loan at the end of two years.

The reason there are two possible amounts listed as being owed at the end of year two in Figure 2.3 is because the amount owed depends on whether the interest rate of 5% is simple interest or compound interest. If it is simple interest, the borrower will be charged \$50.00 for each year he or she does not repay the principal on the loan. If it is compound interest, then the unpaid interest of \$50.00 from the first year is added to the principal amount of \$1,000.00 and the interest for the second year is 5% of \$1,050.00 or \$52.50; therefore, the amount owed at the end of the second year is calculated using Equation 2.5.

Solving for the future amount owed at the end of the second year using Equation 2.5 results in the following:

$$\begin{aligned} F_2 &= \$1,000.00 + \$50.00 + [(\$1,000.00 + \$50.00) \times 0.05] \\ &= \$1,050.00 + \$52.50 \\ &= \$1,102.50 \end{aligned}$$

When performing engineering economic analysis, the interest rate is considered to be compound interest unless stated otherwise.

2.4.2 Using Cash Flow Diagrams to Help Solve Problems

Another example demonstrating the drawing of cash flow diagrams from the lender's point of view is shown in Figure 2.4. The situation represented in this figure is a bank loaning $10,000.00 to an electronics firm to buy parts for a machine it is building for the company. The bank informs the manager of the company it charges 12% interest compounded yearly and the principal and interest are due at the end of three years. The future value shown in this figure includes the interest owed at the end of three years.

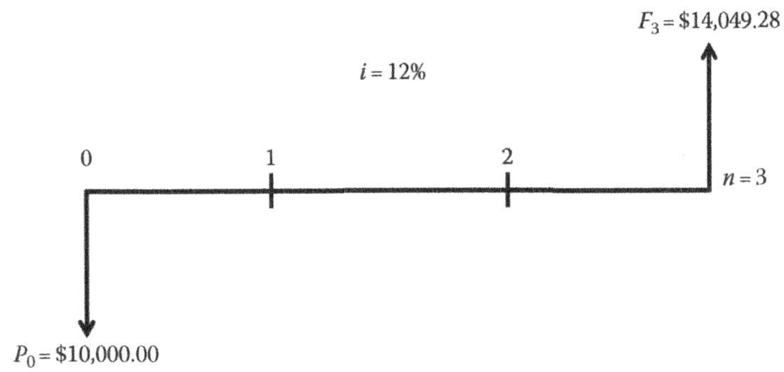

FIGURE 2.4 Cash flow diagram for bank loan to buy parts—lender's perspective.

The calculations for determining the amount owed at the end of year three are the following:

$$\text{Amount owed at the end of year 1} = \$10,000.00 + (\$10,000.00 \times 0.12)$$
$$= \$10,000.00 + \$1,200.00$$
$$= \$11,200.00$$

$$\text{Amount owed at the end of year 2} = \$11,200.00 + (\$11,200.00 \times 0.12)$$
$$= \$11,200.00 + \$1,344.00$$
$$= \$12,544.00$$

$$\text{Amount owed at the end of year 3} = \$12,544.00 + (\$12,544.00 \times 0.12)$$
$$= \$12,544.00 + \$1,505.28$$
$$= \$14,049.28$$

At the end of three years, the borrower pays back the original $10,000.00 in principal along with $4,049.28 in interest.

A second example demonstrating the drawing of cash flow diagrams is shown in Figure 2.5. The cash flow diagram represents a chemical engineer who needs to borrow money to finance an

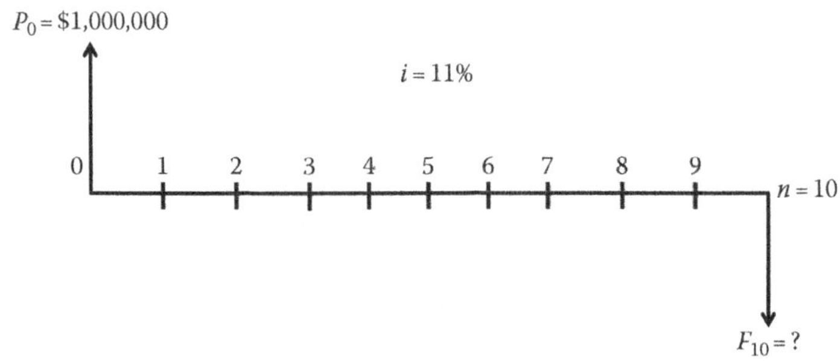

FIGURE 2.5 Cash flow diagram for borrowing funds for an electron microscope—borrower's perspective.

electron microscope. The electron microscope costs \$1,000,000.00. The bank offers to loan the required funds with a payback period of 10 years and an interest rate of 11%.

Another example illustrating the drawing of cash flow diagrams is shown in Figure 2.6. This cash flow diagram represents a bank providing an engineering firm with the option of repaying a loan of \$1,000,000.00 on a yearly basis rather than at the end of the 10 years. The annuity represented by A_1 through A_{10} in Figure 2.6 represents the yearly payments the engineer is responsible for in order to repay the loan by year 10.

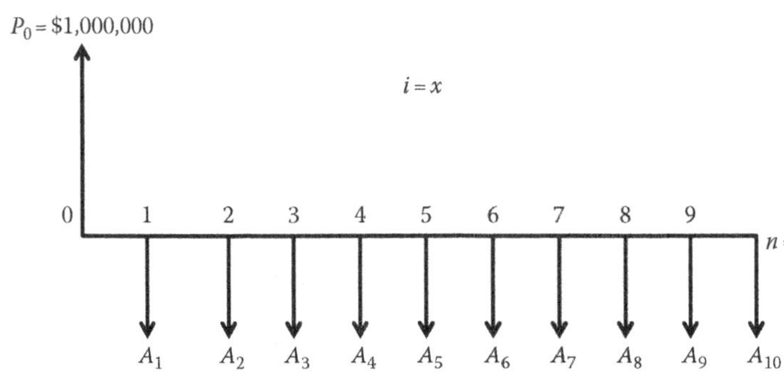

FIGURE 2.6 Cash flow diagram for repaying a loan on a yearly basis—borrower's perspective.

As was demonstrated in the previous examples, cash flow diagrams may be drawn for any type of engineering economic analysis problem. Once a cash flow diagram is drawn, then the engineering economic formulas are developed based on the information and its location in the cash flow diagram. Cash flow diagrams will be used throughout the remainder of this book when solving engineering economic analysis problems.

2.5 SUMMARY

This chapter introduced the concept of the time value of money, which accounts for the interest earned on funds borrowed or loaned to individuals or institutions. This chapter also explained equivalency and how converting alternatives into equivalent terms is required when comparing engineering alternatives on an equal basis when their cash flows occur over different time frames or they have different interest rates. Definitions were provided for simple, compound, nominal, effective, and continuously compounded interest. A definition for rate of return was provided along with an explanation of why rates of return are essential to engineering economic analysis. Definitions were included for present worth, future worth, annuities, salvage value, and sunk cost. The last part of the chapter explained how to draw cash flow diagrams, why they are drawn, and how they are used to help solve engineering economic problems.

KEY TERMS

Annuity
Borrower's point of view
Cash flow diagrams
Compound interest
Compounding period
Continuously compounded interest
Effective interest rate
Equivalency
Future value
Future worth
Greatest equivalent net worth
Interest
Least equivalent cost
Lender's point of view
Nominal interest
Payment period
Present value
Present worth
Present worth factor
Principal
Rate of return
Salvage value
Simple interest
Sunk cost
Time zero

PROBLEMS

2.1 What does having knowledge of the fundamentals of engineering economics allow engineers to do for clients?
2.2 How is interest defined in terms of engineering economics and what is the formula for interest owed on principal at the end of one year?
2.3 What terms associated with loans should borrowers be aware of when borrowing money?
2.4 Explain the difference between simple and compound interest.
2.5 How are nominal interest rates different from annual interest rates?
2.6 Explain how effective interest rates are calculated for use in engineering economic formulas.
2.7 How does continuous compounding affect the amount of interest either received or paid by someone?
2.8 What are rates of return and why are they important in engineering economic analysis?
2.9 Explain the difference between present and future worth.
2.10 What are annuities and how are they used in engineering economics?
2.11 Explain salvage value and its application in engineering economics.
2.12 Discuss sunk costs and why they are hard to quantify in engineering economics.
2.13 Explain how cash flow diagrams are used in engineering economic analysis.

2.14 What at the two perspectives shown in cash flow diagrams?

2.15 Explain what is being shown in Figure 2.6 in the cash flow diagram.

2.16 Explain time value of money in relation to engineering economics.

2.17 Calculate the yearly interest rate if an investment is paid 1.75% interest every two months.

2.18 Calculate the interest rate per interest period if the yearly interest rate is 13% and the number of interest periods per year is three.

2.19 Calculate the number of interest periods per year if the yearly interest rate is 15% and the interest rate per interest period is 2.5%.

2.20 Calculate the yearly interest rate if there are 12 interest periods per year and the interest rate per period is 0.8%.

2.21 Calculate the effective interest rate if the interest rate per compounding period is 1.5% and the number of compounding periods per year is 12.

2.22 Calculate the effective interest rate if the nominal interest rate is 16% and the interest is compounded every two months.

2.23 If a bank pays 9% interest per year and the compounding period is weekly, what is the effective interest rate?

2.24 An engineer locates a bank paying 2% interest per year compounded continuously. If the engineer invests his money in this bank, what is the continuously compounded interest rate?

2.25 An engineering firm borrows funds from a bank charging 9% interest compounded continuously. What is the continuously compounded interest rate?

2.26 If a credit union pays 1% yearly compounded monthly, what is the effective interest rate?

2.27 If an engineer is earning 1.75% interest per year compounded daily, what is the effective interest rate?

2.28 Calculate the nominal interest rate if the period interest rate is 0.5% and there are four compounding periods per year.

2.29 A firm pays year-end dividends to its employees of 1.5% but the funds in the account compound every two months. What is the effective interest rate?

2.30 What is the continuously compounded interest rate if an investor is able to invest her funds at a nominal interest rate of 1.25%?

3 Present Worth, Future Worth, and Unknown Interest Rates

This chapter defines *equivalence* and explains why alternatives should be compared based on their equivalent terms. The single payment compound amount factor (*F*/*P*) is introduced along with the procedure for calculating the future worth of present values. Steps are provided for demonstrating the process for using the interest factor tables in Appendix B for solving engineering economic problems.

A discussion is included on how to calculate the present worth of a future value using the present worth compound amount factor (*P*/*F*) and on how net present worth calculations determine whether an investment results in an equivalent positive or negative present worth. The procedures for calculating the future worth of a present value and the present worth of a future value are explained for situations where the compounding period for the interest is different than the payment period.

The last part of this chapter: (1) covers the process for solving for unknown interest rates using formulas or the interest factor tables in Appendix B and interpolation; and (2) describes the process for calculating unknown number of years using the interest factor tables in Appendix B and interpolation.

3.1 DEFINITION OF EQUIVALENCE

There are two engineering economic formulas available for solving for either the equivalent future worth of a present sum (the present sum plus the interest accrued over time) or the present worth of a future sum with the interest removed from the future value to obtain the present worth of the sum. The reason equivalent future and present worth are calculated is because it is necessary to evaluate potential project alternatives based on equivalent terms, since cash flows occur at different times and projects may have different interest rates. It is not accurate to compare alternatives directly to each other without first converting them into equivalent terms. Engineering economic formulas provide a means for converting payments or disbursements into equivalent present or future worth so that potential alternatives may be compared to each other in equivalent terms to determine which alternative is the most economical alternative.

3.2 FUTURE WORTH: SINGLE PAYMENT COMPOUND AMOUNT FACTOR (*F*/*P*)

The *future worth* formula is the *single payment compound amount factor* (*F*/*P*). This formula considers the effects of compounding interest when determining the future worth of a present value. Therefore, future worth is the sum of the principal plus the compound interest for the number of periods over which interest is charged on the principal. This verbal description of the future worth at the end of the first period is:

$$F_1 = P + P(i)$$

or

$$F_1 = P(1+i) \tag{3.1}$$

where

F_1 is the future worth at the end of year one
P is the principal
i is the interest

In order to calculate the future worth at the end of the second period, the amount of the principal plus interest owed at the end of the first period is multiplied by the interest rate, as shown in Equation 13.2.

$$F_2 = P_0(1+i)(1+i) \tag{3.2}$$

The amount owed at the end of the third period is the amount owed at the end of the second period times the interest rate, as shown in Equation 13.3.

$$F_3 = P_0(1+i)(1+i)(1+i) \tag{3.3}$$

Equation 3.3 is repeated for each period by multiplying the principal or present worth by $(1 + i)$ for the total number of periods resulting in the formula for future worth—the single payment compound amount factor—which is in Equation 13.4.

$$F_n = P_0(1+i)^n \tag{3.4}$$

where
- F_n is the future worth at the end of the nth period
- P_0 is the principal
- i is the interest rate
- n is the number of periods

3.2.1 Example Problems: Solving for Future Worth

This section provides example problems demonstrating the procedure for calculating the future worth of a present value.

Example 3.1

An engineer borrows $1,000.00 at 5% interest for two years. What is the future worth of this transaction at the end of year two? Equation 3.4 is used to calculate the future worth at year two, which represents the principal to be repaid at the end of year two and the interest owed on the principal for two years. Figure 3.1 shows the cash flow diagram for the future worth.

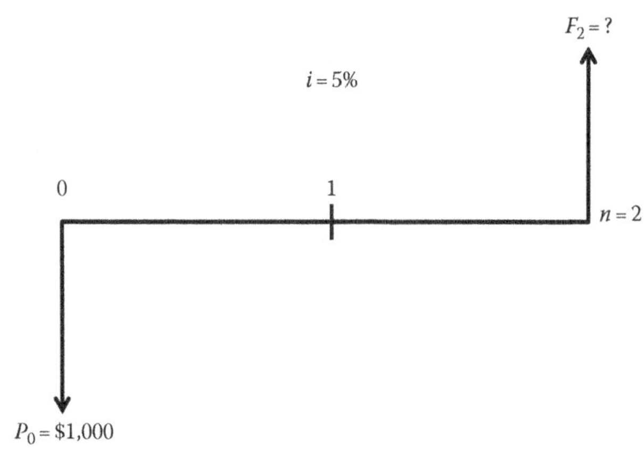

FIGURE 3.1 Cash flow diagram for solving for the future worth for Example 3.1.

Solution

Solve for the future worth at year two using Equation 3.4:

$$\begin{aligned} F_2 &= P_0(1+i)^n \\ &= \$1{,}000.00(1+0.05)^2 \\ &= \$1{,}000.00(1.1025) \\ &= \$1{,}102.50 \end{aligned}$$

At the end of two years, the engineer owes the original principal of $1,000.00 plus $102.50 in interest.

Example 3.2

A bank loans a student $7,000.00 at 8% interest. The student signed a loan agreement indicating he will repay the loan at the end of five years. What is the future worth of this transaction at the end of five years? Figure 3.2 is the cash flow diagram for the future worth of the loan.

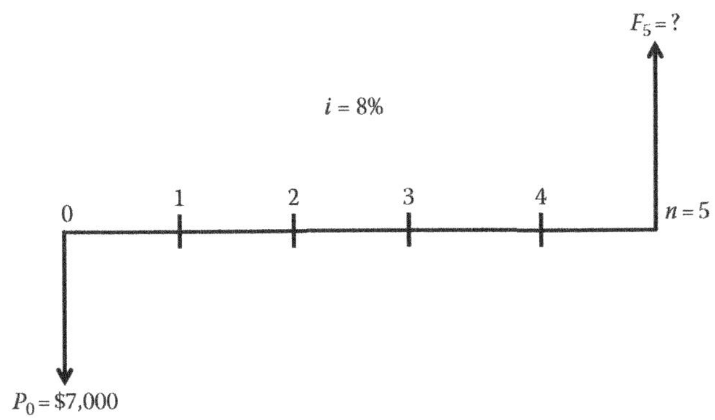

FIGURE 3.2 Cash flow diagram for solving for the future worth of the loan in Example 3.2.

Solution

$$\begin{aligned} F_5 &= P_0(1+i)^n \\ &= \$7{,}000.00(1+0.08)^5 \\ &= \$7{,}000.00(1.4693) \\ &= \$10{,}285.30 \end{aligned}$$

At the end of five years, the student owes the original $7,000.00 borrowed from the bank plus an additional $3,285.30 in interest.

Example 3.3

A bank loans $125,000.00 to an electronics firm at an interest rate of 13.5% for 14 years. How much will the bank be repaid at the end of 14 years? Figure 3.3 is the cash flow diagram for the loan borrowed by the electronics firm.

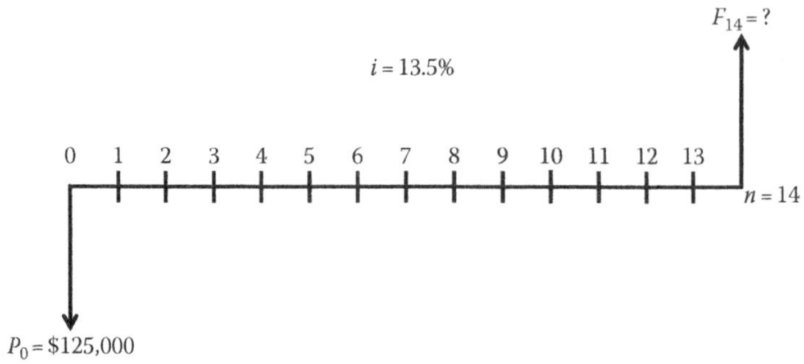

FIGURE 3.3 Cash flow diagram for the loan borrowed by the electronics firm in Example 3.3.

Solution

$$\begin{aligned} F_{14} &= P_0(1+i)^n \\ &= \$125,000.00(1+0.135)^{14} \\ &= \$125,000.00(5.8876) \\ &= \$735,950.00 \end{aligned}$$

Example 3.4

An investment company loans \$292,500.00 to a computer software start-up firm. The investment company charges the computer software firm 4% interest for the loan and the terms of the loan indicate a repayment period of nine years. How much will the investment company be repaid at the end of the nine years? Figure 3.4 is the cash flow diagram for the future worth of the investment company loan.

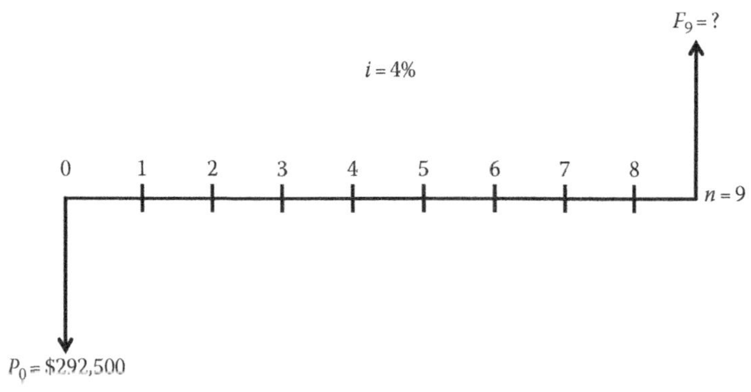

FIGURE 3.4 Cash flow diagram for the loan by the investment company in Example 3.4.

Solution

$$\begin{aligned} F_9 &= P_0(1+i)^n \\ &= \$292,500.00(1+0.04)^9 \\ &= \$292,500.00(1.4233) \\ &= \$416,315.25 \end{aligned}$$

3.2.2 Solving for Future Worth Using Continuous Compounding

In some situations, funds are invested in accounts were the interest is calculated on a continuous basis and this is *continuous compounding*. Continuous compounding is advantageous if someone is receiving interest on an investment since the interest increases faster than nominal or compound interest. Continuous compounding is disadvantageous if someone has to pay the interest being compounded continually. Equation 3.5 is the formula for calculating the future worth in situations where there is continuous compounding, and this formula is the single payment compound amount factor for continuous compounding:

$$F_n = P_0 e^{i_n \times n} \tag{3.5}$$

where

P is the principal

e is 2.718281828… [Euler's number $\left(1+\frac{1}{n}\right)^n$ as n approaches infinity]

i_n is the nominal interest rate

n is the number of periods or years

Example 3.5 solves for the future worth of an investment when the interest is continuously compounded during the life of the investment.

Example 3.5

What is the future worth of $1,000.00 invested for 10 years with an interest rate of 18% compounded continuously? Figure 3.5 is the cash flow diagram for the future worth of the investment.

FIGURE 3.5 Cash flow diagram for the continuously compounded interest in Example 3.5.

Solution

$$F_{10} = Pe^{i_n \times n}$$
$$= \$1,000.00e^{0.18 \times 10}$$
$$= \$1,000.00 \times 6.04965$$
$$= \$6,049.65$$

3.2.3 Using the Interest Factor Tables to Solve for Future Worth

In addition to solving for future worth using the single payment compound amount factor, interest factor tables may be used to calculate the future worth of a present value. The interest factor tables are included in Appendix B. All of the single payment compound amount factors (*F/P*) in Appendix B represent the calculation of $(1 + i)^n$ and they may be used rather than having to calculate this factor using the formula. Equation 3.6 is the formula for calculating the future worth of a present value when using the interest factor tables:

$$F_n = P_0\left(F/P, i, n\right) \tag{3.6}$$

The following steps outline the procedure for solving for the future worth of a present value using the interest factor tables in Appendix B and the information provided in Example 3.4:

1. In Appendix B locate the interest factor table for 4%.
2. Locate the third section on the right-hand side of the interest factor table which is the *future sum factors.*
3. Locate the column under the future sum factors column labeled (*F/P*).
4. In the column labeled (n), which is the number of periods, locate the number 9.
5. The (*F/P*) factor for 9 years ($n = 9$) at 4% is 1.4233, as shown in the (*F/P*) column.
6. Multiple 1.4233 times the present value of \$292,500.00.
7. The resulting future worth from Step 6 is \$416,315.25.

The formulas for the previous steps are the following:

$$F_9 = P_0\left(F/P, i, n\right)$$
$$= \$292{,}500.00\left(F/P, 4, 9\right)$$
$$= \$292{,}500.00(1.4233)$$
$$= \$416{,}315.25$$

In order to solve for future worth using the interest factor tables in Appendix B, the principal (P) must be known along with the interest rate (i) and the number of payment periods (n). With all three of these values, the correct interest factor table is located in Appendix B and the (*F/P*) value is extracted from the table and multiplied by the initial principal.

Example 3.6 is a problem that uses the interest factor tables in Appendix B to solve for future worth.

Example 3.6

An aerospace engineer invests \$10,000.00 and he will be able to withdraw his investment and interest in seven years. If the interest rate is 11%, how much money would he be able to withdraw from the account in seven years? Use the interest factor tables in Appendix B to solve for the future worth. Figure 3.6 is the cash flow diagram for the investment made by the aerospace engineer.

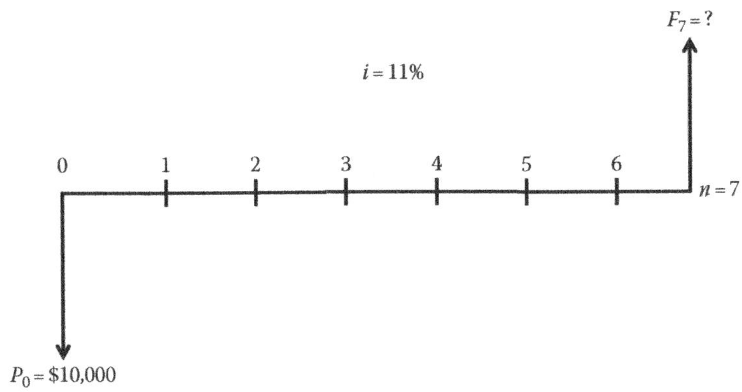

FIGURE 3.6 Cash flow diagram for the investment by the aerospace engineer in Example 3.6.

Solution

$$F_n = P_0(F/P,i,n)$$

The *F/P* factor in Appendix B for $i = 11\%$ and $n = 7$ years is 2.0761.

Multiply the principal of \$10,000.00 by the (*F/P*) factor of 2.0761 to solve for the future worth:

$$\begin{aligned} F_7 &= P_0(F/P,11,7) \\ &= \$10,000.00(2.0761) \\ &= \$20,761.00 \end{aligned}$$

3.3 PRESENT WORTH: PRESENT WORTH COMPOUND AMOUNT FACTOR (*P/F*)

This section explains the process for calculating the present worth of future values and net present worth.

3.3.1 Present Worth of Future Values

The *present worth compound amount factor* (*P/F*) calculates the present worth of a future sum by removing the interest from the future value. The formula for solving for present worth is the reciprocal of the future worth single payment compound amount factor introduced in Section 3.2 and it is Equation 3.7:

Single payment compound amount factor (*F/P*)

$$F_n = P_0(1+i)^n$$

Inverse

Present worth compound amount factor (*P/F*)

$$P_0 = F_n\left[\frac{1}{(1+i)^n}\right] \tag{3.7}$$

where

P_0 is present worth at time zero
F_n is future worth and the end of the nth period
i is interest rate
n is number of periods

The formula for solving for the present worth using the interest factor tables in Appendix B is

$$P_0 = F_n\left(P/F, i, n\right) \tag{3.8}$$

Once again, as with all of the engineering economic factors, the factor being solved for is the numerator in the brackets, and in this case, the factor is (P), the present worth. The denominator is the value being converted to the equivalent value.

Example 3.7 demonstrates calculating the present worth of a future sum using Equations 3.7 and 3.8.

Example 3.7

A municipal engineer deposits funds into an interest-bearing account to pay for a new generator that will cost \$10,000,000.00 in 10 years. The generator will be installed in a wastewater treatment system. If she is able to invest the funds at an interest rate of 6% and she deposits the money into the account now, how much does she need to deposit into the account? Figure 3.7 shows the cash flow diagram representing the investment made by the municipal engineer.

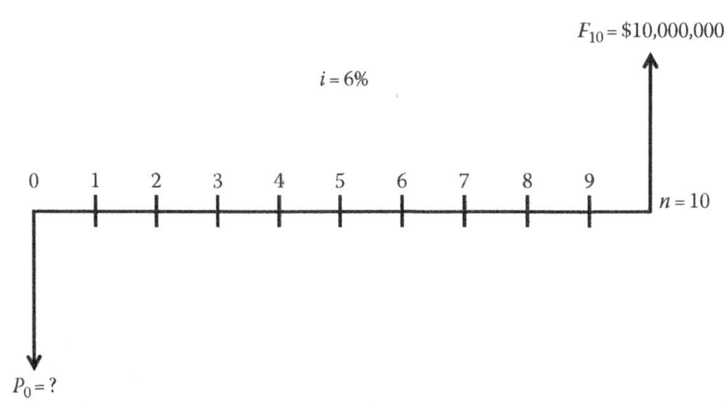

FIGURE 3.7 Cash flow diagram for the new generator in Example 3.7.

Solution

Use Equation 3.7 to calculate the present worth of the desired future sum:

$$P_0 = F_{10}\left[\frac{1}{(1+i)^n}\right]$$
$$= \$10{,}000{,}000.00\left[\frac{1}{(1+0.06)^{10}}\right]$$
$$= \$10{,}000{,}000.00\left(\frac{1}{1.79084}\right)$$
$$= \$10{,}000{,}000.00(0.55839)$$
$$= \$5{,}583{,}900.00$$

Solving for the present worth using the interest factor tables in Appendix B is accomplished by using Equation 3.8:

$$P_0 = F_{10}(P/F,i,n)$$
$$= \$10{,}000{,}000.00(P/F,6,10)$$
$$= \$10{,}000{,}000.00(0.55839)$$
$$= \$5{,}583{,}900.00$$

Example 3.8 provides another problem where the present worth compound amount factor in Equations 3.7 and 3.8 is used for calculating present worth.

Example 3.8

What amount of money could be borrowed from a bank charging 8% interest in order for the bank to be repaid $10,283.00 at the end of five years? Figure 3.8 is the cash flow diagram for the present worth of the borrowed funds.

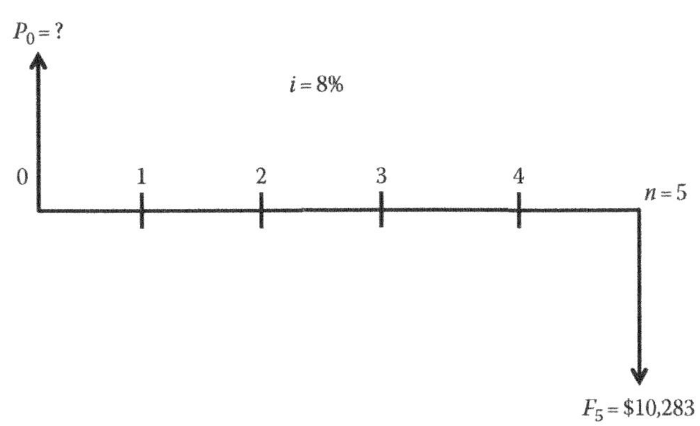

FIGURE 3.8 Cash flow diagram for borrowing funds from the bank in Example 3.8.

Solution

Use Equation 3.7 to calculate the present worth of the future value:

$$P_0 = F_5\left[\frac{1}{(1+i)^n}\right]$$
$$= \$10{,}283.00\left[\frac{1}{(1+0.08)^5}\right]$$
$$= \$10{,}283.00\left(\frac{1}{1.469328}\right)$$
$$= \$10{,}283.00(0.68058)$$
$$= \$6{,}998.40$$

Use Equation 3.8 and the interest factor tables in Appendix B to calculate the present worth of the future value:

$$P_0 = F_5(P/F,i,n)$$
$$= \$10{,}283.00(P/F,8,5)$$
$$= \$10{,}283.00(0.68058)$$
$$= \$6{,}998.40$$

3.3.2 Net Present Worth

In addition to being able to solve for the present worth of a future value using the present worth compound amount factor (*P/F*), this factor is also used when solving for the *net present worth* (NPW) of positive and negative present and future values occurring at different times during the period of analysis. Net present worth is the sum of any positive and negative present values plus or minus the present worth of all future values.

Solving for the net present worth allows a decision maker to determine whether an anticipated investment will have a positive equivalent cash flow rather than the investment having a negative yield. The formula for solving for net present worth is

$$\text{NPW} = \pm P_0 \pm \sum F\left[\frac{1}{(1+i)^n}\right] \tag{3.9}$$

The formula for solving for net present worth using the interest factor tables in Appendix B is

$$\text{NPW} = \pm P_0 \pm \sum F(P/F,i,n) \tag{3.10}$$

Example 3.9 demonstrates calculating the net present worth of a present and future value.

Example 3.9

Determine the net present worth of purchasing a new vehicle for \$30,000.00 and selling it for an estimated salvage value of \$9,000.00 in five years if the interest rate is 12%. Figure 3.9 is the cash flow diagram for purchasing the new vehicle.

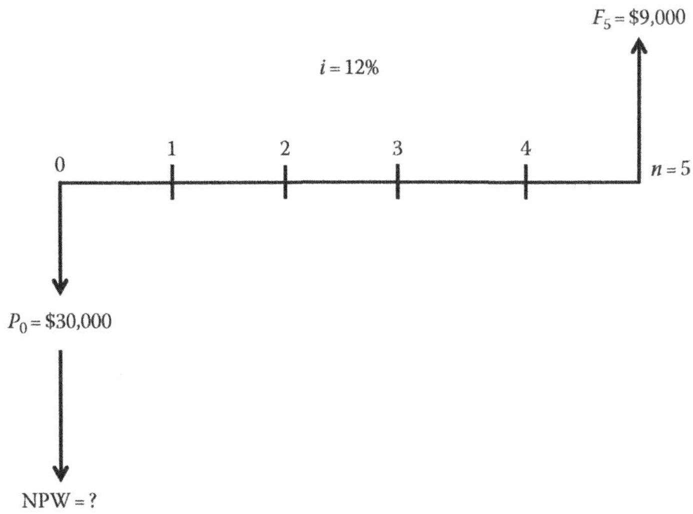

FIGURE 3.9 Cash flow diagram for purchasing the new vehicle in Example 3.9.

Solution

Calculate the net present worth using Equation 3.9:

$$\begin{aligned}
\text{NPW} &= -P_0 + F_5\left[\frac{1}{(1+i)^n}\right] \\
&= -\$30,000.00 + \$9,000.00\left[\frac{1}{(1+0.12)^5}\right] \\
&= -\$30,000.00 + \$9,000.00\left(\frac{1}{1.762342}\right) \\
&= -\$30,000.00 + \$9,000.00(0.56743) \\
&= -\$30,000.00 + \$5,106.87 \\
&= -\$24,893.13
\end{aligned}$$

Calculate the net present worth using the interest factor tables and Equation 3.10:

$$\begin{aligned}
\text{NPW} &= \pm P_0 \pm F_5\,(P/F,i,n) \\
&= -\$30,000.00 + \$9,000.00(P/F,12,5) \\
&= -\$30,000.00 + \$9,000.00(0.56743) \\
&= -\$30,000.00 + \$5,106.87 \\
&= -\$24,893.13
\end{aligned}$$

Another example demonstrating calculating net present worth is Example 3.10.

Example 3.10

The manager of a mechanical engineering firm invests \$23,000.00 in a new product. At the end of three years, the firm receives a return of \$28,500.00 from the investment. If the interest rate is 4%, what is the net present worth of the investment? Figure 3.10 is the cash flow diagram for the mechanical engineering new product.

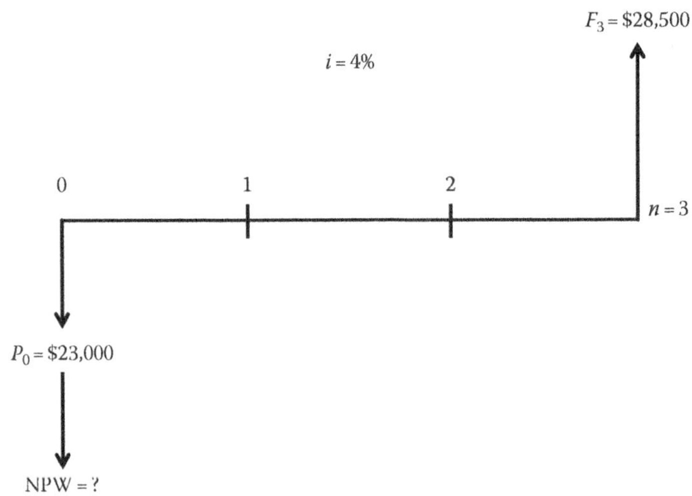

FIGURE 3.10 Cash flow diagram for purchasing the new product in Example 3.10.

Solution

$$\begin{aligned}\text{NPW} &= -P_0 + F_3\left[\frac{1}{(1+i)^n}\right]\\ &= -\$23{,}000.00 + \$28{,}500.00\left[\frac{1}{(1+0.04)^3}\right]\\ &= -\$23{,}000.00 + \$28{,}500.00\left(\frac{1}{1.124864}\right)\\ &= -\$23{,}000.00 + \$28{,}500.00(0.88900)\\ &= -\$23{,}000.00 + \$25{,}336.50\\ &= \$2{,}336.40\end{aligned}$$

Solution using the interest factor tables in Appendix B:

$$\begin{aligned}\text{NPW} &= -P_0 + F_3(P/F, i, n)\\ &= -\$23{,}000.00 + \$28{,}500.00(P/F, 4, 3)\\ &= -\$23{,}000.00 + \$28{,}500.00(0.88900)\\ &= -\$23{,}000.00 + \$25{,}336.50\\ &= \$2{,}336.50\end{aligned}$$

Example 3.11 is another problem solving for net present worth.

Example 3.11

An engineer borrows \$5,500.00 from a bank and has to repay \$6,000.00 at the end of two years. If the interest rate is 2%, what is the net present worth of the amount borrowed from the bank? Figure 3.11 is the cash flow diagram for the loan borrowed by the engineer.

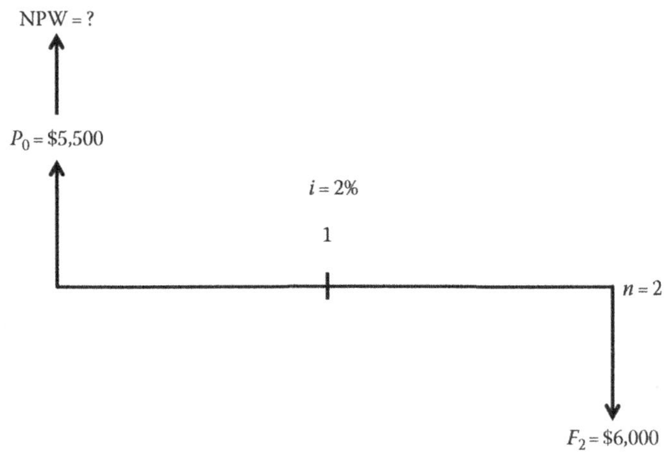

FIGURE 3.11 Cash flow diagram for borrowing money from the bank in Example 3.11.

Solution

$$\begin{aligned}\text{NPW} &= P_0 - F_2\left[\frac{1}{(1+i)^n}\right]\\ &= \$5,500.00 - \$6,000.00\left[\frac{1}{(1+0.02)^2}\right]\\ &= \$5,500.00 - \$6,000.00\left(\frac{1}{1.0404}\right)\\ &= \$5,500.00 - \$6,000.00(0.96116)\\ &= \$5,500.00 - \$5,766.96\\ &= -\$266.96\end{aligned}$$

Solution using the interest factor tables in Appendix B:

$$\begin{aligned}\text{NPW} &= P_0 - F_2\,(P/F,i,n)\\ &= \$5,500.00 - \$6,000.00(P/F,2,2)\\ &= \$5,500.00 - \$6,000.00(0.96116)\\ &= \$5,500.00 - \$5,766.96\\ &= -\$266.96\end{aligned}$$

3.4 COMPOUNDING PERIODS DIFFERENT THAN ONE YEAR

This section addresses situations where the compounding periods are different than one year for future and present worth factors.

3.4.1 Compounding Periods for Interest Different than Payment Period for Future Worth

In all of the previous examples, the compounding period for the interest was yearly since no other compounding period was specified in the problem statements. But in many instances, the compounding period is different than one year, and this has to be accounted for when solving for either future or present worth.

In order to compensate for more than one compounding period each year, the yearly interest rate is divided by the number of compounding periods per year. Examples of this would be four for quarterly, two for semiannually, or 12 for monthly. In addition to converting the interest rate into the interest rate per compounding period, the number of years is multiplied by the number of compounding periods per year such as four for quarterly, two for semiannually, and 12 for monthly. Equation 3.11 shows the formula for solving for the interest rate per compounding period:

$$i = \frac{i_n}{m} \tag{3.11}$$

where

i is the interest rate per compounding period
i_n is the nominal interest rate (interest rate per year)
m is the number of compounding periods per year

Equation 3.12 is the formula for calculating the total number of compounding periods:

$$n = \frac{\text{Number of compounding periods}}{\text{Year}} \times \text{Total number of years} \tag{3.12}$$

The following examples illustrate using these conversion formulas.

Example 3.12 incorporates multiple compounding periods into a future worth problem.

Example 3.12

Find the future worth of $3,000.00 deposited into an interest-bearing account paying 8% per year, compounded quarterly for five years. Figure 3.12 is the cash flow diagram for the future worth of the deposit.

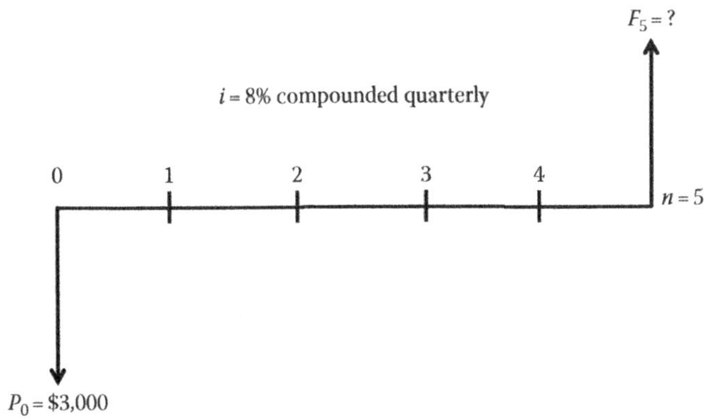

FIGURE 3.12 Cash flow diagram for the funds compounded quarterly in Example 3.12.

Solution

First, solve for the interest rate per compounding period using Equation 3.11:

$$\begin{aligned} i &= \frac{i_n}{m} \\ &= \frac{8\%/\text{year}}{4\ \text{periods/year}} \\ &= 2\%\ \text{per period} \end{aligned}$$

Second, solve for the total number of compounding periods using Equation 3.12:

$$\begin{aligned} n &= \frac{\text{Number of compounding periods}}{\text{Year}} \times \text{Total number of years} \\ &= \frac{4\ \text{periods}}{\text{Year}} \times 5\ \text{years} \\ &= 20\ \text{periods} \end{aligned}$$

Third, calculate the future worth at the end of the total number of years using the interest rate per compounding period and the total number of compounding periods:

$$\begin{aligned} F_5 &= P_0(1+i)^n \\ &= \$3{,}000.00(1+0.02)^{20} \\ &= \$3{,}000.00(1.4859) \\ &= \$4{,}457.70 \end{aligned}$$

This problem could also be solved using the interest factor tables in Appendix B:

$$
\begin{aligned}
F_5 &= P_0\left(F/P,i,n\right)\\
&= \$3{,}000.00\left(F/P,2,20\right)\\
&= \$3{,}000.00(1.4859)\\
&= \$4{,}457.70
\end{aligned}
$$

Example 3.13 demonstrates calculating future worth if the compounding period is daily.

Example 3.13

What is the future worth of an initial investment of \$3,000.00 for five years at an interest rate of 8% compounded daily? Figure 3.13 is the cash flow diagram for the future worth of the investment.

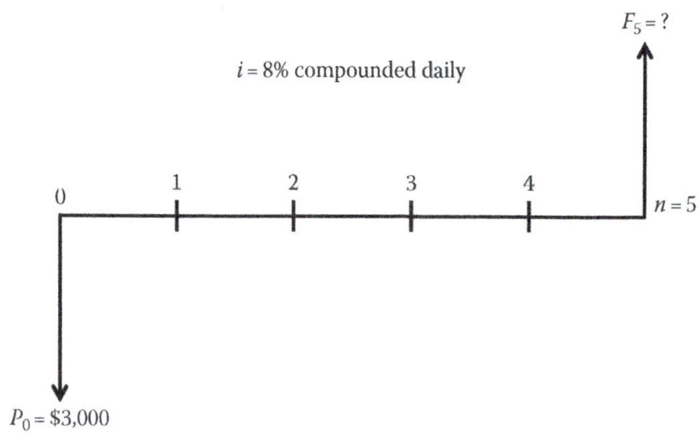

FIGURE 3.13 Cash flow diagram for the funds compounded daily in Example 3.13.

Solution

First, solve for the daily interest rate using Equation 3.11:

$$
\begin{aligned}
i &= \frac{i_n}{m}\\
&= \frac{8\%/\text{year}}{365\ \text{periods/year}}\\
&= 0.021918\%\ \text{per period}
\end{aligned}
$$

Second, solve for (n) using Equation 3.12:

$$
\begin{aligned}
n &= \frac{\text{Number of compounding periods}}{\text{Year}} \times \text{Total number of years}\\
&= \frac{353\ \text{periods}}{\text{Year}} \times 5\ \text{years}\\
&= 1{,}825\ \text{periods}
\end{aligned}
$$

Third, calculate the future worth using the appropriate (i) and (n):

$$F_{1,825} = P_0(1+i)^n$$
$$= \$3,000.00(1+0.00022)^{1,825}$$
$$= \$3,000.00(1.4940)$$
$$= \$4,482.00$$

3.4.2 Compounding Period for Interest Different than Payment Period for Present Worth

When solving for the present worth in situations where the compounding period is different than the payment period, the approach is similar to the method presented in Section 3.4.1 for solving for future worth. The yearly interest rate is divided by the number of compounding periods per year (Equation 3.11). The number of compounding periods per year is multiplied by the total number of years (Equation 3.12). Example 3.14 solves for the present worth when the compounding period is different than the payment period.

Example 3.14

How much money needs to be deposited into a certificate of deposit (CD) account to be able to withdraw $50,000.00 from the account at the end of 10 years if the interest rate is 2% compounded monthly? Figure 3.14 shows the cash flow diagram for the CD.

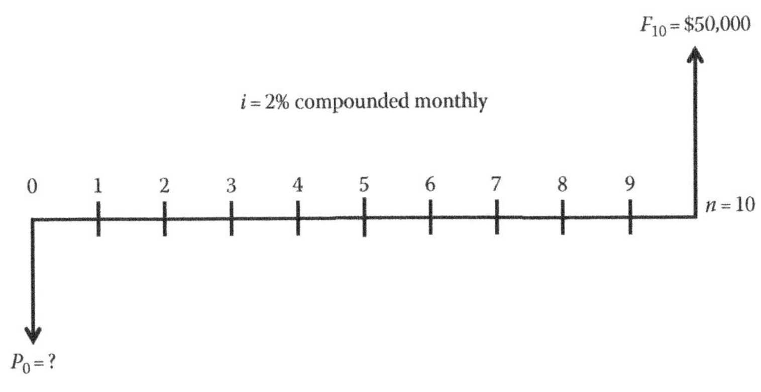

FIGURE 3.14 Cash flow diagram for the certificate of deposit investment in Example 3.14.

Solution

First, use Equation 3.11 to solve for the interest rate per compounding period:

$$i = \frac{i_n}{m}$$
$$= \frac{2\%/\text{year}}{12\ \text{periods/year}}$$
$$= 0.1667\%\ \text{per period}$$

Second, use Equation 3.12 to solve for the total number of compounding periods:

$$
\begin{aligned}
n &= \frac{\text{Number of compounding periods}}{\text{Year}} \times \text{Total number of years} \\
&= \frac{12 \text{ periods}}{\text{Year}} \times 10 \text{ years} \\
&= 120 \text{ periods}
\end{aligned}
$$

Third, calculate the present worth using the appropriate interest rate and number of compounding periods:

$$
\begin{aligned}
P_0 &= F_{120}\left[\frac{1}{(1+i)^n}\right] \\
&= \$50{,}000.00\left[\frac{1}{(1+0.001667)^{120}}\right] \\
&= \$50{,}000.00\left(\frac{1}{1.221248}\right) \\
&= \$50{,}000.00(0.8188) \\
&= \$40{,}940.00
\end{aligned}
$$

Example 3.15 provides another problem solving for the present worth with a compounded interest rate.

Example 3.15

If the interest rate in Example 3.10 of 4% was compounded quarterly, what would be the net present worth of the investment? Figure 3.15 is the cash flow diagram for the present worth of the investment.

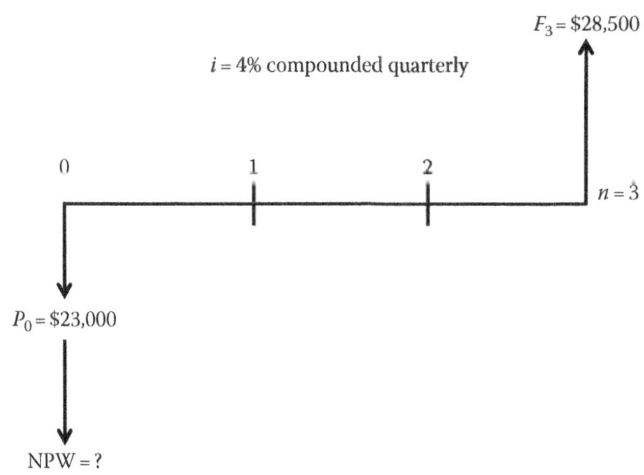

FIGURE 3.15 Cash flow diagram for funds compounded quarterly in Example 3.15.

Solution

First, use Equation 3.11 to solve for the quarterly interest rate:

$$i = \frac{i_n}{m}$$

$$i = \frac{4\%/\text{year}}{4\ \text{periods/year}}$$

$$= 1\%\ \text{per period}$$

Second, use Equation 3.12 to calculate the total number of compounding periods:

$$n = \frac{\text{Number of compounding periods}}{\text{Year}} \times \text{Total number of years}$$

$$n = \frac{4\ \text{periods}}{\text{Year}} \times 3\ \text{years}$$

$$= 12\ \text{periods}$$

Third, use the quarterly interest rate and the total number of compounding periods per year to calculate the net present worth using Equation 3.9:

$$\text{NPW} = -P + F_3\left[\frac{1}{(1+i)^n}\right]$$

$$= -\$23{,}000.00 + \$28{,}500.00\left[\frac{1}{(1+0.01)^{12}}\right]$$

$$= -\$23{,}000.00 + \$28{,}500.00\left(\frac{1}{1.126825}\right)$$

$$= -\$23{,}000.00 + \$28{,}500.00(0.8875)$$

$$= -\$23{,}000.00 + \$25{,}293.75$$

$$= \$2{,}293.75$$

3.5 SOLVING FOR UNKNOWN INTEREST RATES: RATE OF RETURN

When individuals or managers of firms make decisions on whether to invest in a project, a major factor in the decision process is the *rate of return* (*ROR*) on the investment. The definition for ROR was provided in Section 2.2.6 and it is the following:

$$\text{ROR (percent)} = \frac{\text{Total amount of money received} - \text{Original investment}}{\text{Original investment}} \times 100\%$$

or

$$\text{ROR (percent)} = \frac{\text{Profit}}{\text{Original investment}} \times 100\%$$

In order to incorporate the time value of money into ROR calculations, the calculations are based on the formulas for solving for present worth (*P*/*F*) or future worth (*F*/*P*). The ROR is the interest rate where the present worth and future worth are equal.

3.5.1 Solving for Unknown Rates of Return Using Formulas

The derivation of the formula for solving for the *unknown ROR* (i) is shown in steps 1 through 3 and Equation 3.13 is the formula for solving for unknown interest rates (ROR):

1. $F_n = P_0(1+i)^n$

2. $\sqrt[n]{\frac{F_n}{P_0}} = 1+i$

3. $i = \sqrt[n]{\frac{F_n}{P_0}} - 1$ (3.13)

Example 3.16 demonstrates the process for using Equation 3.13 to calculate an unknown interest rate.

Example 3.16

A nuclear engineer invests $55,000.00 in 2017 and his investment will be worth $65,000.00 in 2021. What is the rate of return on this investment? Figure 3.16 is the cash flow diagram for the investment of the nuclear engineer.

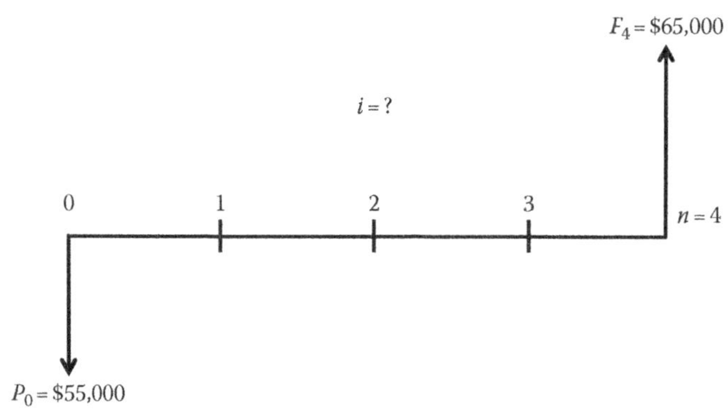

FIGURE 3.16 Cash flow diagram for the unknown rate of return in Example 3.16.

Solution

Use Equation 3.13 to solve for the unknown ROR:

$$i = \sqrt[n]{\frac{F_n}{P_0}} - 1$$

$$= \sqrt[4]{\frac{\$65{,}000.00}{\$55{,}000.00}} - 1$$

$$= \sqrt[4]{1.181818} - 1$$

$$= 1.042647 - 1$$

$$= 4.26\%$$

3.5.2 Solving for Unknown Rates of Return Using the Interest Factor Tables

An unknown ROR may also be calculated by using the interest factor tables in Appendix B and *interpolation*. The formula for interpolating the ROR is

$$i = c + \left(\frac{a}{b}\right)d \tag{3.14}$$

The format for interpolation is shown in Table 3.1.

TABLE 3.1
Table for Developing Interpolation Problems for Unknown Rate of Return

	i	*F/P*		
d [	c	*A*]	*a*	]
	Unknown *i*	*B*]		*b*
	e	*C*		]

Note: $a = A - B$; $b = A - C$; $d = c - e$.

For the data in Example 3.16, the ROR is calculated in Example 3.17 using the interest factor tables in Appendix B.

Example 3.17

Calculate the rate of return in Example 3.16 using the interest factor tables. The cash flow diagram for Example 3.17 is in Figure 3.16.

Solution

From Example 3.16, the single payment compound amount factor (*F/P*) is 1.181818 and $n = 4$.

Locating the (*F/P*) factor closest to 1.181818 in Appendix B at $n = 4$ indicates the interest rate is between 4% and 5%. Use interpolation to solve for the interest rate.

First, set up the interest table in Table 3.2 using the format from Table 3.1.

TABLE 3.2
Solving for the Unknown Interest Rate for *F/P* of 1.181819 and *n* = 4 Years

	i	*F/P*		
d [	4	1.1698]	*a*	]
	Unknown *i*	1.1818]		*b*
	5	1.2155		]

Second, use Equation 3.14 to solve for (i):

$$\begin{aligned} i &= c + \left(\frac{a}{b}\right)d \\ &= 4 + \left(\frac{1.1818 - 1.1698}{1.2155 - 1.1698}\right)(5 - 4) \\ &= 4 + \left(\frac{0.0120}{0.4570}\right) \times 1 \\ &= 4 + (0.026258 \times 1) \\ &= 4 + 0.26248 \\ &= 4.26\% \end{aligned}$$

The process for solving for an unknown ROR shown in Example 3.17 could be used to solve for the ROR for any type of problem using the single payment compound amount factor (*F/P*).

3.6 SOLVING FOR UNKNOWN NUMBER OF PERIODS

In some instances, a decision maker needs to solve for the number of periods (or years) over which the investor or firm should hold onto an investment. To calculate the unknown number of periods, the format in Table 3.3 and the interest factor tables in Appendix B are used along with interpolation formula in Equation 3.15.

TABLE 3.3
Format for Developing Interpolation Problems for Unknown Period

	n	*P/F*		
d [	c	A]	*a*]	*b*
	Unknown *n*	B		
	e	*C*		

Note: $a = A - B$; $b = A - C$; $d = c - e$.

$$n = c + \left(\frac{a}{b}\right)d \tag{3.15}$$

Example 3.18 provides an investment scenario where the number of years is unknown and it is solved using Equation 3.15.

Example 3.18

How long would it take for $1,000.00 to double in value if the interest rate is 5%?

Solution

Use the single payment present worth factor (*P/F*) and the interest factor tables in Appendix B to solve for the unknown number of years.

First, calculate the (*P*/*F*) factor:

$$P_0 = F_n\left(P/F,i,N\right)$$

$$\$1{,}000.00 = \$2{,}000.00\left(P/F,5,N\right)$$

$$\left(P/F,5,N\right) = 0.50$$

According to the interest factor table for 5% in Appendix B, the (*P*/*F*) factor is between 14 and 15 years. Table 3.4 is the interpolation table for unknown years. Use Equation 3.15 to solve for the number of years:

TABLE 3.4
Table for Solving for Unknown Years of (*P*/*F*) of 0.50 at 5% Interest

	n	(*P*/*F*)	
d [	14	0.50507	] *a*] *b*
	Unknown *n*	0.50000	
]	15	0.48102	]

$$n = c + \left(\frac{a}{b}\right)d$$

$$= 14 + \left(\frac{0.50507 - 0.50}{0.50507 - 0.48102}\right)(15-14)$$

$$= 14 + \left(\frac{0.00507}{0.02405}\right)\times 1$$

$$= 14 + 0.210811$$

$$= 14.21 \text{ years}$$

Appendix C includes spreadsheet formulas for calculating present and future worth.

3.7 SUMMARY

This chapter provided a definition for equivalence and explained why it is necessary to compare alternatives based on equivalent terms. It explained why equivalent terms are calculated to compare alternatives occurring over different time frames and having different interest rates. The single payment compound amount factor (*F*/*P*) was introduced along with an explanation of the procedure for calculating the future worth of present values. Steps were included demonstrating the process for using the interest factor tables in Appendix B for solving for the future worth of a present value.

The process for calculating the present worth of a future value using the present worth compound amount factor (*P*/*F*) was discussed and an explanation was provided on how net present worth calculations determine whether an investment results in an equivalent positive or negative present worth. The procedures for calculating the future worth of a present value and the present worth of a future value were provided for situations where the compounding period for the interest is different than the payment period.

The last part of this chapter: (1) explained the process for solving for unknown interest rates using formulas or the interest factor tables in Appendix B and interpolation; and (2) covered the process for calculating unknown number of years using the interest factor tables in Appendix B and interpolation.

KEY TERMS

Continuous compounding
Equivalence
Future worth Interpolation
Net present worth
Present worth compound amount factor
Rate of return
Single payment compound amount factor
Unknown rate of return

PROBLEMS

3.1 Explain equivalence and how it relates to engineering economics.

3.2 What is the factor for solving for the future worth of a present value called and what is the formula?

3.3 What is the formula for solving for the future worth of a present value if the interest rate is continuously compounding?

3.4 How are the interest factor tables in Appendix B used to solve for future worth?

3.5 What is the factor called for calculating the present worth of a future value and what is the formula?

3.6 Explain the process for calculating the net present worth of present and future values.

3.7 Explain how to solve for the future worth if compounding periods are different than payment periods and what is the formula for solving for the interest rate per compounding period.

3.8 Explain why time value of money is considered in engineering economic problems.

3.9 What is the formula for solving for unknown interest rates?

3.10 What is the formula for solving for unknown interest rates using interpolation?

3.11 What is the formula for solving for unknown number of years using interpolation?

3.12 A firm manufacturing nuclear submarines has an opportunity to purchase a new type of turbine. They may either purchase the turbine now for $10,000,000.00 or at year three. If the interest rate is 6%, what is the equivalent future worth at year three of the turbine?

3.13 A vehicle manufacturer enters into a contract with a parts manufacturer to purchase parts for $75,000,000.00 when the parts will be ready at year two. What is the present worth of the contract if the interest rate is 5%?

3.14 How much could a manufacturer of thermal expansion joints afford to spend to purchase equipment in seven years rather than purchasing the equipment now for $675,000.00 if the interest rate is 8%?

3.15 An electrical engineering company could reduce the amount of recalled electrical parts if it purchases a new piece of equipment to manufacture the parts. The engineer assigned to the project needs to determine whether the present worth of the future cost savings of $327,000.00 at year four justifies the cost of purchasing the equipment now for $297,000.00. Calculate the present worth of the future cost savings to determine whether it will be more than the cost of the equipment if the interest rate is 7%. Solve using the net present worth formula.

3.16 A manufacturing firm plans to purchase a computer network control (CDC) machine for \$650,000.00. If the firm makes a profit from the products of the machine of \$265,000.00 at years three and four and the firm will be able to sell the machine at year four for \$125,000.00 (salvage value), is the firm able to recover the cost of the machine if the interest rate is 4%? Solve using the net present worth formula.

3.17 An agriculture company plans on buying a new tractor for \$1,125,000.00. The company is able to use the tractor for 12 years and then they will sell it for a salvage value of \$750,000.00. What is the net present worth of the tractor if the interest rate is 4%?

3.18 A software company purchases a new computer for \$176,000.00. What is the future worth of the computer at year three if the interest rate is 3%?

3.19 An investment pays 5% compounded monthly. What is the net present worth of the investment if \$52,000.00 is invested now and it is worth \$63,500.00 in four years?

3.20 Calculate the future worth of \$30,000.00 deposited into an interest-bearing account paying 4% interest per year, compounded quarterly for five years.

3.21 A petroleum engineer is calculating the rate of return on a potential purchase for the company before presenting it to his boss for consideration. If the initial cost is \$267,000.00 and it will result in a projected profit of \$301,000.00 after six years, what is the rate of return on the investment? Solve using the rate of return formula.

3.22 Solve for the rate of return in Problem 3.21 using interpolation and the interest factor tables in Appendix B.

3.23 A systems engineer invests \$65,000.00 for his company into an account and the money is worth \$90,000.00 at 10 years. What is the rate of return on the investment? Solve using the rate of return formula.

3.24 Solve for the rate of return in Problem 3.22 using the interest factor tables in Appendix B and $n = 4$ years.

3.25 A computer consulting company purchases computer parts to build a new computer system. If the parts cost \$35,500.00 and the computer will sell for \$47,500.00 in two years, what is the rate of return on the investment? Solve using the rate of return equation.

3.26 Solve for the rate of return in Problem 3.25 using the interest factor tables in Appendix B.

3.27 An electrical engineer is calculating the number of years it would take an investment of \$10,500.00 to triple if the interest rate is 9%. Solve for the unknown number of years using interpolation and the interest factor tables in Appendix B.

3.28 A systems engineer installs a new piece of equipment costing \$372,000.00 into a new processing plant, and by installing the new equipment, the company is able to sell the processing plant for \$512,000.00. If the interest rate is 6%, how many years do they have to wait before selling the plant in order for the selling price to cover the cost of the new equipment? Solve using interpolation and the interest factor tables in Appendix B.

3.29 An environmental engineering firm is financing a site reclamation project costing \$15,700,000.00. They are able to borrow the money for the project at 12% interest compounded quarterly and repay the loan at the end of seven years. How much will the firm have to repay at the end of seven years?

3.30 A student borrows \$2,500.00 to purchase a new computer. If she borrows the money at 9% interest compounded semiannually and has to repay the loan in four years, how much will she have to repay?

4 Annuities
Uniform Series

This chapter introduces annuities, which are uniform series of payments or disbursements deposited or withdrawn at equal intervals such as yearly, monthly, or daily. This chapter also explains the difference between problem time zero (PTZ) and equation time zero (ETZ). The formulas for the uniform series compound amount factor and the uniform series present worth factor are introduced, and a case study is included on social security income that illustrates using the uniform series present worth factor. The uniform series sinking fund factor and the uniform capital recovery factor are covered along with the process for calculating the remaining balances on loans and balloon payments. The last section of this chapter demonstrates the process for calculating the present worth of infinite uniform series and the infinite uniform series of a present worth.

Uniform series (*annuities*) are (n) end-of-period payments or disbursements compounded at an interest rate of (i) on the balance in an account at the end of each period. As each uniform series amount is deposited into, or withdrawn from, an interest-bearing account, the account draws interest at rate (i). The account continues to draw interest at this rate as each new uniform payment is deposited, or disbursement is withdrawn, in subsequent time periods.

Cash flow diagrams representing uniform series of payments or disbursements are always drawn with the annuity at the end of each period unless stated otherwise. There is also an annuity included at the end of the last time period representing the deposit or disbursement for the last period. The uniform series formulas, and the interest factor tables in Appendix B, were developed for uniform series to be considered at the end of each time period. A sample cash flow diagram for a uniform series is shown in Figure 4.1.

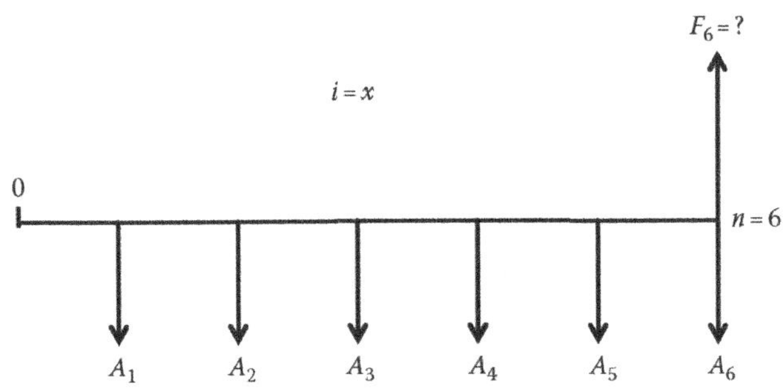

FIGURE 4.1 Sample cash flow diagram for a uniform series of individual payments.

Since a uniform series may start at any point in time, not just at time zero, there are two points in time referred to as time zero:

1. *Equation time zero* (ETZ): With reference to a particular payment or disbursement, the time at which $n = 0$. The ETZ for a payment or disbursement is one time period prior to the start of the uniform series. The number of periods (n) for the ETZ is the ending period minus the ETZ.
2. *Problem Time Zero* (PTZ): The present time according to the problem statement. If a uniform series starts at year one, then the ETZ = PTZ since both occur at the same time.

Figure 4.2 shows a cash flow diagram with an ETZ in a different location than the PTZ. The PTZ is at time zero and the ETZ is at year two.

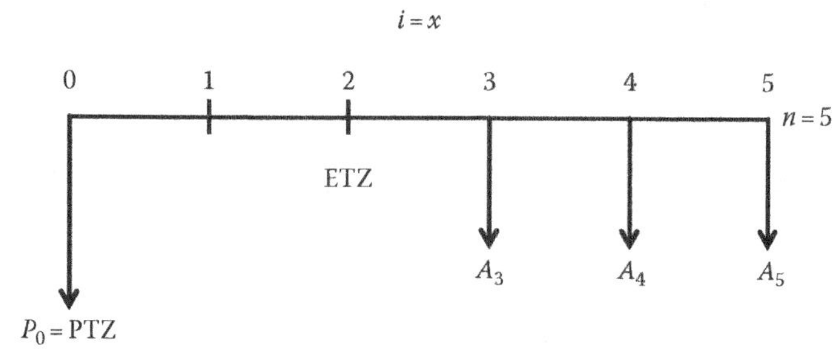

FIGURE 4.2 Cash flow diagram for ETZ ≠ PTZ.

For the uniform series shown in Figure 4.2, the number of periods is calculated using the following equation:

$$\begin{aligned} n &= \text{Last payment period} - \text{Equation time zero (ETZ)} \\ &= 5-2 \\ &= 3 \end{aligned} \tag{4.1}$$

Another method for determining the total number of periods is to count the number of periodic payments and this number is the number of periods used in the formulas.

The first type of uniform series to be discussed in this chapter is the uniform series compound amount factor.

4.1 UNIFORM SERIES COMPOUND AMOUNT FACTOR: FUTURE WORTH OF ANNUITIES (*F/A*)

The *uniform series compound amount factor* (*F/A*) is the formula for solving for the future worth of a uniform series of payments or disbursements. When calculating the future worth of a uniform series, the future value occurs at the same time as the last periodic deposit or disbursement. Figure 4.3 shows the same cash flow diagram as the cash flow diagram in Figure 4.1, but the periodic payments are shown as a rectangular box with the uniform series value listed for the rectangular box rather than showing an arrow for each yearly annuity.

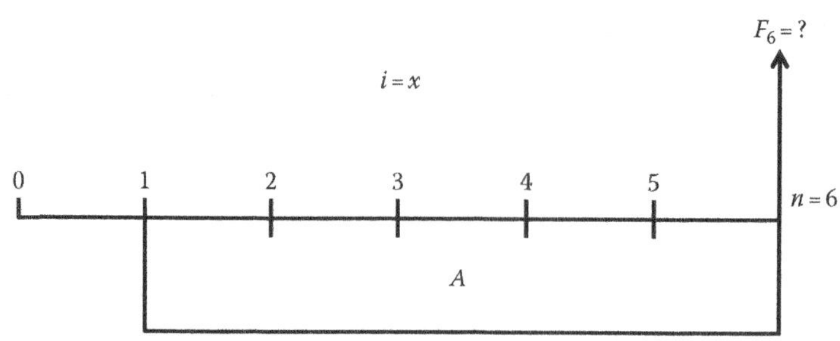

FIGURE 4.3 Cash flow diagram for a uniform series of payments.

Equation 4.2 is the formula for the uniform series compound amount factor. The derivation of Equation 4.2 is provided in Appendix D:

$$F_n = A\left[\frac{(1+i)^n - 1}{i}\right] \tag{4.2}$$

The interest factor table formula for the uniform series compound amount factor is Equation 4.3.

$$F_n = A(F/A, i, n) \tag{4.3}$$

Example 4.1 demonstrates using the uniform series compound amount factor formula for calculating the future worth of a uniform annual series.

Example 4.1

The manager of a mechanical engineering firm is investing $150,000.00 each year for nine years. He will be able to withdraw the funds deposited plus 8% interest at the end of the nine years. What will the investment be worth at the end of nine years? Figure 4.4 is the cash flow diagram for the investment made by the mechanical engineer.

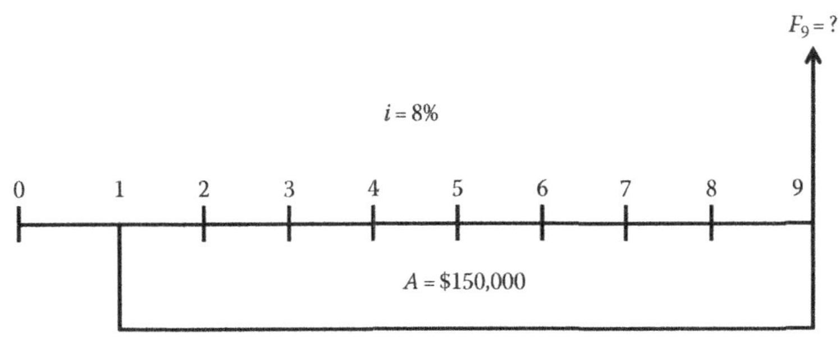

FIGURE 4.4 Cash flow diagram for the mechanical engineering investment in Example 4.1.

Solution

Use Equation 4.2 to calculate the future worth of the uniform series:

$$\begin{aligned} F_9 &= A\left[\frac{(1+i)^n - 1}{i}\right] \\ &= \$150{,}000.00\left[\frac{(1+0.08)^9 - 1}{0.08}\right] \\ &= \$150{,}000.00\left(\frac{0.999005}{0.08}\right) \\ &= \$150{,}000.00(12.487) \\ &= \$1{,}873{,}050.00 \end{aligned}$$

This problem could also be solved using the interest factor tables in Appendix B by locating the (*F*/*A*) factor in the table for $i = 8\%$ and $n = 9$ and multiplying the (*F*/*A*) factor by the uniform series amount using Equation 4.3:

$$
\begin{aligned}
F_9 &= A(F/A,i,n) \\
&= \$150{,}000.00(F/A,8,9) \\
&= \$150{,}000.00(12.487) \\
&= \$1{,}873{,}050.00
\end{aligned}
$$

Example 4.2 provides another uniform series compound amount factor (*F*/*A*) problem that calculates the future worth of a uniform series.

Example 4.2

A student borrows $2,500.00 per year for eight years at 3% interest. How much will she have to repay at the end of eight years? Figure 4.5 is the cash flow diagram for Example 4.2.

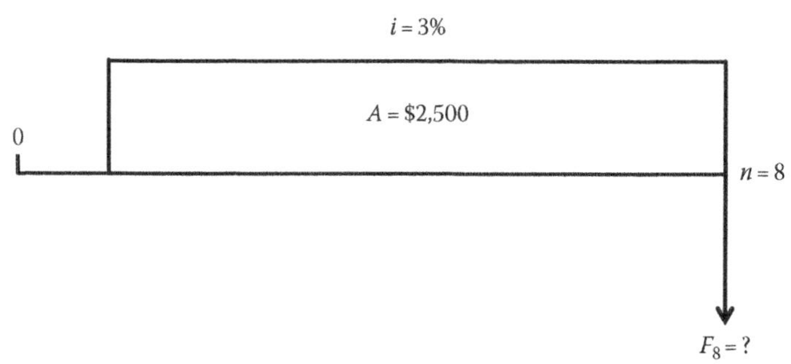

FIGURE 4.5 Cash flow diagram for borrowing funds in Example 4.2.

Solution

$$
\begin{aligned}
F_8 &= A\left[\frac{(1+i)^n - 1}{i}\right] \\
&= \$2{,}500.00\left[\frac{(1+0.03)^8 - 1}{0.03}\right] \\
&= \$2{,}500.00\left(\frac{0.22677}{0.03}\right) \\
&= \$2{,}500.00(8.8923) \\
&= \$22{,}230.75
\end{aligned}
$$

Using the interest factor tables in Appendix B to solve for the future worth of the uniform series results in the following:

$$\begin{aligned} F_8 &= A(F/A,i,n) \\ &= \$2{,}500.00(F/A,3,8) \\ &= \$2{,}500.00(8.8923) \\ &= \$22{,}230.75 \end{aligned}$$

4.2 UNIFORM SERIES PRESENT WORTH FACTOR: PRESENT WORTH OF ANNUITIES (*P/A*)

The *uniform series present worth factor* (*P/A*) was developed to calculate the present worth of a uniform series of payments or disbursements. The first payment or disbursement starts one year after the ETZ and the last one occurs at the end of the last time period. Figure 4.6 is a sample cash flow diagram for the uniform series present worth factor.

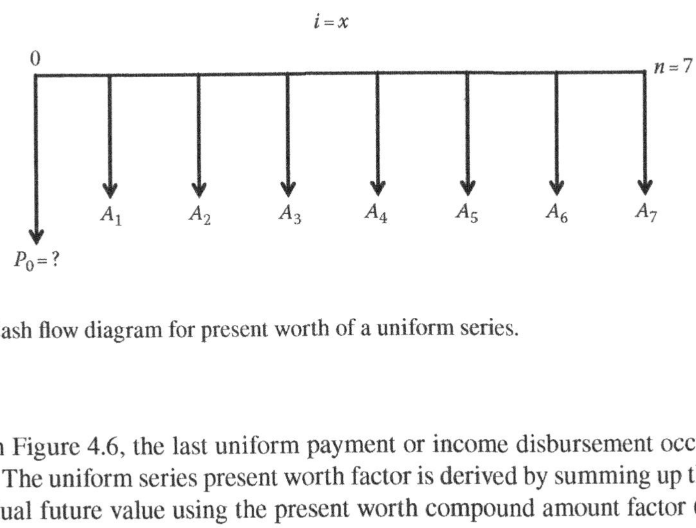

FIGURE 4.6 Cash flow diagram for present worth of a uniform series.

As shown in Figure 4.6, the last uniform payment or income disbursement occurs at the end of the last period. The uniform series present worth factor is derived by summing up the present worth of each individual future value using the present worth compound amount factor (*P/F*) introduced in Section 3.3, as shown in Equation 4.4:

$$P_0 = F_1\left(\frac{1}{(1+i)^1}\right) + F_2\left(\frac{1}{(1+i)^2}\right) + F_3\left(\frac{1}{(1+i)^3}\right) + \cdots + F_n\left(\frac{1}{(1+i)^n}\right) \tag{4.4}$$

Substituting (*A*) for (*F*) in Equation 4.4 results in Equation 4.5:

$$P_0 = A\left(\frac{1}{(1+i)^1}\right) + A\left(\frac{1}{(1+i)^2}\right) + A\left(\frac{1}{(1+i)^3}\right) + \cdots + A\left(\frac{1}{(1+i)^n}\right) \tag{4.5}$$

Appendix D continues the derivation for the uniform series present worth factor and the resulting formula is Equation 4.6.

$$P_0 = A\left[\frac{(1+i)^n - 1}{i(1+i)^n}\right] \tag{4.6}$$

The formula for the uniform series present worth factor (*P/A*) used with the interest factor tables in Appendix B is Equation 4.7.

$$P_0 = A(P/A, i, n) \tag{4.7}$$

The uniform series present worth factor in Equation 4.6 calculates the present worth of any uniform series whether it starts at PTZ or ETZ. Example 4.3 illustrates using Equation 4.6 to calculate the present worth of a uniform series.

Example 4.3

Calculate the present worth of the amount of money that needs to be deposited into an interest-bearing account so an annuity of $2,000.00 could be withdrawn for the next 10 years if the interest rate is 4%. Figure 4.7 is the cash flow diagram for Example 4.3.

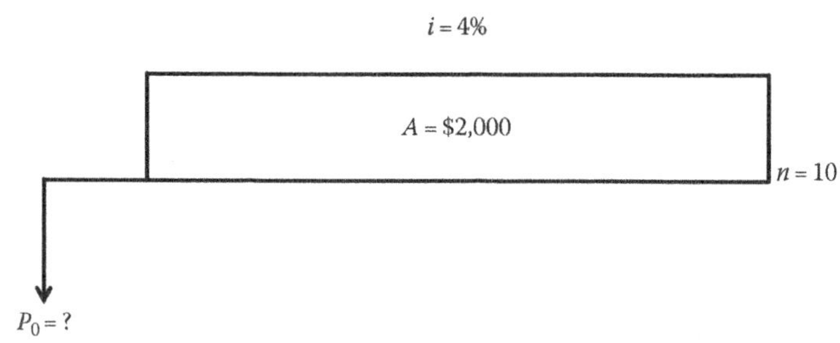

FIGURE 4.7 Cash flow diagram for the present worth of a uniform series in Example 4.3.

Solution

Calculate the present worth of the uniform series (*A*) using Equation 4.6:

$$P_0 = A\left[\frac{(1+i)^n - 1}{i(1+i)^n}\right]$$

$$= \$2{,}000.00\left[\frac{(1+0.04)^{10} - 1}{0.04(1+0.04)^{10}}\right]$$

$$= \$2{,}000.00\left(\frac{0.480244}{0.059210}\right)$$

$$= \$2{,}000.00(8.1109)$$

$$= \$16{,}221.80$$

This problem may also be solved using Equation 4.6 and the interest factor tables in Appendix B.

Locate the interest factor table in Appendix B for $i = 5\%$ and $n = 10$ and use the appropriate (P/A) factor in Equation 4.7:

$$P_0 = A(P/A,i,n)$$
$$= \$2{,}000.00(P/A,4,10)$$
$$= \$2{,}000.00(8.1109)$$
$$= \$16{,}221.80$$

Example 4.4 provides another problem that calculates the present worth of a uniform series.

Example 4.4

A software engineer borrows money to pay her employees while they are working at her start-up company. The bank offers an interest rate on the loan of 7% and the loan will be repaid in equal annual installments over seven years at \$70,000.00 per year. How much is the engineer able to borrow with the terms offered by the bank? Figure 4.8 is the cash flow diagram for the loan borrowed by the software engineer.

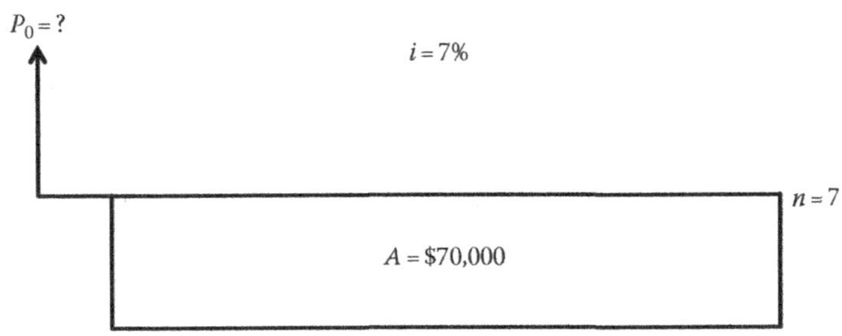

FIGURE 4.8 Cash flow diagram for the software engineer borrowing funds in Example 4.4.

Solution

$$P_0 = A\left[\frac{(1+i)^n - 1}{i(1+i)^n}\right]$$
$$= \$70{,}000.00\left[\frac{(1+0.07)^7 - 1}{0.07(1+0.07)^7}\right]$$
$$= \$70{,}000.00\left(\frac{0.605781}{0.112405}\right)$$
$$= \$70{,}000.00(5.3893)$$
$$= \$377{,}251.00$$

Solving Example 4.4 with Equation 4.7 and the interest factor tables in Appendix B yields the following:

$$
\begin{aligned}
P_0 &= A(P/A,i,n) \\
&= \$70{,}000.00(P/A,7,7) \\
&= \$70{,}000.00(5.3893) \\
&= \$377{,}251.00
\end{aligned}
$$

Case Study 4.1 illustrates the procedure for calculating the uniform series present worth factor by providing different options for yearly social security income.

Case Study 4.1 Social Security Income Calculations

An engineer has the option of retiring and collecting social security when he is 62, 66.5, or 70 years old. If he decides to start his social security income at age 62 or 66.5, he will only be entitled to a reduced amount rather than the entire amount he would receive if he retires at age 70. Table 4.1 contains the amounts the engineer would receive on a monthly basis for each of the three different starting ages.

Part A—If the engineer estimates he will live to be 85 years old, what would be the present worth for each of the social security age alternatives using an interest rate of 4%? Figures 4.9 through 4.11 are the cash flow diagrams for each of the social security options for 85 years.

Part B—What would be the present worth of the three options if the engineer only lives to 77 years old? Figures 4.12 through 4.14 are the cash flow diagrams for each of the social security options for 77 years.

TABLE 4.1
Social Security Monthly Payments at Ages 62, 66.5, and 70 Years

Age	Monthly Income	Number of Years to Age 85	Number of Years to Age 77
62	$1,799.00	23	15
66.5	$2,412.00	18.5	10.5
70	$3,151.00	15	7

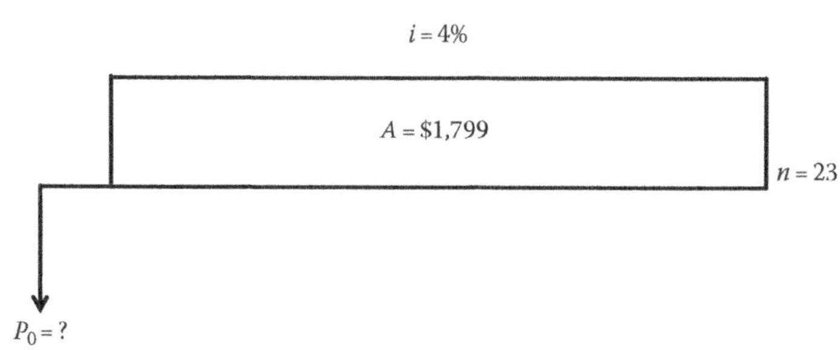

FIGURE 4.9 Cash flow diagram for social security starting at age 62 until age 85.

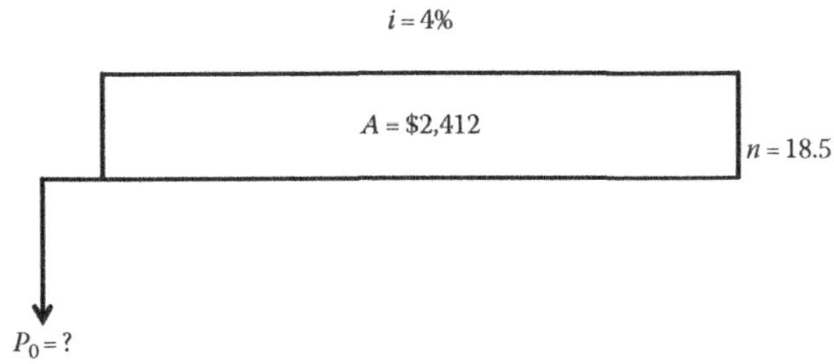

FIGURE 4.10 Cash flow diagram for social security starting at age 66.5 until age 85.

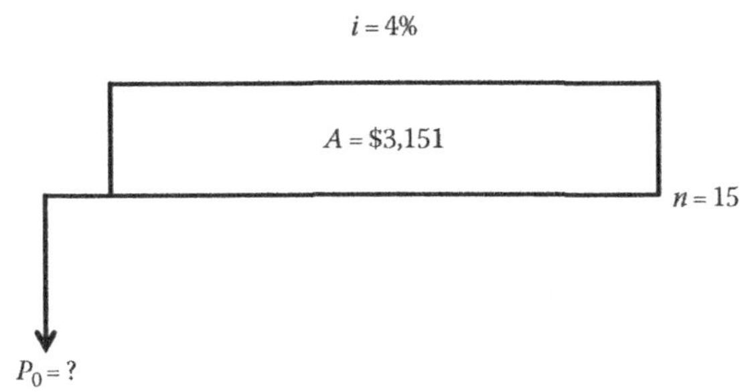

FIGURE 4.11 Cash flow diagram for social security starting at age 70 until age 85.

SOLUTION TO PART A—SOCIAL SECURITY MONTHLY WITHDRAWALS UNTIL AGE 85

First, solve for the total number of monthly periods and the monthly interest rate for age 62 until 85:

$$n = \frac{\text{Number of compounding periods}}{\text{Year}} \times \text{Total number of years}$$

$$n = \frac{12 \text{ periods}}{\text{Year}} \times 23 \text{ years}$$

$$= 276 \text{ periods}$$

$$i = \frac{i_n}{m}$$

$$i = \frac{4\%/\text{year}}{12 \text{ periods/year}}$$

$$= 0.33\% \text{ per period}$$

Second, calculate the present worth of the social security income for age 62 until 85 using Equation 4.6:

$$P_0 = A\left[\frac{(1+i)^n - 1}{i(1+i)^n}\right]$$

$$= \$1{,}799.00\left[\frac{(1+0.0033)^{276} - 1}{0.0033(1+0.0033)^{276}}\right]$$

$$= \$1{,}799.00\left(\frac{1.482585}{0.008193}\right)$$

$$= \$1{,}799.00(180.9575)$$

$$= \$325{,}542.54 \text{ age 62 until 85}$$

Third, calculate the total number of periods and the period interest rate for age 66.4 until 85:

$$n = \frac{\text{Number of compounding periods}}{\text{Year}} \times \text{Total number of years}$$

$$n = \frac{12 \text{ periods}}{\text{Year}} \times 23 \text{ years}$$

$$n = 222 \text{ periods}$$

$$i = \frac{i_n}{m}$$

$$i = \frac{4\%/\text{year}}{12 \text{ periods/year}}$$

$$= 0.33\% \text{ per period}$$

Fourth, calculate the present worth of the social security income for age 66.5 until 85:

$$P_0 = A\left[\frac{(1+i)^n - 1}{i(1+i)^n}\right]$$

$$= \$2{,}412.00\left[\frac{(1+0.0033)^{222} - 1}{0.0033(1+0.0033)^{222}}\right]$$

$$= \$2{,}412.00\left(\frac{1.077975}{0.006857}\right)$$

$$= \$2{,}412.00(157.2080)$$

$$= \$379{,}185.70 \text{ age } 66.5 \text{ until } 85$$

Fifth, calculate the total number of periods and the period interest rate for age 70 until 85:

$$n = \frac{\text{Number of compounding periods}}{\text{Year}} \times \text{Total number of years}$$

$$= \frac{12 \text{ periods}}{\text{Year}} \times 15 \text{ years}$$

$$= 180 \text{ periods}$$

$$i = \frac{i_n}{m}$$

$$i = \frac{4\%/\text{year}}{12 \text{ periods/year}}$$

$$= 0.33\% \text{ per period}$$

Sixth, calculate the present worth of the social security income for age 70 until 85:

$$P_0 = A\left[\frac{(1+i)^n - 1}{i(1+i)^n}\right]$$

$$= \$3{,}151.00\left[\frac{(1+0.0033)^{180} - 1}{0.0033(1+0.0033)^{180}}\right]$$

$$= \$3{,}151.00\left(\frac{0.809448}{0.005971}\right)$$

$$= \$3{,}151.00(135.5632)$$

$$= \$427{,}159.64 \text{ age } 70 \text{ until } 85$$

SOLUTION TO PART B—SOCIAL SECURITY MONTHLY WITHDRAWALS UNTIL AGE 77

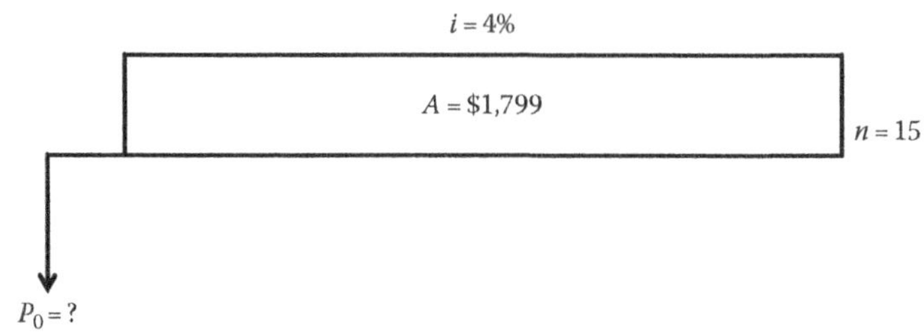

FIGURE 4.12 Cash flow diagram for social security starting at age 62 until age 77.

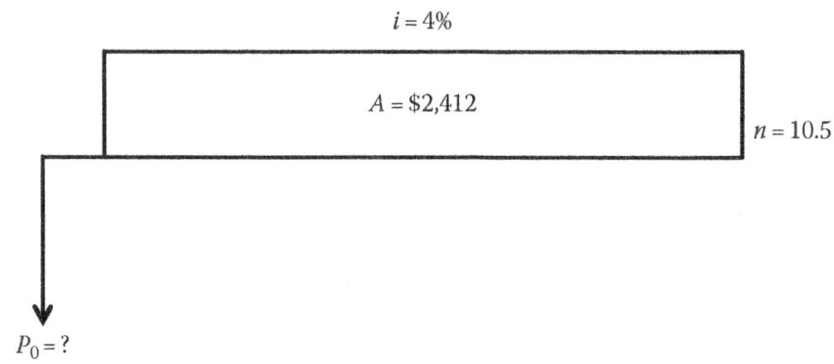

FIGURE 4.13 Cash flow diagram for social security starting at age 66.5 until age 77.

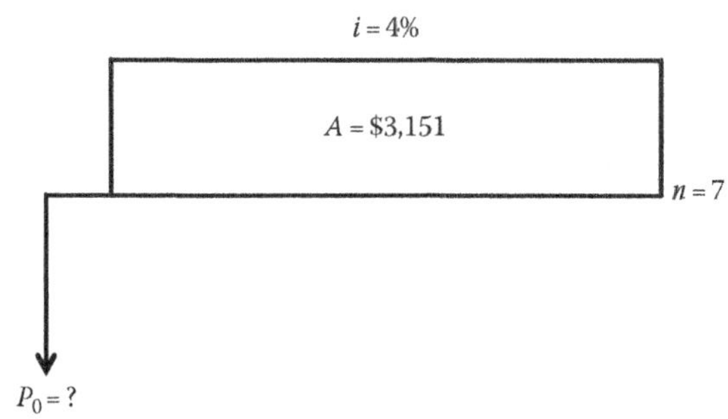

FIGURE 4.14 Cash flow diagram for social security starting at age 70 until age 77.

First, solve for the total number of periods and the period interest rate from age 62 until 77:

$$n = \frac{\text{Number of compounding periods}}{\text{Year}} \times \text{Total number of years}$$

$$= \frac{12 \text{ periods}}{\text{Year}} \times 15 \text{ years}$$

$$= 180 \text{ periods}$$

$$i = \frac{i_n}{m}$$

$$i = \frac{4\%/\text{year}}{12 \text{ periods/year}}$$

$$= 0.33\% \text{ per period}$$

Second, calculate the present worth of the social security income for age 62 until 77 using Equation 4.6:

$$P_0 = A\left[\frac{(1+i)^n - 1}{i(1+i)^n}\right]$$

$$= \$1{,}799.00\left[\frac{(1+0.0033)^{180} - 1}{0.0033(1+0.0033)^{180}}\right]$$

$$= \$1{,}799.00\left(\frac{0.809448}{0.005971}\right)$$

$$= \$1{,}799.00(135.5632)$$

$$= \$243{,}878.20 \text{ age 62 until 77}$$

Third, solve for the total number of periods and the period interest rate for age 66.5 until 77:

$$n = \frac{\text{Number of compounding periods}}{\text{Year}} \times \text{Total number of years}$$

$$= \frac{12 \text{ periods}}{\text{Year}} \times 10.5 \text{ years}$$

$$= 126 \text{ periods}$$

$$i = \frac{i_n}{m}$$

$$i = \frac{4\%/\text{year}}{12 \text{ periods/year}}$$

$$= 0.33\% \text{ per period}$$

Fourth, calculate the present worth of the social security income for age 66.5 until 77:

$$P_0 = A\left[\frac{(1+i)^n - 1}{i(1+i)^n}\right]$$

$$= \$2{,}412.00\left[\frac{(1+0.0033)^{126} - 1}{0.0033(1+0.0033)^{126}}\right]$$

$$= \$2{,}412.00\left(\frac{0.514546}{0.004998}\right)$$

$$= \$2{,}412.00(102.9504)$$

$$= \$248{,}316.36 \text{ age } 66.5 \text{ until } 77$$

Fifth, solve for the total number of periods and the period interest rate for age 70 until 77:

$$n = \frac{\text{Number of compounding periods}}{\text{Year}} \times \text{Total number of years}$$

$$= \frac{12 \text{ periods}}{\text{Year}} \times 7 \text{ years}$$

$$= 84 \text{ periods}$$

$$i = \frac{i_n}{m}$$

$$i = \frac{4\%/\text{year}}{12 \text{ periods/year}}$$

$$= 0.33\% \text{ per period}$$

Sixth, calculate the present worth of the social security income for age 70 until 77:

$$P_0 = A\left[\frac{(1+i)^n - 1}{i(1+i)^n}\right]$$

$$= \$3{,}151.00\left[\frac{(1+0.0033)^{84} - 1}{0.0033(1+0.0033)^{84}}\right]$$

$$= \$3{,}151.00\left(\frac{0.318828}{0.004352}\right)$$

$$= \$3{,}151.00(73.2601)$$

$$= \$230{,}842.57 \text{ age } 70 \text{ until } 77$$

Table 4.2 shows the results for all of the social security options in Parts A and B.

TABLE 4.2
Results for Social Security Options

Retirement Age	Total Income if Live until Age 85	Total Income if Live until Age 77
62.0	\$325,542.54	\$243,878.20
66.5	\$379,185.70	\$248,316.36
70.0	\$427,159.64	\$230,842.57

4.3 UNIFORM SERIES SINKING FUND FACTOR: ANNUITY OF A FUTURE VALUE (*A/F*)

In addition to being able to solve for the future worth of a present value and the present worth of a uniform series, there is also a formula for calculating what a uniform series would be based on its future worth. This formula is the *uniform series sinking fund factor* (*A/F*). The uniform series sinking fund factor is Equation 4.8 and it is the inverse of the uniform series compound amount factor (*F/A*) that was introduced in Section 4.1.

Uniform series compound amount factor (*F/A*)

$$F_n = A\left[\frac{(1+i)^n - 1}{i}\right]$$

Inverse

Uniform series sinking fund factor (*A/F*)

$$A = F_n\left[\frac{i}{(1+i)^n - 1}\right] \quad (4.8)$$

Equation 4.9 is the formula for solving for the future worth of a uniform series when using the interest factor tables in Appendix B:

$$A = F_n\left(A/F, i, n\right) \quad (4.9)$$

The uniform series sinking fund factor (*A/F*) determines a uniform series based on its future worth when the number of periods and the interest rate per period are known. An example of when this formula would be used is when an organization needs to determine the amount to invest each year to be able to withdraw a certain amount of money after a predefined period of time. Another example is when an individual is able to withdraw funds from an account each year and repays a predefined amount at the end of a certain period of time. In Example 4.5, the uniform series sinking fund factor is used to determine the amount of a uniform series based on its future worth.

Example 4.5

The manager of a small agricultural engineering firm realizes that he will have to replace a seeding machine in 20 years. The manager would like to know how much he should deposit each year in an interest-bearing account earning 6% a year to have the \$250,000.00 required for the seeding machine at the end of 20 years. Figure 4.15 is the cash flow diagram for the replacement seeding machine.

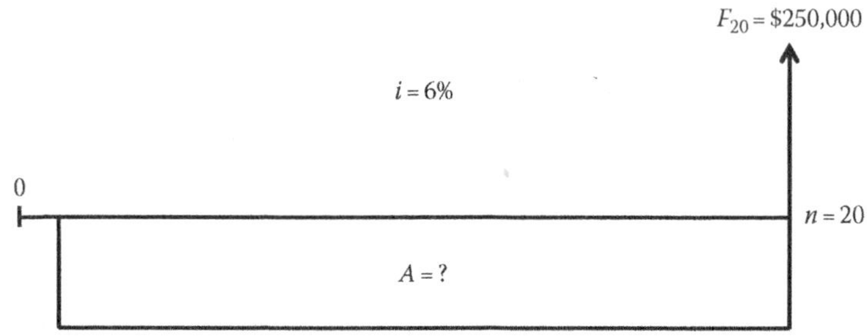

FIGURE 4.15 Cash flow diagram for the replacement seeding machine in Example 4.5.

Solution

Calculate the equivalent uniform annual series of the future value using Equation 4.8:

$$A = F_{20}\left[\frac{i}{(1+i)^n - 1}\right]$$

$$= \$250{,}000.00\left[\frac{0.06}{(1+0.06)^{20} - 1}\right]$$

$$= \$250{,}000.00\left(\frac{0.06}{2.207135}\right)$$

$$= \$250{,}000.00(0.02718)$$

$$= \$6{,}795.00$$

Calculate the equivalent uniform annual series of the future value using Equation 4.9 and the interest factor tables in Appendix B:

$$A = F_{20}\left(A/F, i, n\right)$$

$$= \$250{,}000.00\left(A/F, 6, 20\right)$$

$$= \$250{,}000.00(0.02718)$$

$$= \$6{,}795.00$$

Example 4.6 provides another uniform series sinking fund factor problem that calculates the equivalent uniform annual amount of a future worth.

Example 4.6

A process engineer has been asked by her boss to determine the amount of money the company could withdraw from a project so the funds are available for another project for five years if the second project would be able to return $1,500,000.00 to the first project at the end of the five years. The interest rate is 4%. Figure 4.16 is the cash flow diagram for the second project.

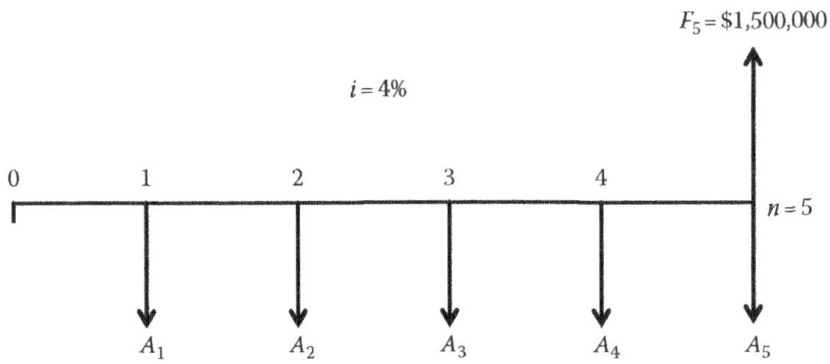

FIGURE 4.16 Cash flow diagram for the second process engineering project in Example 4.6.

Solution

$$
\begin{aligned}
A &= F_5\left[\frac{i}{(1+i)^n - 1}\right] \\
&= \$1{,}500{,}000.00\left[\frac{0.04}{(1+0.04)^5 - 1}\right] \\
&= \$1{,}500{,}000.00\left(\frac{0.04}{0.216653}\right) \\
&= \$1{,}500{,}000.00(0.18463) \\
&= \$276{,}945.00
\end{aligned}
$$

Using the interest factor tables in Appendix B results in the following solution:

$$
\begin{aligned}
A &= F_5(A/F, i, n) \\
&= \$1{,}500{,}000.00(A/F, 4, 5) \\
&= \$1{,}500{,}000.00(0.18463) \\
&= \$276{,}945.00
\end{aligned}
$$

4.4 UNIFORM SERIES CAPITAL RECOVERY FACTOR: ANNUITY OF A PRESENT WORTH (*A*/*P*)

Along with being able to solve for an equivalent uniform series of a future value using the uniform series sinking fund factor (*A*/*F*), as explained in Section 4.3, there is a *uniform series capital recovery factor* (*A*/*P*) for calculating the uniform series equivalent to a present value. The formula for the uniform series capital recovery factor is Equation 4.10, and it is the inverse of the uniform series present worth factor.

Uniform series present worth factor (P/A)

$$P_0 = A\left[\frac{(1+i)^n - 1}{i(1+i)^n}\right]$$

Inverse

Uniform series capital recovery factor (A/P)

$$A = P_0\left[\frac{i(1+i)^n}{(1+i)^n - 1}\right] \tag{4.10}$$

The interest factor table formula for solving for an equivalent uniform series of a present value is given in Equation 4.11.

$$A = P_0\left(A/P, i, n\right) \tag{4.11}$$

The uniform series capital recovery factor (*A*/*P*) determines the amount of money that could be withdrawn from a fund each year for a predetermined amount of time based on an initial deposit into the account. It is also used when funds are initially withdrawn from an account to calculate the amount of money repaid each year to replenish the account.

Example 4.7 demonstrates using the uniform series capital recovery factor to calculate an equivalent uniform series of a present value.

Example 4.7

An industrial engineer inherits $575,000.00 from his father, but instead of inheriting the money outright, his father has invested the funds in an account paying 3% interest and the son is able to withdraw a yearly annuity from the account for 10 years. Determine how much the engineer is able to withdraw each year from the account. Figure 4.17 is the cash flow diagram for the inheritance.

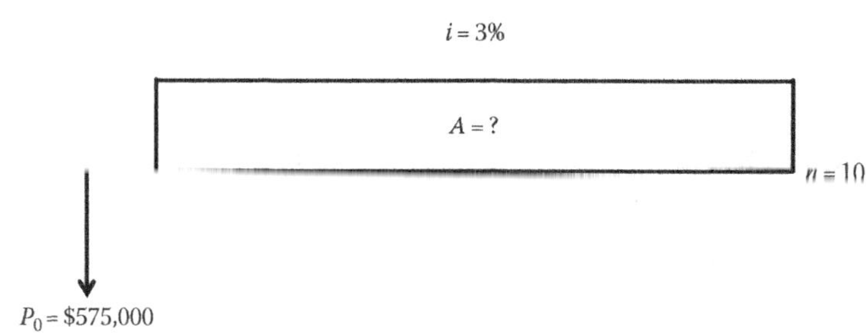

FIGURE 4.17 Cash flow diagram for the uniform series of withdrawals in Example 4.7.

Solution

Use Equation 4.10 to solve for the equivalent uniform series of the present value:

$$A = P_0\left[\frac{i(1+i)^n}{(1+i)^n - 1}\right]$$

$$= \$575{,}000.00\left[\frac{0.03(1+0.03)^{10}}{(1+0.03)^{10} - 1}\right]$$

$$= \$575{,}000.00\left(\frac{0.040317}{0.343916}\right)$$

$$= \$575{,}000.00(0.11723)$$

$$= \$67{,}407.25$$

Using Equation 4.11 and the interest factor tables in Appendix B to solve this problem results in the following solution:

$$A = P_0\left(A/P, i, n\right)$$

$$= \$575{,}000.00\left(A/P, 3, 10\right)$$

$$= \$575{,}000.00(0.11723)$$

$$= \$67{,}407.25$$

Example 4.8 uses the uniform series capital recovery factor (*A*/*P*) to convert a present value into an equivalent uniform series.

Example 4.8

A manufacturing engineer recommends to his boss that the company purchase a new computer numerical control (CNC) machine. The cost of the machine is \$500,000.00. The company has to borrow the money to purchase the machine at an interest rate of 6% and repay the loan over five years. How much will the company have to repay each year? Figure 4.18 is the cash flow diagram for the new CNC machine.

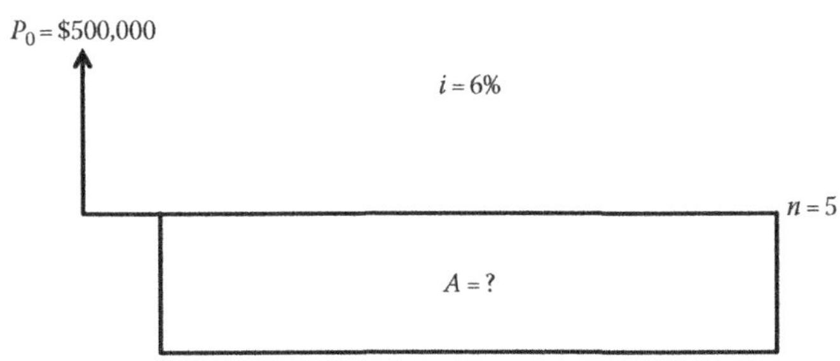

FIGURE 4.18 Cash flow diagram for the new computer numerical control machine in Example 4.8.

Solution

$$A = P_0\left[\frac{i(1+i)^n}{(1+i)^n - 1}\right]$$

$$= \$500{,}000.00\left[\frac{0.06(1+0.06)^5}{(1+0.06)^5 - 1}\right]$$

$$= \$500{,}000.00\left(\frac{0.080294}{0.338226}\right)$$

$$= \$500{,}000.00(0.23740)$$

$$= \$118{,}700.00$$

Using the interest factor tables in Appendix B to solve for the equivalent uniform series of the present value yields the following solution:

$$A = P_0\left(A/P, i, n\right)$$

$$= \$500{,}000.00\left(A/P, 6, 5\right)$$

$$= \$500{,}000.00(0.23740)$$

$$= \$118{,}700.00$$

4.4.1 Solving for Remaining Balances

In addition to the uniform series capital recovery factor (*A*/*P*) determining the equivalent uniform series of a present value based on the interest rate, the amount of the annuity going toward repaying part of the principal each period could also be calculated along with what portion of the annuity is interest each payment period. To calculate what portion of an annuity payment is interest, first, the interest rate per compounding period is calculated using Equation 3.11:

$$i = \frac{i_n}{m}$$

The interest owed each period declines and the amount of the annuity repaying the principal increases over time. The amount of the annuity repaying the principal each period is the payment amount minus the interest for the period. Equation 4.12 is the formula for solving for the interest for each period:

$$\text{Interest per period} = \text{Remaining principal} \times \text{Period interest rate} \tag{4.12}$$

The payment (annuity) minus the interest for each period is the amount of the principal paid in that period. The principal repaid in each period is subtracted from the remaining principal at the beginning of each period to calculate the remaining balance for each period. Equations 4.13 and 4.14 are the formulas for calculating the principal paid and the remaining balance owed at the end of each period:

$$\text{Principal paid per period} = \text{Payment} - \text{Interest paid in that period} \tag{4.13}$$

$$\text{Remaining balance} = \text{Starting principal for period} - \text{Principal paid that period} \tag{4.14}$$

An efficient method for tracking payments and balances is to enter the data and formulas into a spreadsheet in the format shown in Table 4.3.

TABLE 4.3
Format for a Spreadsheet for Tracking Payments and Balances on Loans

Period	Remaining Principal Balance	Payment This Period	Interest Paid This Period	Principal Paid This Period	New Principal Balance
1	Original principal	$A = P_0\left[\frac{i(1+i)^n}{(1+i)^n-1}\right]$	Original Principal × Period interest rate	Payment – Interest paid this period	Remaining principal balance – Principal paid this period
2	New principal balance from previous year	$A = P_0\left[\frac{i(1+i)^n}{(1+i)^n-1}\right]$	Remaining principal balance × Period interest rate	Payment – Interest paid this period	Remaining principal balance – Principal paid this period
3	New principal balance from previous year	$A = P_0\left[\frac{i(1+i)^n}{(1+i)^n-1}\right]$	Remaining principal balance × Period interest rate	Payment – Interest paid this period	Remaining principal balance – Principal paid this period

In Table 4.3, the first remaining principal balance is the original principal and the new principal balance is the remaining principal balance at the beginning of the period minus the principal paid in the period. The new principal balance at the end of each period becomes the remaining principal balance for the next period. Case Study 4.2 demonstrates the procedure for determining remaining balances on loans.

Case Study 4.2 Solving for Remaining Balances on Loans

A petroleum engineer needs to purchase new equipment when he is working in the oil fields. The equipment cost $50,000.00 and he has to borrow the funds to pay for the equipment. He agrees to repay the loan over four years by making monthly payments at a yearly interest rate of 10%. Calculate the monthly payments and fill in the spreadsheet shown in Table 4.3 for the first three payment periods. Figure 4.19 is the cash flow diagram for purchasing the oil field equipment.

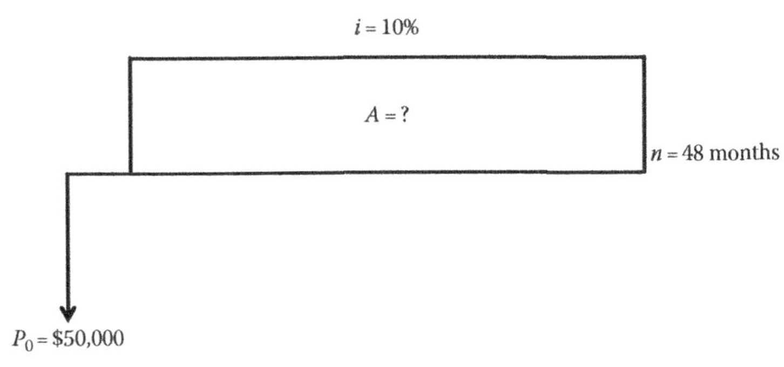

FIGURE 4.19 Cash flow diagram for the new oil field equipment in Case Study 4.2.

Solution

Procedure

First, calculate the period interest rate and the total number of periods.

Second, calculate the amount of the monthly annuity using the uniform series capital recovery factor formula Equation 4.8.

Third, use Equations 4.13 and 4.14 to perform the calculations for the values required in the spreadsheet format shown in Table 4.3.

Part I

$$n = \frac{\text{Number of compounding periods}}{\text{Year}} \times \text{Total number of years}$$

$$= \frac{12 \text{ periods}}{\text{Year}} \times 4 \text{ years}$$

$$= 48 \text{ periods}$$

$$i = \frac{i_n}{m}$$

$$i = \frac{10\%/\text{year}}{12 \text{ periods/year}}$$

$$= 0.83\% \text{ per period}$$

Part II

Calculate the monthly equivalent uniform series of the present value:

$$A = P_0\left[\frac{i(1+i)^n}{(1+i)^n - 1}\right]$$

$$= \$50{,}000.00\left[\frac{0.0083(1+0.0083)^{48}}{(1+0.0083)^{48} - 1}\right]$$

$$= \$50{,}000.00\left(\frac{0.012342}{0.486993}\right)$$

$$= \$50{,}000.00(0.025343)$$

$$= \$1{,}267.16$$

Part III

The spreadsheet in Table 4.4 is developed using Equations 4.13 and 4.14 and the format provided in Table 4.3.

TABLE 4.4
Spreadsheet for Solving for Payments and Remaining Balances

Period	Remaining Principal Balance	Payment This Period	Interest Paid This Period	Principal Paid This Period	New Principal Balance
1	\$50,000.00	\$1,267.16	\$50,000.00 × 0.0083 = \$415.00	\$1,267.16–\$415.00 = \$852.16	\$50,000.00–\$852.16 = \$49,147.84
2	\$49,147.84	\$1,267.16	\$49,147.84 × 0.0083 = \$407.93	\$1,267.16–\$407.93 = \$859.23	\$49,147.84–\$859.23 = \$488,288.61
3	\$48,288.61	\$1,267.16	\$49,288.61 × 0.0083 = \$409.10	\$1,267.16–\$409.10 =\$858.06	\$48,288.61–\$858.06 = \$47,430.55

4.4.2 Amount of a Final Payment in a Series: Balloon Payments

In some instances, a lending institution will set a loan payment at an artificially low amount not equal to what is required to repay the loan during the loan repayment period. In this case, the total amount paid over the term of the loan is not adequate to repay the loan balance. There will be a balance remaining at the end of the loan period that has to be repaid in one lump sum and this is a *balloon payment.* To calculate the value of the last payment (balloon payment), subtract one period from the loan repayment period and use $(n-1)$ as the number of periods to calculate the present worth of the $(n-1)$ payments. Once the present worth of the $(n-1)$ number of payments is determined, subtract this value from the original principal and then calculate the future worth of the original principal using the total number of periods (n) minus the present worth of the amount repaid. Equation 4.15 calculates the present worth of the remaining balance.

$$\begin{aligned}\text{Present worth of the remaining balance} &= \text{Present worth of the principal}\\ &\quad - \text{Present worth of the amount paid in } n-1 \text{ periods}\end{aligned} \tag{4.15}$$

Example 4.9 demonstrates the procedure for determining a final balloon payment.

Example 4.9

Calculate the amount of the final payment if \$50,000.00 is borrowed by a biomedical engineering company to expand the business and the bank sets the amount of the repayment schedule at \$8,000.00 per year for 10 years with an interest rate of 10%. Figure 4.20 is the cash flow diagram for the final payment for the biomedical engineering company loan.

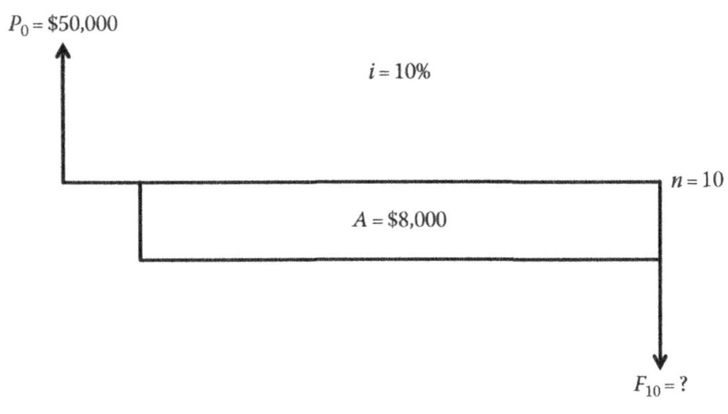

FIGURE 4.20 Cash flow diagram for expanding the biomedical engineering firm in Example 4.9.

Solution

Solve for $n-1$:

$$\begin{aligned} n-1 &= 10-1 \\ &= 9 \end{aligned}$$

Calculate the present worth of $(n-1)$ payments:

$$P_0 = A\left[\frac{(1+i)^n - 1}{i(1+i)^n}\right]$$

$$= \$8{,}000.00\left[\frac{(1+0.10)^9 - 1}{0.10(1+0.10)^9}\right]$$

$$= \$8{,}000.00\left(\frac{1.357948}{0.235795}\right)$$

$$= \$8{,}000.00(5.7590)$$

$$= \$46{,}072.00$$

Calculate the remaining balance at time zero using Equation 4.15:

Present worth of the remaining balance = Present worth of the principal

−Present worth of the amount paid in $n-1$ periods

Present worth of the remaining balance = \$50,000.00 − \$46,072.00

= \$3,928.00

Calculate the future worth at year 10 of the remaining balance:

$$F_{10} = P_0(1+i)^n$$

$$= \$3{,}928.00(1+0.10)^{10}$$

$$= \$3{,}928.00(2.5937)$$

$$= \$10{,}188.05$$

4.5 PRESENT WORTH OF AN INFINITE UNIFORM SERIES

This section explains the process for calculating the present worth of a uniform series that continues forever (perpetual life), which is an *infinite uniform series.*

To calculate the present worth of an infinite uniform series, the annuity amount is divided by the interest rate. Equation 4.16 is the formula for solving for the present worth of an infinite uniform series:

$$P_0 = \frac{A}{i} \tag{4.16}$$

Example 4.10 demonstrates solving for the present worth of an infinite uniform series.

Example 4.10

What is the principal amount that would have to be deposited now if the interest rate is 5% to be able to withdraw \$5000.00 forever? Figure 4.21 is the cash flow diagram for the infinite series.

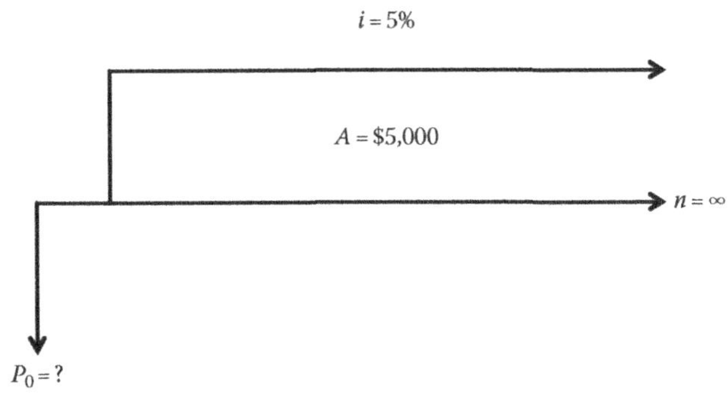

FIGURE 4.21 Cash flow diagram for the infinite uniform series in Example 4.10.

Solution

Use Equation 4.15 to solve for the present worth of the infinite uniform series:

$$P_0 = \frac{A}{i}$$

$$= \frac{\$5{,}000.00}{0.05}$$

$$= \$100{,}000.00$$

4.6 INFINITE UNIFORM SERIES OF PRESENT WORTH

To calculate an *equivalent infinite uniform series* of a present value, the formula in Equation 4.16 is inverted and it becomes

$$A = P_0(i) \tag{4.17}$$

Whatever amount is deposited into a *perpetual life account* (an account where money may be withdrawn forever) is multiplied by the interest rate to determine the amount that may be withdrawn from the account each year forever without depleting the principal. Example 4.11 calculates a perpetual life annuity of a present value.

Example 4.11

An investor deposits \$20,000.00 into an account paying 6% interest. How much may the investor withdraw from the account every year forever? Figure 4.22 is the cash flow diagram for the infinite series withdrawal.

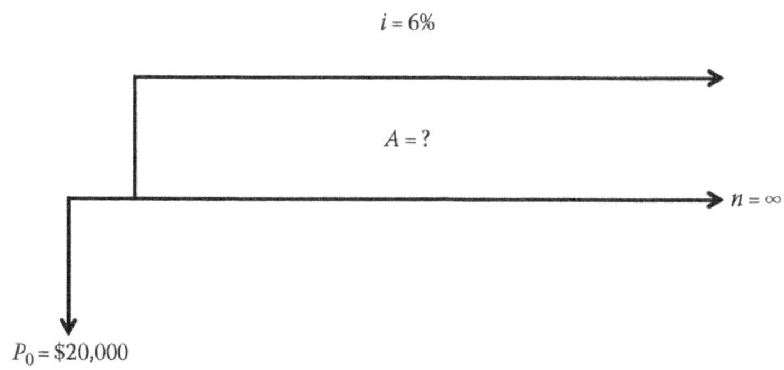

FIGURE 4.22 Cash flow diagram for the infinite uniform series in Example 4.11.

Solution

Use Equation 4.16 to solve for the equivalent infinite uniform annual series of the present worth:

$$A = P_0(i)$$
$$= \$20{,}000.00(0.06)$$
$$= \$1{,}200.00$$

If an infinite uniform series is not yearly but it repeats as a uniform withdrawal or disbursement on a set schedule then an equivalent yearly uniform series is determined by distributing each future amount over the years between amounts. For example, if there is a uniform amount being paid out every 10 years forever, the amount is converted into a yearly uniform annual series over the prior 10 years. Since the amount repeats every 10 years, it only has to be converted into an equivalent uniform annual series over the prior 10 years one time. Once it has been converted into an equivalent uniform annual series, the present worth of the uniform series is calculated using Equation 4.16. When drawing cash flow diagrams for this type of repeating infinite series, a bar is inserted above the annuity to indicate the series repeats forever.

Example 4.12 demonstrates solving for the present worth of a repetitive infinite uniform series that does not occur annually.

Example 4.12

What amount of money needs to be deposited into an account to be able to withdraw $10,000 every 10 years if the interest rate is 5%. Figure 4.23 is the cash flow diagram for the repetitive infinite series.

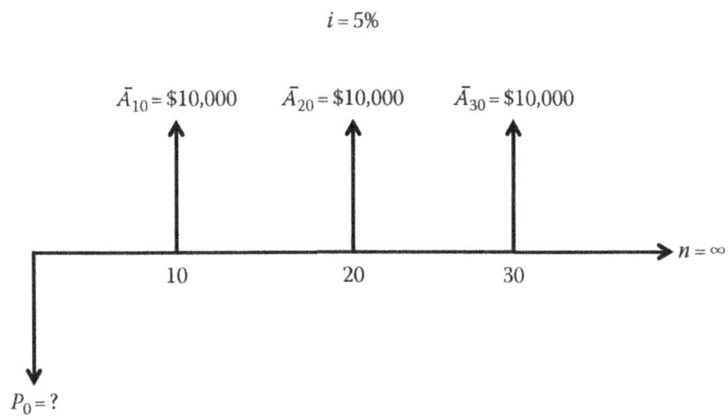

FIGURE 4.23 Cash flow diagram for the repetitive infinite uniform series in Example 4.12.

Solution

First, calculate the equivalent uniform annual series for the $10,000.00 occurring every 10 years by converting the future value of $10,000.00 into a uniform series over 10 years:

$$A = F_{10}\left[\frac{i}{(1+i)^n - 1}\right]$$
$$= \$10,000.00\left[\frac{0.05}{(1+0.05)^{10} - 1}\right]$$
$$= \$10,000.00\left(\frac{0.05}{0.628895}\right)$$
$$= \$10,000.00(0.07950)$$
$$= \$795.00$$

Second, calculate the present worth of the equivalent infinite uniform annual series:

$$P_0 = \frac{A}{i}$$
$$= \frac{\$795.00}{0.05}$$
$$= \$15,900.00$$

In some instances, funds may not be deposited into an account immediately or they may be deposited immediately and not withdrawn for several years. In either of these cases, the PTZ will be different than the ETZ. In addition to a different ETZ than PTZ, a problem may also include an equivalent infinite uniform series as part of the problem. Example 4.13 demonstrates solving a problem with both of these situations.

Example 4.13

An investment firm is offering an annuity where investors are able to withdraw $50,000.00 every five years forever starting in year eight. If the interest rate is 10%, what is the present worth of the equivalent infinite uniform series? Figure 4.24 is the cash flow diagram for the repetitive infinite series withdrawals.

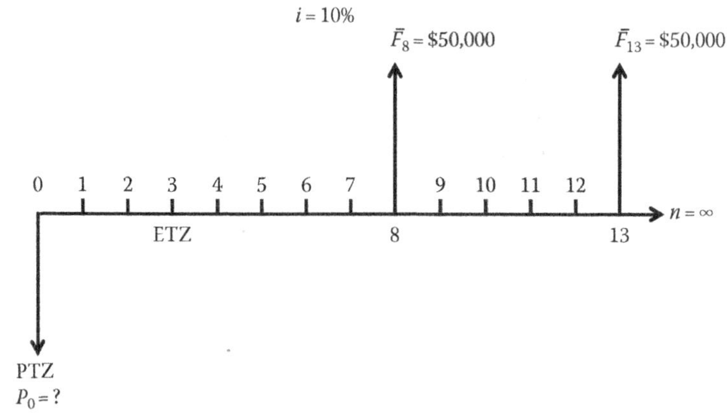

FIGURE 4.24 Cash flow diagram for the infinite uniform series in Example 4.13.

Solution

Procedure

First, calculate the equivalent uniform annual series of the future value over five years. This uniform series will be repeated indefinitely.

Second, calculate the present worth of the infinite uniform annual series at year three, the ETZ.

Third, calculate the present worth at time zero of the future worth at year three (this is the present worth of the infinite uniform annual series).

Part I

$$
\begin{aligned}
A &= F_5\left[\frac{i}{(1+i)^n - 1}\right] \\
&= \$50{,}000.00\left[\frac{0.10}{(1+0.10)^5 - 1}\right] \\
&= \$50{,}000.00\left(\frac{0.10}{0.610510}\right) \\
&= \$50{,}000.00(0.16380) \\
&= \$8{,}190.00
\end{aligned}
$$

Part II

$$
\begin{aligned}
P_3 &= A(P/A, i, n) \\
&= \$8{,}190.00(P/A, 10, \infty) \\
&= \frac{A}{i} \\
&= \frac{\$8{,}190.00}{0.10} \\
&= \$81{,}900.00
\end{aligned}
$$

Part III

$$P_3 = F_3$$

$$P_0 = F_3\left(\frac{1}{(1+i)^n}\right)$$

$$= \$81{,}890.00\left(\frac{1}{(1+0.10)^3}\right)$$

$$= \$81{,}890.00\left(\frac{1}{1.3310}\right)$$

$$= \$81{,}890.00(0.7513)$$

$$= \$61{,}523.96$$

Appendix C includes spreadsheet formulas for calculating the present and future worth of uniform series and converting present and future values into equivalent uniform series.

4.7 SUMMARY

This chapter explained the processes used for solving for the future and present worth of uniform series and for the equivalent uniform series of future and present values. Equation time zero and problem time zero were introduced, as they relate to the location of where a problem is solving for the initial time of a uniform series. The uniform series compound amount factor (*F*/*A*) was discussed, which solves for the future value of a uniform series of payments or disbursements.

The uniform series present worth factor (*P*/*A*) and its role in the process of calculating the present worth of a uniform series of payments or disbursements was explained and a case study was provided to demonstrate the calculations required for solving for the present worth of three social security options occurring over different time periods. The uniform series sinking fund factor (*A*/*F*) was introduced, which is used for calculating the equivalent uniform series of a future value.

The uniform series capital recovery factor (*A*/*P*) was covered and calculations for solving for the equivalent uniform series of a present value were reviewed to demonstrate their use in engineering economic analysis. In addition, a spreadsheet format for solving for remaining balances on loans at the end of each period was provided along with an explanation on how to calculate balloon payments and the final payment in a series when it is a balloon payment. The last section of this chapter presented formulas for solving for the present worth of an infinite uniform series and for the equivalent infinite uniform series of a present value.

KEY TERMS

Annuities
Balloon payment
Equation time zero
Equivalent infinite uniform series
Infinite series
Perpetual life account
Problem time zero
Uniform series
Uniform series capital recovery factor

Uniform series compound amount factor
Uniform series formulas
Uniform series present worth factor
Uniform series sinking fund factor

PROBLEMS

4.1 A manager at a mold and die company software design firm is purchasing a new communication system. If the company makes payments of $15,000.00 per year for four years, what amount will the company have paid for the system by year four if the interest rate is 7%? Solve this problem using formulas.

4.2 Solve Problem 4.1 using the interest factor tables in Appendix B.

4.3 A small petroleum company needs to replace the flow meters at one well sight. The company borrows funds to pay for the flow meters and repays $65,000.00 per year for six years. The interest rate is 12%. What amount will they have repaid at six years? Solve this problem using formulas.

4.4 Solve Problem 4.3 using the interest factor tables in Appendix B.

4.5 Solve Problem 4.3 but use a compounding period for the interest rate of quarterly. Use formulas to solve this problem.

4.6 Solve Problem 4.5 using the interest factor tables in Appendix B.

4.7 A manufacturing firm is able to invest $127,000.00 per year for 10 years at 2.5% interest. How much would the investment be worth now? Solve this problem using formulas.

4.8 Solve Problem 4.7 using the interest factor tables in Appendix B.

4.9 An engineer plans on investing $5,000.00 per year for the next 36 years at 1.5% interest. What amount of money will she have in her investment when she retires at 36 years? Solve using the interest factor tables in Appendix B.

4.10 If the engineer in Problem 4.9 finds a different account paying 4% interest and she still deposits $5,000.00 per year into the new account, how much will she have in the account at the end of 36 years? Solve using the interest factor tables in Appendix B.

4.11 What would be the present worth of the $5,000.00 invested every year at 4% in Problem 4.10 for 36 years? Solve this problem using the interest factor tables in Appendix B.

4.12 An engineer plans on purchasing a new fleet of five trucks. If the engineer is able to afford to pay $5,000.00 per month for 60 months at an interest rate of 0.05% per month, what amount are they able to finance? Solve this problem using the interest factor tables in Appendix B.

4.13 A new sports arena costs $350,000,000.00 and it will generate profits of $35,000,000.00 per year. If the money to build the arena is borrowed at 8% interest and repaid at $35,000.000.00 per year, how many years will it take to repay the loan?

4.14 How much should someone be willing to pay to purchase a note that pays dividends of $6,000.00 per year for 10 years if the interest rate is 2.5%? Solve using the interest factor tables.

4.15 If a person purchases a note for $50,000.00 and it pays 3% interest, how much would they receive in dividends each year for 10 years? Solve using the interest factor tables.

4.16 If the manager of an engineering consulting firm needs to have $1,500,000.00 in 10 years, what amount does he need to invest each year at 6% interest? Solve using the interest factor tables.

4.17 A manager of a process engineering company is considering the purchase of 10 3D printers costing $5,100.00 each. How much needs to be saved each year to recover the investment over six years at an interest rate of 3%? Solve using the interest factor tables.

4.18 The manager of a materials engineering firm purchased cell phones for its 1,000 employees at a cost of $500.00 per cell phone. What amount does the company have to repay each year over four years at an interest rate of 7% to repay the loan taken out to pay for the cell phones? Solve using the interest factor tables.

4.19 An engineer graduates from college with a student loan debt of $87,000.00. If she has to repay the debt yearly at an interest rate of 5% over 20 years, what amount will she be repaying each year? Solve using the interest factor tables.

4.20 If the engineer in Problem 4.19 repays the student loan debt at $87,000.00 yearly for 20 years with an interest rate of 5%, how much will she have paid at 20 years? Solve using the interest factor tables.

4.21 If an engineer repays his student loan debt by paying $300.00 per month at an annual interest rate of 5% compounded monthly, what amount will he have repaid at 20 years? Solve using the formulas.

4.22 If an engineer repays $600.00 per month on his student loan debt at 5% interest per year compounded monthly for 20 years, how much will he have repaid at 20 years? Solve using formulas.

4.23 A civil engineer passed the professional engineers examination and as a result he received a raise of $3,000.00 per year. If he works for 36 more years, what is the present worth of the raise at an interest rate of 2%?

4.24 What is the future worth of the $3,000.00 raise for passing the professional engineer's examination after 36 years at an annual interest rate of 2%?

4.25 A process engineer invests $2,000,000.00 at an interest rate of 4% per year in an account, what amount would she be able to withdraw from the account each year forever to pay maintenance costs on their equipment?

4.26 A county needs to have $200,000.00 a year to be able to pay for maintenance on a small road system. How much would the county need to invest now into an account paying 4% interest per year to be able to withdraw $200,000.00 a year forever?

4.27 A communications company has to repair their microwave towers at a cost of $65,000.00 each repair. If the first repair begins at year 10 and repairs are required every four years forever, what amount does the firm need to invest now at 5% to cover the cost of the repairs forever?

4.28 How much does a gas company need to deposit now into an interest-bearing account paying 3% interest per year to be able to withdraw $125,000.00 per year starting at year six and be able to withdraw $125,000.00 forever?

4.29 A construction firm plans on withdrawing funds every year starting at year eight. If they deposit $9,800,000.00 now, what amount could they withdraw each year forever if the interest rate is 6%?

4.30 What is the current value of a uniform series of $1,000.00 forever at an interest rate of 13%?

5 Arithmetic and Geometric Gradients

This chapter discusses arithmetic and geometric gradients and explains their purpose and the procedures for incorporating gradients into engineering economic analysis. Arithmetic gradients are defined, and the process for calculating the future worth of arithmetic gradients is presented along with example problems and a case study demonstrating using the future worth gradient factor (*F/G*). The procedure for determining the present worth of arithmetic gradients using the present worth gradient factor (*P/G*) is presented along with the process for solving for decreasing values of uniform gradients series and increasing and decreasing gradients.

This chapter introduces the method for calculating an equivalent uniform annual series from a uniform gradient series and provides a case study illustrating the process for converting uniform gradient series into equivalent uniform annual series using the annual cost gradient factor (*A/G*). Noncontinuous arithmetic gradient series are defined and three methods are included for calculating the present worth of noncontinuous gradient series.

Perpetual life gradient series are covered in this chapter along with the formula for calculating the present worth of an infinite gradient and the infinite gradient of a present value. Geometric gradients are introduced in the last section of this chapter and they are gradients increasing by a uniform rate. Three formulas for calculating the present worth of geometric gradients are included for use when the rate of increase equals, is greater than, or is less than the interest rate.

5.1 DEFINITION OF ARITHMETIC GRADIENTS

In personal financial matters, and when working in an organization, there are always situations where periodic payments or disbursements are not uniform during each period, but they increase or decrease by an arithmetic amount each period. Payment or disbursement streams that do not steadily increase or decrease but increase or decrease in set increments are *arithmetic gradients.*

Arithmetic gradients are defined as either a progressive increase or decrease in the flow of funds at the end of each period for (n) periods. The amount the flow of funds increases or decreases is a constant amount in each period. Figure 5.1 shows a cash flow diagram for an arithmetic gradient.

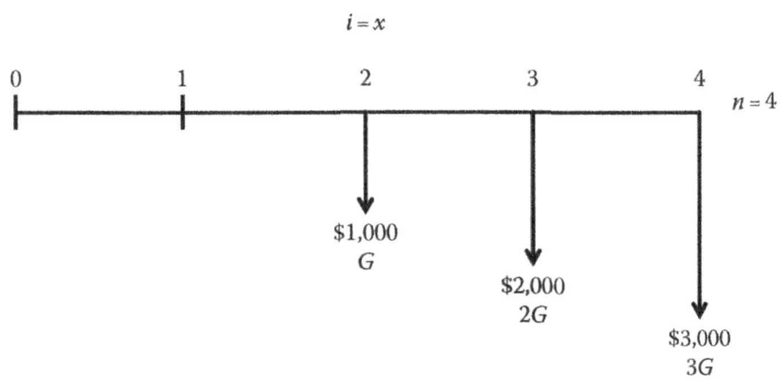

FIGURE 5.1 Cash flow diagram for an arithmetic gradient.

In Figure 5.1, (G) equals the annual arithmetic change in the magnitude of receipts or disbursements. For an arithmetic gradient, the number of payment or disbursement periods is $(n-1)$ since the arithmetic gradients start two years after equation time zero (ETZ). In Figure 5.1, if a line is drawn from the beginning of the gradient to the end at year four, it demonstrates that the gradient actually starts increasing in year 1; therefore, ETZ for the gradient is year zero and the number of periods is the number of gradients plus 1. Figure 5.2 represents the same gradient shown in Figure 5.1, but it uses a line to represent the gradient instead of separate, increasing arrows.

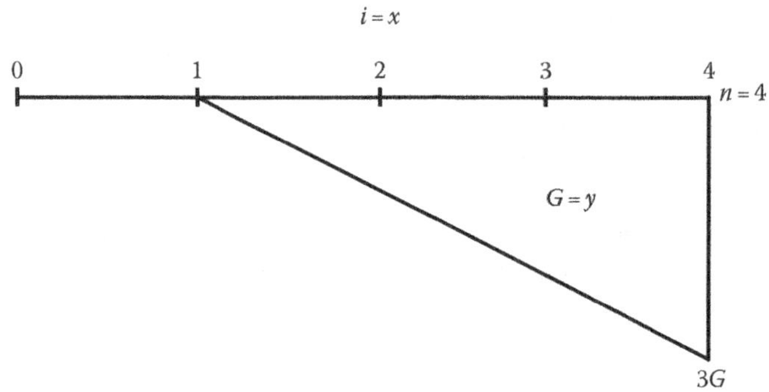

FIGURE 5.2 Cash flow diagram for a triangular arithmetic gradient instead of arrows.

Arithmetic gradients may also be visualized as increasing or decreasing uniform annual series, as shown in Figure 5.3.

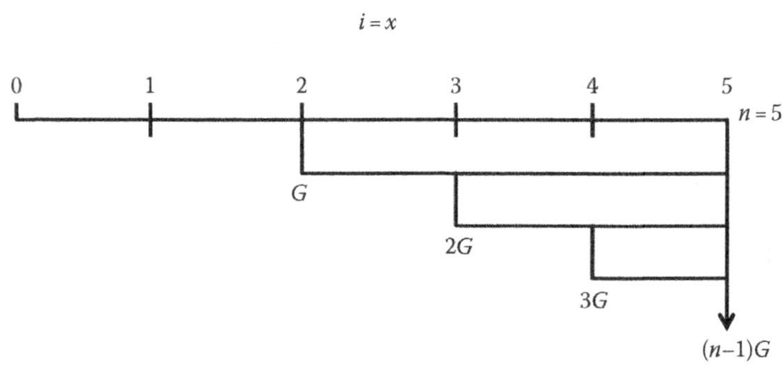

FIGURE 5.3 Cash flow diagram for an arithmetic gradient decomposed into annuities.

Gradients may be positive or negative amounts, but they increase or decrease in absolute value over time. For gradients, the increasing increment is a constant amount and its progression is described as being arithmetic. Situations where gradients appear include the following:

- Reducing expenses each year in order to increase profits
- Electrical companies with increasing costs to maintain transmission lines
- Federal government agency decreasing items in its budget every year to try and balance the budget

- Increasing costs of maintaining agricultural equipment
- Increasing maintenance costs for a piece of equipment or a facility
- Maintenance costs of a vehicle increasing yearly as the vehicle ages
- Manufacturing plant budgeting for increasing expenses on a yearly basis
- Municipality with increasing costs to maintain a roadway system
- Petroleum firm budgeting for increasing costs for maintaining oil field operations
- Trust fund where a person is adding money to the trust fund by increasing their deposit arithmetically each year

5.2 FUTURE WORTH OF ARITHMETIC GRADIENTS (*F/G*)

The derivations for the arithmetic gradient equations are provided in Appendix D. One of the formulas resulting from the derivations is the equation for solving for the future worth of an arithmetic gradient, the *future worth gradient factor* (*F/G*). Equation 5.1 is the formula for calculating the future worth of an arithmetic gradient:

$$F_n = G\left[\left(\frac{1}{i}\right)\left(\frac{(1+i)^n - 1}{i} - n\right)\right] \tag{5.1}$$

Equation 5.2 is the formula for solving for the future worth of a gradient using the interest factor tables in Appendix B:

$$F_n = G(F/G, i, n) \tag{5.2}$$

The formulas for calculating the future and present worth of arithmetic gradients, and for calculating equivalent arithmetic gradients for present and future values, were developed to account for gradients starting one time period after the first period; therefore, when using any of the arithmetic gradients formulas, the total number of time periods used in the formulas is (n + 1) where (n) is the number of gradients.

Example 5.1 demonstrates calculating the future worth of an arithmetic gradient.

Example 5.1

A construction firm invests its profits over five years. It makes an investment at year two of \$50,000.00 that increases by \$50,000.00 up until year five. If the interest rate is 6%, what is the future worth of the invested profits? Figure 5.4 is the cash flow diagram for the gradient investment.

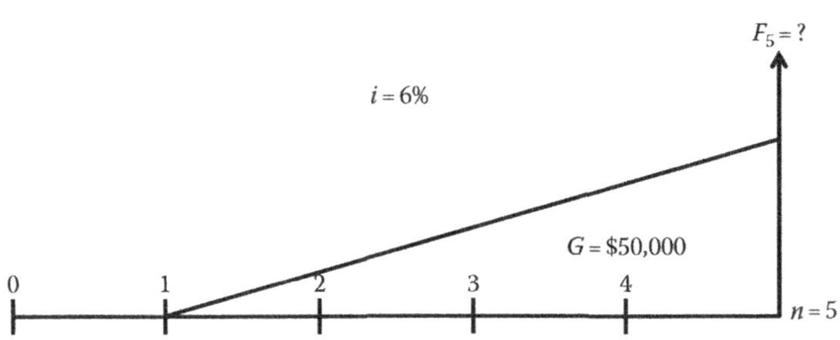

FIGURE 5.4 Cash flow diagram for the profits invested by the construction firm in Example 5.1.

Solution

Use Equation 5.1 to calculate the future worth of the gradient:

$$
\begin{aligned}
F_5 &= G\left[\left(\frac{1}{i}\right)\frac{(1+i)^n-1}{i}-n\right] \\
&= \$50,000.00\left[\left(\frac{1}{i}\right)\frac{(1+0.06)^5-1}{0.06}-5\right] \\
&= \$50,000.00\left[\left(\frac{1}{0.06}\right)\left(\frac{0.338226}{0.06}-5\right)\right] \\
&= \$50,000.00\left[16.667\times(5.637093-5)\right] \\
&= \$50,000.00(16.667\times 0.637093) \\
&= \$50,000.00(10.618) \\
&= \$530,900.00
\end{aligned}
$$

Solve for the future worth of the gradient using Equation 5.2 and the interest factor tables in Appendix B:

$$
\begin{aligned}
F_n &= G(F/G,i,n) = \$50,000.00(F/G,6,5) = \$50,000.00(10.618) \\
&= \$530,900.00
\end{aligned}
$$

Case Study 5.1 demonstrates using the future worth gradient factor and the uniform series compound amount factor to calculate the future worth of an arithmetic gradient and an annuity.

Case Study 5.1 Future Worth of a Gradient and an Annuity

A mechanical engineer plans on completing his Ph.D. in four years. During the four years of his education, he will be earning \$20,000.00 per year as a research assistant. His first year is paid for by a scholarship, but his expenses for subsequent years will be \$10,000.00 for the second year, \$20,000.00 for the third year, and \$30,000.00 for the fourth year. The engineering student would like to have \$50,000.00 remaining when he graduates in four years so he needs to determine whether he will have enough funds remaining with his projected expenses if the interest rate is 10%.

Part I—Solve for the future worth of the income and expenses at the end of four years. Figure 5.5 is the cash flow diagram for funding the Ph.D.

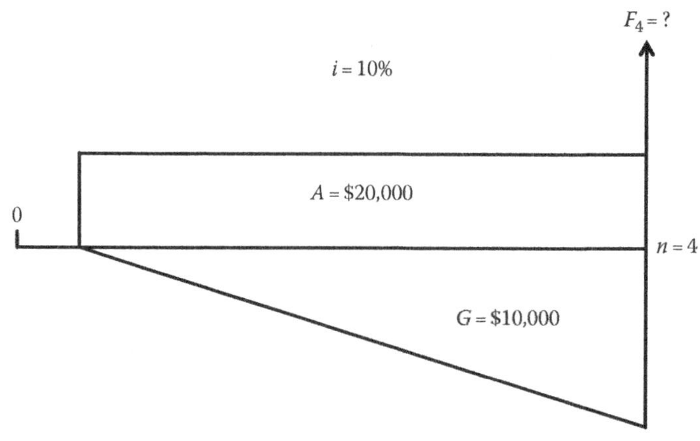

FIGURE 5.5 Cash flow diagram for funding the Ph.D. in Case Study 5.1 Part A.

Part II—If the student will not have \$50,000.00 remaining at the end of four years with his projected expenses, would cutting his expenses by \$5,000.00 a year result in his having \$50,000.00 at the end of four years? Solve for the future worth of his income and new projected expenses to determine whether he will have \$50,000.00 remaining at the end of four years. Figure 5.6 is the cash flow diagram for the option of reducing expenses by \$5,000.00 a year.

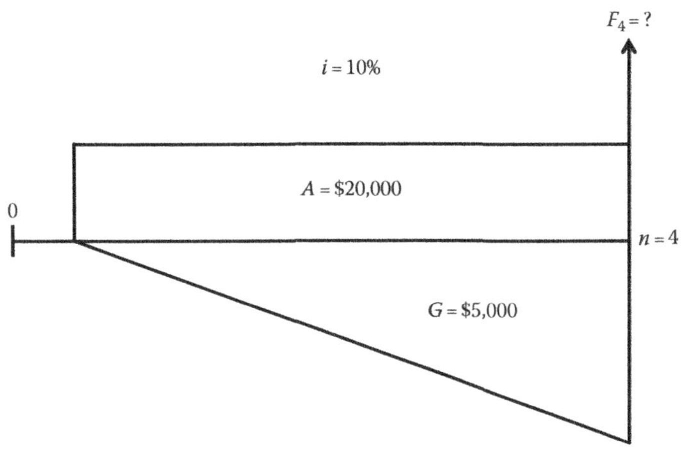

FIGURE 5.6 Cash flow diagram for funding the Ph.D. in Case Study 5.1 Part B.

Solution

Part I

First, calculate the future worth of the uniform annual series of \$20,000.00:

$$
\begin{aligned}
F_4 &= A\left[\frac{(1+i)^n - 1}{i}\right] \\
&= \$20,000.00\left[\frac{(1+0.10)^4 - 1}{0.10}\right] \\
&= \$20,000.00\left(\frac{0.46410}{0.10}\right) \\
&= \$20,000.00(4.6410) \\
&= \$92,820.00
\end{aligned}
$$

Second, calculate the future worth of the $10,000.00 arithmetic gradient using Equation 5.1:

$$F_4 = G\left[\left(\frac{1}{i}\right)\left(\frac{(1+i)^n - 1}{i} - n\right)\right]$$

$$= \$10{,}000.00\left[\left(\frac{1}{i}\right)\left(\frac{(1+0.10)^4 - 1}{0.10} - 4\right)\right]$$

$$= \$10{,}000.00\left[\left(\frac{1}{0.10}\right)\left(\frac{0.46410}{0.10} - 4\right)\right]$$

$$= \$10{,}000.00\left[10 \times (4.6410 - 4)\right]$$

$$= \$10{,}000.00(6.410)$$

$$= \$64{,}100.00$$

Third, solve for the total future worth by subtracting the future worth of the arithmetic gradient from the future worth of the annuity:

$$F_4 = F_A - F_G$$

$$= \$92{,}820.00 - \$64{,}100.00$$

$$= \$28{,}720.00$$

Therefore, the engineer will not have $50,000.00 when he graduates in 4 years.

The future worth of the gradient could also be calculated using Equation 5.2 and the interest factor tables in Appendix B:

$$F_4 = G(F/G, i, n)$$

$$= \$10{,}000.00(F/G, 10, 4)$$

$$= \$10{,}000.00(6.4100)$$

$$= \$64{,}100.00$$

Part II

The uniform annual series is the same as for Part I; therefore, its future worth is still \$92,820.00. First, calculate for the future worth of the revised gradient of \$5,000.00:

$$F_4 = G\left[\left(\frac{1}{i}\right)\left(\frac{(1+i)^n - 1}{i} - n\right)\right]$$

$$= \$5{,}000.00\left[\left(\frac{1}{i}\right)\left(\frac{(1+0.10)^4 - 1}{0.10} - 4\right)\right]$$

$$= \$5{,}000.00\left[\left(\frac{1}{0.10}\right)\left(\frac{0.46410}{0.10} - 4\right)\right]$$

$$= \$5{,}000.00\left[10 \times (4.6410 - 4)\right]$$

$$= \$5{,}000.00(6.410)$$

$$= \$32{,}050.00$$

Second, solve for the total future worth by subtracting the future worth of the gradient from the future worth of the annuity:

$$F_4 = F_A - F_G$$

$$= \$92{,}820.00 - \$32{,}050.00$$

$$= \$69{,}770.00$$

Therefore, the engineer will have over \$50,000.00 left at the end of four years if he reduces his expenses to a \$5,000.00 arithmetic gradient starting in year two.

The future worth of the gradient could also be calculated using Equation 5.2 and the interest factor tables in Appendix B:

$$F_4 = G(F/G, i, n)$$

$$= \$5{,}000.00(F/G, 10, 4)$$

$$= \$5{,}000.00(6.4100)$$

$$= \$32{,}050.00$$

5.3 PRESENT WORTH OF ARITHMETIC GRADIENTS (*P/G*)

The present worth of a gradient series could also be viewed as a series of increasing or decreasing future values; therefore, the formula for the present worth of a gradient is based on summing up all of the future values and the derivation of the ***present worth gradient factor*** formula (*P/G*) is shown in Appendix D. The formula resulting from the derivation is the present worth gradient factor shown in the following equation:

$$P_0 = \frac{G}{i}\left[\frac{(1+i)^n - 1}{i(1+i)^n} - \frac{n}{(1+i)^n}\right] \tag{5.3}$$

Equation 5.4 is the formula for solving for the present worth of an arithmetic gradient using the interest factor tables in Appendix B:

$$P_0 = G(P/G,i,n) \tag{5.4}$$

Since the present worth gradient factor is a lengthy equation, the calculations for solving the present worth of an arithmetic gradient are usually performed using the interest factor table formula shown in Equation 5.4. The number of periods used in the present worth gradient factor formula and the interest factor table formula are the number of gradient payments or disbursements plus one, which indicates that the ETZ is two time periods prior to the first gradient. Figure 5.7 shows a cash flow diagram for solving for the present worth of an arithmetic gradient.

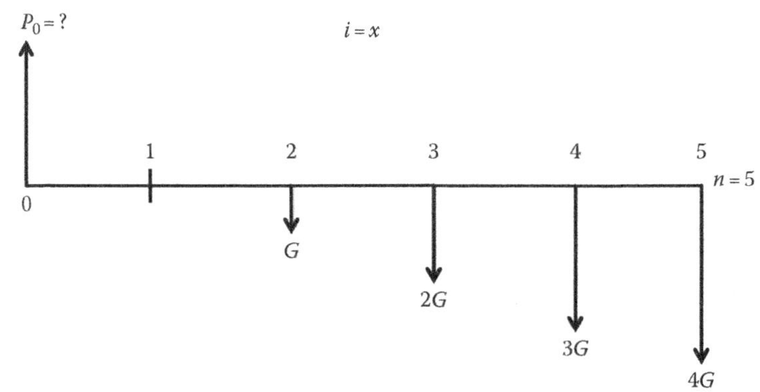

FIGURE 5.7 Cash flow diagram for solving for the present worth of an arithmetic gradient.

The present worth of an arithmetic gradient is calculated to determine the amount of money that needs to be deposited or withdrawn at time zero to cover the value of the increasing or decreasing payments or disbursements in the arithmetic gradient. If an arithmetic gradient increases or decreases and it has an initial value that is a uniform series, the present worth of the uniform series and the gradient is calculated using Equation 5.5.

$$P_0 = \pm A(P/A,i,n) \pm G(P/G,i,n) \tag{5.5}$$

Example 5.2 uses the present worth gradient factor formula to calculate the present worth of an arithmetic gradient.

Example 5.2

A nuclear engineer has to determine the amount of money her firm may borrow to purchase a new machine if the firm is able to start repaying the loan by paying $10,000.00 in year two increasing by $10,000.00 a year until year six with an interest rate of 10%. Figure 5.8 is the cash flow diagram for the new machine.

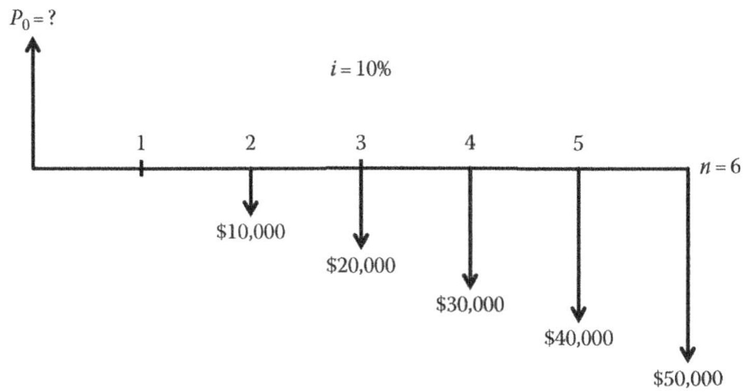

FIGURE 5.8 Cash flow diagram for purchasing the new machine in Example 5.2.

Solution

Calculate for the present worth of the arithmetic gradient using Equation 5.3:

$$\begin{aligned} P_0 &= \frac{G}{i}\left[\frac{(1+i)^n - 1}{i(1+i)^n} - \frac{n}{(1+i)^n}\right] \\ &= \frac{\$10{,}000.00}{0.10}\left[\frac{(1+0.10)^6 - 1}{0.10(1+0.10)^6} - \frac{6}{(1+0.10)^6}\right] \\ &= \$100{,}000.00\left(\frac{0.771561}{0.177156} - \frac{6}{1.771561}\right) \\ &= \$100{,}000.00(4.355263 - 3.386844) \\ &= \$100{,}000.00(0.96842) \\ &= \$96{,}842.00 \end{aligned}$$

Solve for the present worth of the arithmetic gradient using Equation 5.4 and the interest factor tables in Appendix B:

$$\begin{aligned} P_0 &= G(P/G,i,n) \\ &= \$10{,}000.00(P/G,10,6) \\ &= \$10{,}000.00(9.6842) \\ &= \$96{,}842.00 \end{aligned}$$

Example 5.3 provides a problem solving for the present worth of an arithmetic gradient and a uniform annual series.

Example 5.3

A construction firm pays \$5,000.00 for the first year of a maintenance service contract that increases by \$1,000.00 in each subsequent year until year six. If the interest rate is 6%, what would be the present worth of the uniform annual series and the gradient? Figure 5.9 is the cash flow diagram for the maintenance service contract.

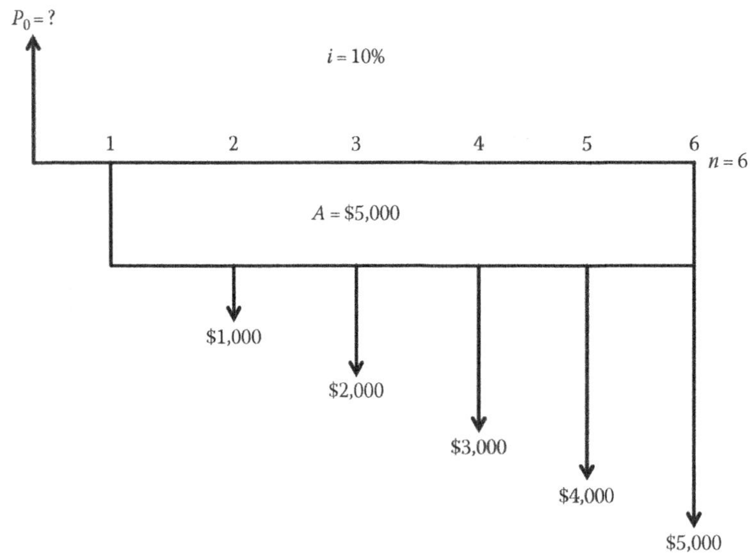

FIGURE 5.9 Cash flow diagram for the maintenance service contract in Example 5.3.

Solution

This problem is solved using the interest factor tables in Appendix B and Equation 5.5:

$$
\begin{aligned}
P_0 &= A(P/A,i,n)+G(P/G,i,n)\\
&= \$5,000.00(P/A,10,6)+\$1,000.00(P/G,10,6)\\
&= \$5,000.00(4.3552)+\$1,000.00(9.6842)\\
&= \$21,776.00+\$9,684.20\\
&= \$31,460.20
\end{aligned}
$$

5.3.1 Decreasing Value of Uniform Gradients

In the previous gradient examples, the gradients were always increasing in value with time. In some instances, a gradient will start at its highest value and decrease over time to its lowest value. In some cases, a gradient is a value subtracted from a uniform annual series represented by the highest value; therefore, *decreasing uniform gradients* are always negative values. Figure 5.10 shows a sample cash flow diagram for a decreasing uniform gradient and Figure 5.11 shows the same cash flow diagram as Figure 5.10, but the decreasing uniform gradient is represented as a uniform series with a decreasing uniform gradient.

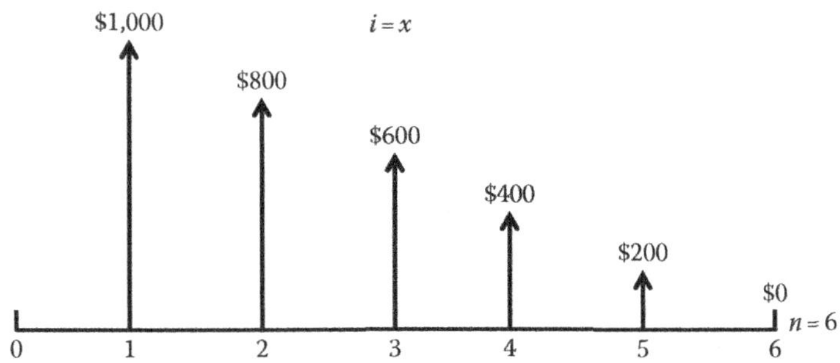

FIGURE 5.10 Cash flow diagram for a decreasing uniform gradient.

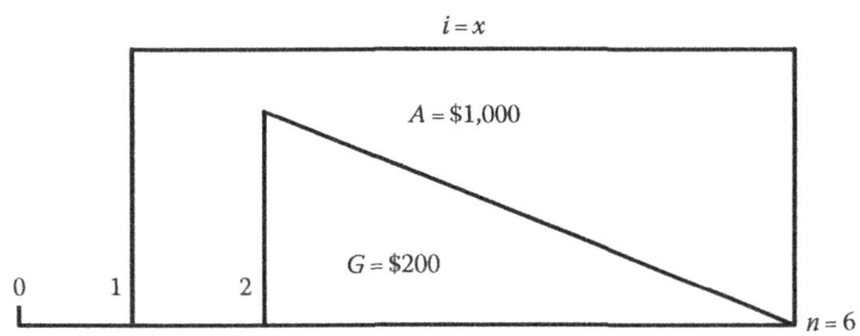

FIGURE 5.11 Cash flow diagram for a uniform series with a decreasing uniform gradient.

Example 5.4 provides calculations for determining the present worth of the uniform annual series minus the gradient in Figures 5.10 and 5.11.

Example 5.4

Calculate the present worth of the gradient series in Figure 5.10 of $1,000.00 starting at year one and decreasing by $200.00 per year until year six using an interest rate of 7%.

Solution

Calculate the present worth of the uniform annual series minus the gradient (see Figure 5.11) using Equation 5.5:

$$\begin{aligned} P_0 &= A(P/A,i,n)+G(P/G,i,n) \\ &= \$1,000.00(P/A,7,6)-\$200.00(P/G,7,6) \\ &= \$1,000.00(4.7665)-\$200.00(10.978) \\ &= \$4,776.50-\$2,195.60 \\ &= \$2,580.90 \end{aligned}$$

5.3.2 Increasing and Decreasing Values of Uniform Gradients

When calculating the future or present worth of uniform gradients, the uniform gradients could either increase or decrease or there could be situations where a problem has an increasing gradient and a separate decreasing gradient. If both types of gradients are included in a problem, the present worth for each uniform gradient series is solved for individually and then summed up to determine the total present worth of all of the uniform gradient series. Case Study 5.2 contains a situation where there is both an increasing and a decreasing uniform gradient series.

Case Study 5.2 Increasing and Decreasing Gradients

An aerospace engineer recommends to his boss that if they start investing funds this year in the amount of $20,000.00 and increase the deposits each year by $5,000.00 for four years for a total of five years, they should have enough money invested to start withdrawing $50,000.00 a year at year six decreasing by $10,000.00 a year until year 10. Calculate the present worth of both gradients to determine if the deposits will be sufficient to cover all of the withdrawals. Use an interest rate of 7%. Figure 5.12 is the cash flow diagram for the aerospace engineering firm investment scheme.

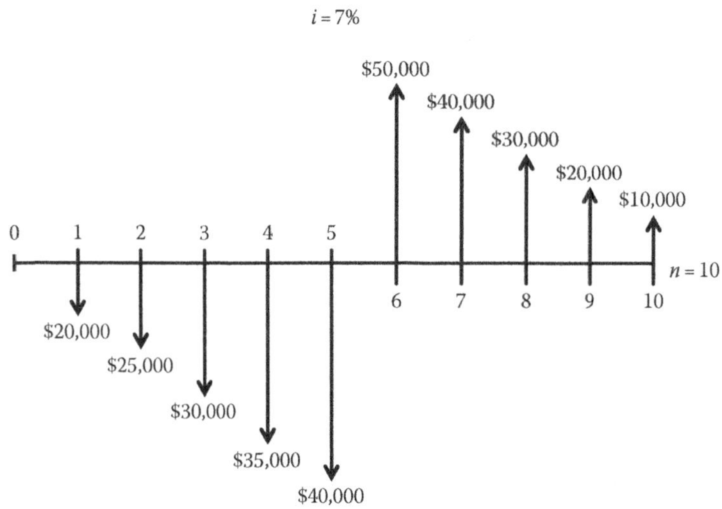

FIGURE 5.12 Cash flow diagram for the aerospace engineering firm investments in Case Study 5.2.

Solution

Procedure

First, calculate the present worth of the uniform annual series of $20,000.00 per year for five years plus the present worth of the gradient of $5,000.00 per year for five years using Equation 5.5.

Second, calculate the present worth of the uniform series starting in year six with the ETZ at year five and calculate the present worth of the decreasing uniform gradient series starting at year seven with the ETZ at year five using Equation 5.5.

Third, calculate the present worth at the PTZ of zero for the future worth of the second uniform series and the uniform gradient at year five.

Fourth, sum up the present worth at time zero of the two uniform annual series and the two uniform gradients:

Part I

$$P_0 = A(P/A,i,n) + G(P/G,i,n)$$
$$= -\$20{,}000.00(P/A,7,5) - \$5{,}000.00(P/G,7,5)$$
$$= -\$20{,}000.00(4.1002) - \$5{,}000.00(7.6466)$$
$$= -\$82{,}004.00 - \$38{,}233.00$$
$$= -\$120{,}237.00$$

Part II

$$P_5 = A(P/A,i,n) + G(P/G,i,n)$$
$$= \$50{,}000.00(P/A,7,5) - \$10{,}000.00(P/G,7,5)$$
$$= \$50{,}000.00(4.1002) - \$10{,}000.00(7.6466)$$
$$= \$205{,}010.00 - \$76{,}466.00$$
$$= \$128{,}544.00$$

Part III

$$P_5 = F_5$$
$$P_0 = F_5(P/F,i,n)$$
$$= \$128{,}544.00(P/F,7,5)$$
$$= \$128{,}544.00(0.71299)$$
$$= \$91{,}650.59$$

Part IV

$$P_{Total} = -P_0 \text{ of investments} + P_0 \text{ of withdrawals}$$
$$= -\$120{,}237.00 + \$91{,}650.59$$
$$= -\ \$28{,}586.41$$

Therefore, the investment stream is sufficient to cover the future withdrawal stream.

ALTERNATIVE SOLUTION

First, calculate the future worth of the \$20,000.00 uniform annual series and the \$5,000.00 gradient at year five.

Second, calculate the present worth of the \$50,000.00 uniform annual series and the \$10,000.00 gradient at year five.

Third, compare the future worth at year five of the first two series to the present worth of the second two series at year five.

Part I

$$F_5 = A(F/A,i,n) + G(F/G,i,n) = -\$20{,}000.00(F/A,7,5) - \$5{,}000.00(F/G,7,5)$$
$$= -\$20{,}000.00(5.7507) - \$5{,}000.00(10.724)$$
$$= -\$115{,}014.00 - \$53{,}620.00$$
$$= -\$168{,}634.00$$

Part II

$$P_5 = A(P/A,i,n) - G(P/G,i,n) = \$50{,}000.00(P/A,7,5) - \$10{,}000.00(P/G,7,5)$$
$$= \$50{,}000.00(4.1002) - \$10{,}000.00(7.6466)$$
$$= \$205{,}010.00 - \$76{,}466.00$$
$$= \$128{,}544.00$$

Part III

$$P_5 + F_5 = -\$168{,}634.00 + \$128{,}544.00 = -\$40{,}090.00$$

Therefore, the investment stream is sufficient to cover the future withdrawal stream.

5.4 EQUIVALENT UNIFORM GRADIENT FOR UNIFORM ANNUAL SERIES (*A/G*)

It is also possible to convert uniform gradient series into equivalent annual series, and the factor for performing this conversion is the *annual cost gradient factor* (*A/G*). In some factor tables, there are no uniform gradient present worth factors (*P/G*) or future sum gradient factors (*F/G*), and if this is the case, the uniform gradient series needs to first be converted into an equivalent uniform annual series using the annual cost gradient factor and then the equivalent uniform annual series is converted into a present worth using the uniform series present worth factor (*P/A*) or into a future worth using the uniform series compound amount factor (*F/A*). Appendix D provides the derivation of the formula for the uniform gradient annual series factor and the resulting formula for converting a uniform gradient series into an equivalent uniform annual series is the Equation 5.6.

$$A = G\left[\frac{1}{i} - \frac{n}{(1+i)^n - 1}\right] \tag{5.6}$$

Equation 5.7 is the formula for solving for the equivalent uniform annual series of a uniform gradient series using the interest factor tables in Appendix B:

$$A = G(A/G,i,n) \tag{5.7}$$

Case Study 5.3 provides a problem converting a uniform gradient into a uniform gradient annual series.

Case Study 5.3 Converting Uniform Gradients into Uniform Series

An electrical engineer works for an electrical distribution company. Each year, the company raises electricity rates by 10% per year. The initial rate is \$15,000.00 for the first year, the rate increases by 10% per year for five years, and then the rates are reevaluated and a new rate scheme is determined for the next five years. The electrical engineer is approached by a manager from a firm called Electroplus, Inc. with a request that they would prefer to pay a uniform amount for five years equal to the rate increasing by 10% per year. If the electric rate for Electroplus for year one were \$15,000.00, what would be the equivalent uniform annual series they would pay rather than having their rate increase by 10% per year? Use an interest rate of 3%. Figure 5.13 is the cash flow diagram for the electrical distribution services.

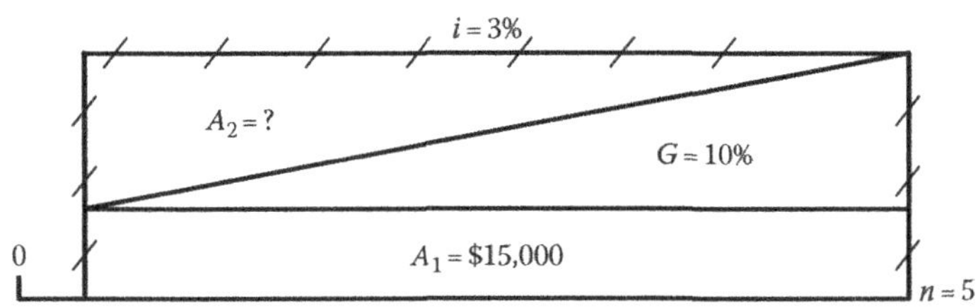

FIGURE 5.13 Cash flow diagram for the electrical distribution services in Example 5.7.

Solution

Procedure

First, calculate what the monetary amount of the gradient increase would be if the electric rates start at \$15,000.00 the first year and increase by 10% each year.

Second, calculate the equivalent uniform annual series for the uniform gradient series.

Third, add the equivalent uniform annual series of the gradient to the uniform annual series of \$15,000.00.

Part I

The amount of the gradient increase starting at year two is calculated using the following formula:

$$\text{Gradient increase} = \text{Initial amount} \times \text{Percentage increase per year}$$

$$\text{Gradient increase for years two through five} = \$15{,}000.00 \times 0.10 = \$1{,}500.00$$

Part II

$$\begin{aligned} A_2 &= G(A/G,i,n) \\ &= \$1{,}500.00(A/G,3,5) \\ &= \$1{,}500.00(1.9409) \\ &= \$2{,}911.35 \end{aligned}$$

Part III

$$A_{total} = A_1 + A_2$$
$$= \$15,000.00 + \$2911.35$$
$$= \$17,911.35$$

5.5 NONCONTINUOUS ARITHMETIC GRADIENT SERIES

Arithmetic gradients may also have their payments or disbursements start at a future period, as was demonstrated in Example 5.6, and this type of series is a *noncontinuous arithmetic gradients series.* Arithmetic gradients starting at an ETZ different from the problem time zero (PTZ) are addressed in one of the following three ways:

1. Calculate the present worth of the uniform gradient at the ETZ and then convert that value into a present worth at the PTZ by treating it as a future worth.
2. Convert the uniform gradient series into an equivalent uniform annual series, then solve for the present worth of the uniform annual series at the ETZ, and convert that value into a present worth at the PTZ by treating it as a future worth.
3. Calculate the future worth of the uniform gradient series and then calculate the present worth of the future worth at PTZ.

All three of these methods require two or three sets of calculations and they all yield the same answer; therefore, all three methods are used for calculating the present worth of a uniform gradient with an ETZ different from the PTZ. Example 5.5 provides calculations for all three of the methods for solving for the present worth of a noncontinuous arithmetic gradient.

Example 5.5

A systems engineer has convinced her boss that if the company makes an investment now in a new time saving product, the company will see an increase in profits of $10,000.00 starting in year six and the profits will continue to increase until year 10 by $10,000.00 per year. The interest rate is 10%. How much would the company need to invest now to realize the projected increase in profits? This problem is solved using all three of the methods for calculating the present worth of a noncontinuous series. Figure 5.14 is the cash flow diagram for the systems engineering firm product.

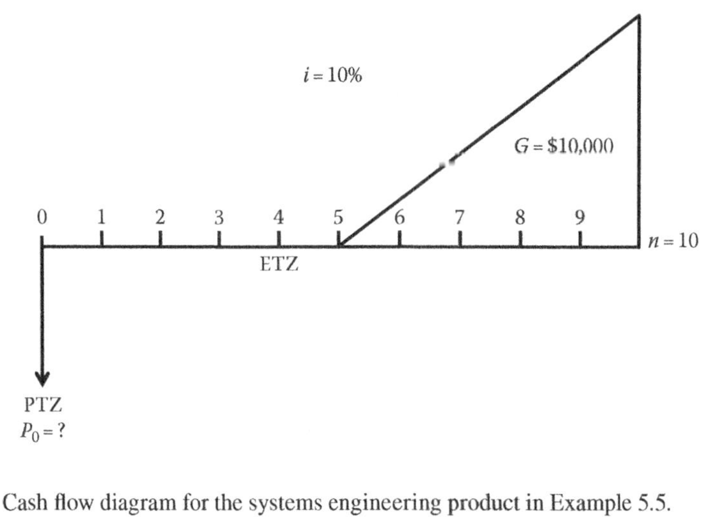

FIGURE 5.14 Cash flow diagram for the systems engineering product in Example 5.5.

Solution

For the first two methods, the ETZ is at year four, two years prior to the first increase in the profits.
Solve for the number of periods using the ETZ at four years:

$$\begin{aligned} n &= \text{Number of periods} - \text{ETZ} \\ &= 10 - 4 \\ &= 6 \end{aligned}$$

Method I

Calculate the present worth of the uniform gradient series at the ETZ at year four:

$$\begin{aligned} P_4 &= G(P/G,i,n) \\ &= \$10{,}000.00(P/G,10,6) \\ &= \$10{,}000.00(9.6842) \\ &= \$96{,}842.00 \end{aligned}$$

Second, calculate the present worth at the PTZ of year zero of the future value at year four:

$$\begin{aligned} P_0 &= F_4(P/F,i,n) \\ &= \$96{,}842.00(P/F,10,4) \\ &= \$96{,}842.00(0.68301) \\ &= \$66{,}144.05 \end{aligned}$$

Method II

First, convert the uniform gradient series into a uniform annual series at the ETZ of year four:

$$\begin{aligned} A_{ETZ\text{ of }4} &= G(A/G,i,n) \\ &= \$10{,}000.00(A/G,10,6) \\ &= \$10{,}000.00(2.2235) \\ &= \$22{,}235.00 \end{aligned}$$

Second, convert the uniform annual series into a present worth at year four:

$$\begin{aligned} P_4 &= A(P/A,i,n) \\ &= \$22{,}235.00(P/A,10,6) \\ &= \$22{,}235.00(4.3552) \\ &= \$96{,}837.87 \end{aligned}$$

Third, solve for the present worth of the future value at year four at the PTZ of year zero:

$$\begin{aligned} P_4 &= F_4 \\ P_0 &= F_4(P/F,i,n) \\ P_0 &= \$96{,}837.87(P/F,10,4) \\ &= \$96{,}837.87(0.68301) \\ &= \$66{,}141.23 \end{aligned}$$

Method III

First, calculate the future worth of the uniform gradient series at year 10:

$$
\begin{aligned}
F_{10} &= G(F/G,i,n) \\
&= \$10,000.00(F/G,10,6) \\
&= \$10,000.00(17.156) \\
&= \$171,560.00
\end{aligned}
$$

Second, calculate the present worth of the future value at year 10 at the PTZ at year zero:

$$
\begin{aligned}
P_0 &= F_{10}(P/F,i,n) \\
&= \$171,560.00(P/F,10,10) \\
&= \$171,560.00(0.38554) \\
&= \$66,143.24
\end{aligned}
$$

5.6 PERPETUAL LIFE GRADIENT SERIES: INFINITE SERIES

The formula for the present worth of an infinite uniform series was introduced in Section 4.5 and it is the following:

$$P_0 = \frac{A}{i}$$

In addition to infinite uniform series, there also could be gradients with a perpetual life such as the situation of increasing maintenance costs for a municipality or a trust fund where the person withdrawing funds is allowed to withdraw the funds in an increasing amount indefinitely. The formula for solving for the present worth of the perpetual life of a gradient is given in Equation 5.8.

$$P_0 = \frac{G}{i^2} \tag{5.8}$$

Example 5.6 is a problem where Equation 5.8 is used to calculate the present worth of a perpetual life gradient.

Example 5.6

A nuclear engineer realizes that the new nuclear power plant he is working on will have a maintenance cost starting to increase in year two by $10 million per year for the perpetual life of the plant. He recommends to his boss that the firm deposit an amount of money into an interest-bearing account now to cover the increasing maintenance costs. If the money is invested at 4%, what amount of money needs to be deposited now to cover the future maintenance cost? Figure 5.15 is the cash flow diagram for the maintenance costs.

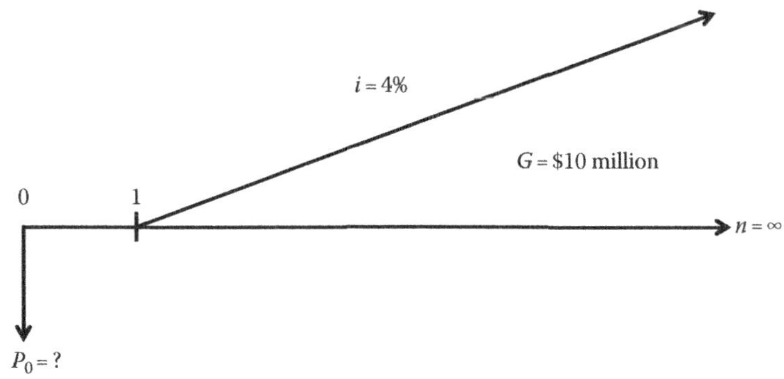

FIGURE 5.15 Cash flow diagram for the nuclear power plant maintenance cost in Example 5.6.

Solution

Equation 5.8 is used to solve for the present worth of the perpetual life gradient.

$$\begin{aligned} P_0 &= \frac{G}{i^2} \\ &= \frac{\$10{,}000{,}000.00}{0.04^2} \\ &= \frac{\$10{,}000{,}000.00}{0.0016} \\ &= \$6{,}250{,}000{,}000.00 \end{aligned}$$

The inverse of the formula for solving for the present worth of an infinite gradient series would be solving for the infinite gradient of a present worth and the formula for this is:

$$G = P_0\left(i^2\right) \tag{5.9}$$

Example 5.7 demonstrates calculating the infinite gradient of a present value.

Example 5.7

If a firm deposits $10,500,000.00 in an interest-bearing account, what amount of money may be withdrawn from this account in an increasing infinite gradient if the interest rate is 8%? Figure 5.16 is the cash flow diagram for the infinite uniform gradient series.

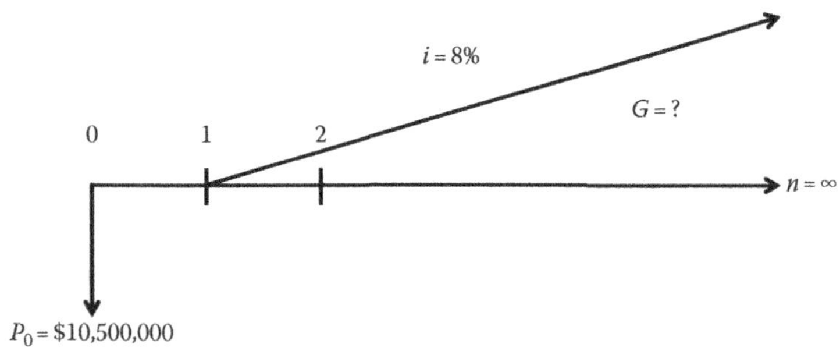

FIGURE 5.16 Cash flow diagram for the infinite uniform gradient series in Example 5.7.

Solution

Equation 5.9 is used to solve for the infinite uniform gradient series of the present value:

$$G = P_0(i^2)$$
$$= \$10{,}500{,}000.00(0.08^2)$$
$$= \$10{,}500{,}000.00(0.0064)$$
$$= \$67{,}200.00$$

5.7 GEOMETRIC GRADIENTS

In addition to calculating the present worth of arithmetic gradients representing a uniform series of increasing or decreasing amounts, there are also situations where gradients increase or decrease by a constant percentage. In Case Study 5.3, the electrical rates increasing by 10% per year represent a *geometric gradient.* Geometric gradients increase or decrease by a constant percentage but the amount of the increase or decrease is variable. When calculating geometric gradients, an initial amount (C) is included in the calculations and (C) represents the first payment of the increasing series. The first payment is the *base payment.*

For geometric gradients, there is a *rate of growth* or decline, and this growth rate is represented by the symbol (r) to distinguish it from the interest rate (i). A growth rate of 9% means $r = 0.09$ and a declining growth rate of 5% indicates $r = -0.05$. There are three formulas for solving for the present worth of a geometric gradient and the three formulas are based on whether

1. $r > i$
2. $r < i$
3. $r = i$

The three formulas for calculating the present worth of geometric gradients are the following equations:

1. When $r > i$, then

$$w = \frac{1+r}{1+i} - 1, \quad P_0 = \frac{C}{1+i}(F/A,i,n)$$
$$= \frac{C}{1+i}\left[\frac{(1+w)^n - 1}{w}\right] \qquad P/C \Rightarrow \infty \qquad (5.10)$$

2. When $r < i$, then

$$w = \frac{1+i}{1+r} - 1, \quad \text{and} \quad P_0 = \frac{C}{1+r}(P/A,i,n)$$
$$= \frac{C}{1+r}\left[\frac{(1+w)^n - 1}{w(1+w)^n}\right] \qquad P/C \Rightarrow \frac{1}{(1+r)w} \qquad (5.11)$$

3. When $r = i$, then

$$P_0 = \frac{C\times n}{(1+r)} = \frac{C\times n}{(1+i)} \qquad P/C \Rightarrow \infty \qquad (5.12)$$

Figure 5.17 is the cash flow diagram for geometric gradients.

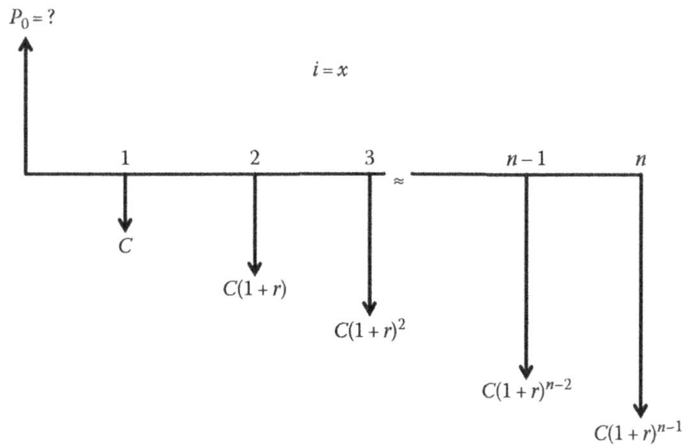

FIGURE 5.17 Cash flow diagram for geometric gradients.

Example 5.8 demonstrates using two of the geometric gradient equations to solve for the different situations for (r).

Example 5.8

The initial amount of a geometric gradient series is \$600.00, the number of years is 25, the interest rate is 10%, the rate of growth is 10% for Part I, and 12% for Part II. What is the future worth of the gradient? Figure 5.18 is the cash flow diagram for Example 5.7 for the growth rate of 10% and Figure 5.19 is the cash flow diagram for the growth rate of 12%.

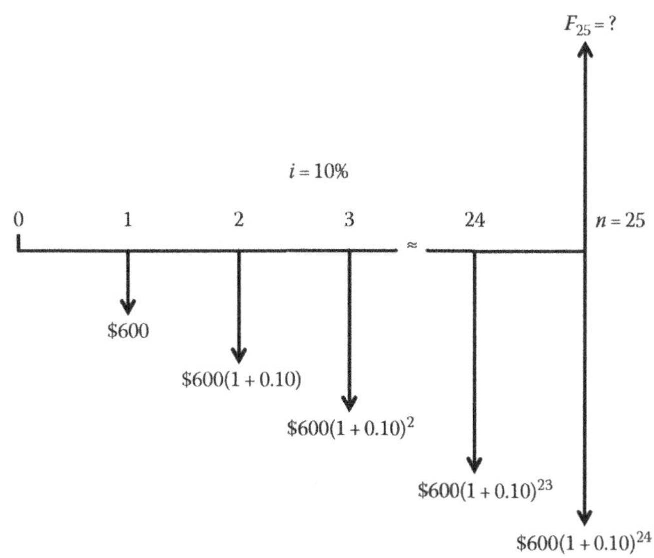

FIGURE 5.18 Cash flow diagram for Example 5.9 for a growth rate of 10%.

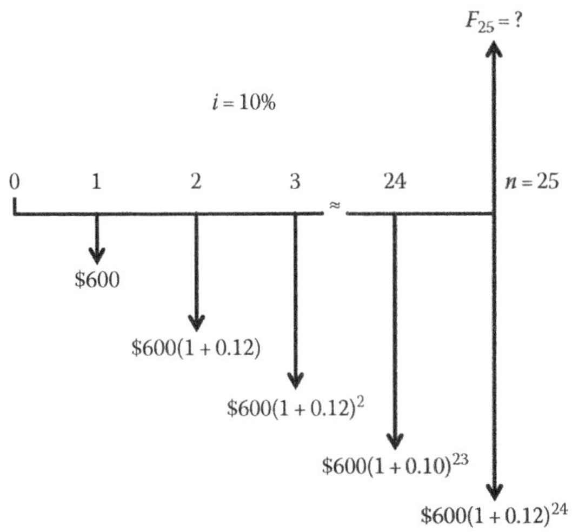

FIGURE 5.19 Cash flow diagram for Example 5.9 for a growth rate of 12%.

Solution

Part I

$r = i$

Use Equation 5.10 to calculate the present worth of the geometric gradient at time 0 and then calculate the future worth at year 25:

$$P_0 = \frac{C \times n}{(1+r)} = \frac{C \times n}{(1+i)}$$

$$= \frac{\$600.00 \times 25}{(1+0.10)}$$

$$= \frac{\$15,000.00}{1.10}$$

$$= \$13,636.36$$

$$F_{25} = P(F/P, i, n)$$

$$= \$13,636.36(F/P, 10, 25)$$

$$= \$13,636.36(10.834)$$

$$= \$147,736.32$$

Part II

$r > i$

Use Equation 5.9 to calculate the present worth of the geometric gradient series at time 0 and then calculate the future worth at 25 years:

$$w = \frac{1+r}{1+i} - 1$$

$$w = \frac{1+0.12}{1+0.10} - 1$$

$$= \frac{1.12}{1.10} - 1$$

$$= 0.018182$$

$$P_0 = \frac{C}{1+i}\left[\frac{(1+w)^n - 1}{w}\right]$$

$$= \frac{\$600.00}{(1+0.10)}\left[\frac{(1+0.018182)^{25} - 1}{0.018182}\right]$$

$$= \$545.45\left(\frac{0.569045}{0.018182}\right)$$

$$= \$545.45(31.297)$$

$$= \$17,070.95$$

$$F_{25} = P(F/P,i,n)$$

$$= \$17,070.95(F/P,10,25)$$

$$= \$17,071.04(10.834)$$

$$= \$184,946.66$$

Example 5.9 provides a problem where the rate of growth is lesser than the interest rate.

Example 5.9

A student is leasing a vehicle. The lease costs \$2000.00 this year and it will increase by 4% a year. If the interest rate is 10% a year, what is the present worth of the leased vehicle payments over four years? Figure 5.20 is the cash flow diagram for the leased vehicle.

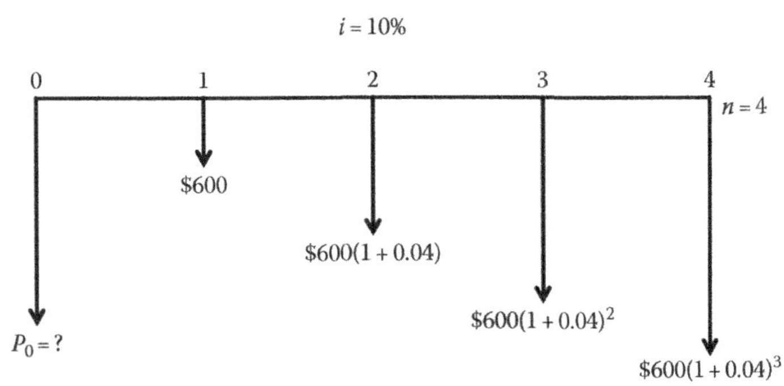

FIGURE 5.20 Cash flow diagram for leasing the vehicle in Example 5.8.

Solution

For $r < i$

Use Equation 5.10 to calculate the present worth of the geometric gradient:

$$w = \frac{1+i}{1+r} - 1$$

$$= \frac{1+0.10}{1+0.04} - 1$$

$$= \frac{1.10}{1.04} - 1$$

$$= 0.057692$$

$$P_0 = \frac{C}{1+r}\left[\frac{(1+w)^n - 1}{w(1+w)^n}\right]$$

$$= \frac{\$2,000.00}{(1+0.04)}\left[\frac{(1+0.057692)^4 - 1}{0.057692(1+0.057692)^4}\right]$$

$$= \frac{\$2,000.00}{(1.04)}\left(\frac{0.251517}{0.072203}\right)$$

$$= \$1,923.08(3.4835)$$

$$= \$6,699.05$$

5.8 SUMMARY

This chapter discussed and defined arithmetic gradients and explained the process for calculating the future worth of arithmetic gradients. Example problems and a case study demonstrating the process for using the future worth gradient factor (*F*/*G*) were included in this chapter. The calculations for determining the present worth of arithmetic gradients using the present worth gradient factor (*P*/*G*) were covered along with the process for solving for decreasing values of uniform gradients series and increasing and decreasing gradients.

This chapter also demonstrated the method for calculating an equivalent uniform annual series from a uniform gradient series and provided a case study explaining the process for converting a uniform gradient series into an equivalent uniform annual series using the annual cost gradient factor (*A*/*G*). Noncontinuous arithmetic gradient series were explained and three methods were provided for solving for the present worth of noncontinuous gradient series.

This chapter addressed perpetual life gradient series and included the formula for calculating the present worth of an infinite gradient and the infinite gradient of a present value. Section 5.7 introduced geometric gradients, which are gradients increasing by a uniform rate, and the last section of this chapter also provided three formulas for calculating the present worth of geometric gradients for when the rate of increase equals, is greater than, or is less than the interest rate.

KEY TERMS

Annual cost gradient factor
Arithmetic gradient
Base payment

Decreasing uniform gradients
Future worth gradient factor
Geometric gradient
Noncontinuous arithmetic gradient series
Present worth gradient factor
Rate of growth
Uniform gradient annual series factor

PROBLEMS

5.1 A county engineer plans on implementing a road improvement project over the next six years. The first improvements will be implemented at year two and cost $200,000.00. The cost of the improvements will increase by $50,000.00 per year starting at year two until year six. If the interest rate is 3%, what is the present worth of the road improvement project? Solve using the interest factor tables in Appendix B.

5.2 A manufacturing plant is budgeting for increasing expenses and determines the costs will be $10,000.00 starting at year two and increasing by $10,000.00 per year until year 10. What is the future worth of the expenses if the interest rate is 2%? Use the future worth gradient factor formula to solve this problem.

5.3 Solve Problem 5.2 using the interest factor tables in Appendix B.

5.4 A mining company is spending funds to improve the efficiency of one of their mines. The company will spend $100,000.00 the first year and the amount will increase by $10,000.00 per year for the next four years. What would be the future worth of the expenditures after five years if the interest rate is 7%? Use formulas to solve this problem.

5.5 Solve for the present worth of the scenario in Problem 5.4 using the interest factor tables in Appendix B.

5.6 A communications provider expects to make a profit of $10,500,000.00 the first year of operations. The manager of the company estimates profits will decrease by $400,000.00 each year after the first year until year five when the firm will install new equipment. What is the future worth of the profits at year five if the interest rate is 6%? Solve using the interest factor tables in Appendix B.

5.7 Solve for the present worth of the scenario in Problem 5.6 using the interest factor tables in Appendix B.

5.8 A chemical engineering firm needs to invest funds over the next five years in preparation for covering increasing operating costs. The firm plans on investing $275,000.00 the first year and to increase its investment by $50,000.00 per year until year seven. The firm will start withdrawing funds at year eight. At year eight, the firm will withdraw $250,000.00 and the amount they withdraw will increase by $100,000.00 each year until year 14. If the interest rate is 8%, will the firm have enough funds deposited to be able to withdraw the proposed amounts?

5.9 A geotechnical engineering firm has been awarded a job to design the foundation of an office building. The project is postponed and will not start for four more years. The firm needs to estimate the present worth of the costs of the project. At year four, the cost will be $175,000.00 decreasing by $50,000.00 per year for the next three years. If the interest rate is 4%, what is the present worth of the projected cost of the project? Solve by calculating the present worth using the interest factor tables in Appendix B.

5.10 An engineer plans on depositing money into a savings account each year starting with $500.00 at year two and increasing the amount by $500.00 per year until the end of year 20. If the interest rate is 0.75% per year, how much will the engineer have in his account at the end of 20 years?

5.11 What would be the present worth of the $500.00 gradient in Problem 5.10?

5.12 A rural electric cooperative company is budgeting funds to cover electrical transmission line repairs over the next 10 years. An electrical engineer working for the company estimates a yearly cost of $375,000.00 increasing by $50,000.00 per year starting at year four. If the interest rate is 9%, what amount of money needs to be deposited now to cover the transmission line repairs over the next 10 years?

5.13 An agricultural engineering firm has purchased a new machine in anticipation of its increasing profits starting at year four by $20,000.00 increasing by $20,000.00 for five more years. Using an interest rate of 13%, what is the future worth of the increased profits?

5.14 Solve for the present worth of the increased profits in Problem 5.13.

5.15 An engineer sets up an investment to pay for vehicle repairs. She plans on investing $200.00 the first year increasing by $200.00 per year for four more years. She anticipates repairs of $500.00 per year starting at year six through year 10. Will there be enough funds in the account to cover the estimated repairs from year six through year 10 if the interest rate is 1%?

5.16 A mechanical engineering firm has projected expenses of $4,500,000.00 per year increasing by $275,000.00 per year for five more years. What is the equivalent uniform annual series for the expenses if the interest rate is 7%?

5.17 A biomedical firm plans on investing a gradient series of $10,000.00 in a perpetual life investment starting at year two. If the interest rate is 2.5%, what is the present worth of the investment?

5.18 An engineer deposits $3,000.00 into a savings account. He will increase the amount in the account by 5% a year for 10 years. If the interest rate is 2% per year, what is the present worth of the amount in the account?

5.19 An aerospace engineering firm needs to maintain a piece of equipment costing $62,000.00 a year increasing by 4% per year for the next eight years. Using an interest rate of 6%, calculate the present worth of the maintenance costs.

5.20 A systems engineering firm installed a new process for a manufacturing company. The new process will save the manufacturing company $100,000.00 a year increasing by 2% per year for 12 years. Using an interest rate of 2%, calculate the present worth of the savings.

6 Multiple Factors in Engineering Economic Problems

This chapter demonstrates using the single payment compound amount factor (*F*/*P*), single payment present worth factor (*P*/*F*), uniform series compound amount factor (*F*/*A*), uniform series present worth factor (*P*/*A*), uniform series sinking fund factor (*A*/*F*), uniform series capital recovery factor (*A*/*P*), uniform gradient present worth factor (*P*/*G*), uniform gradient future worth factor (*F*/*G*), and the annual cost gradient factor (*A*/*G*)—all of which were introduced in Chapters 3 through 5—to solve problems with multiple payment and disbursement streams.

6.1 COMBINING FACTORS TO SOLVE FOR THE FUTURE OR PRESENT WORTH OF DIFFERENT SERIES

When payments or disbursement series start at a time not equal to problem time zero (*PTZ*), *multiple factors* are combined to solve for the present or future worth of the series. Figures 6.1 through 6.6 show sample cash flow diagrams for problems requiring multiple factors.

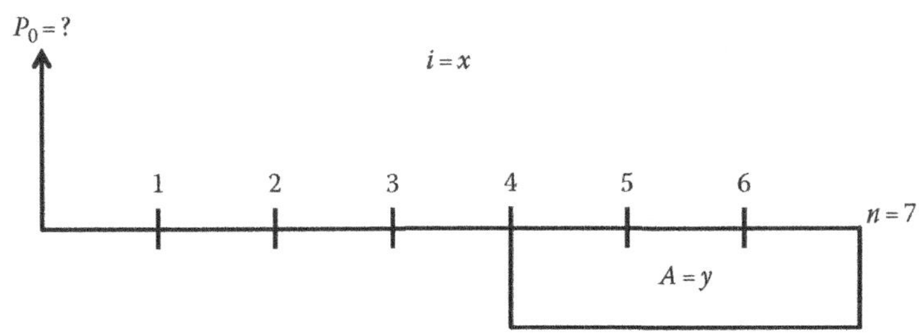

FIGURE 6.1 Sample A—cash flow diagram of problem requiring multiple factors.

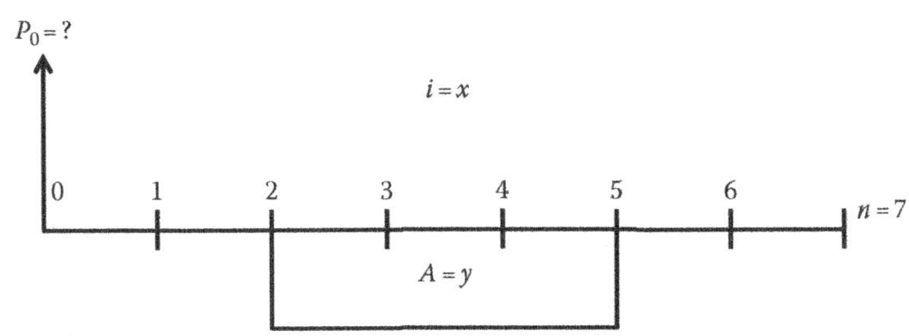

FIGURE 6.2 Sample B—cash flow diagram of problem requiring multiple factors.

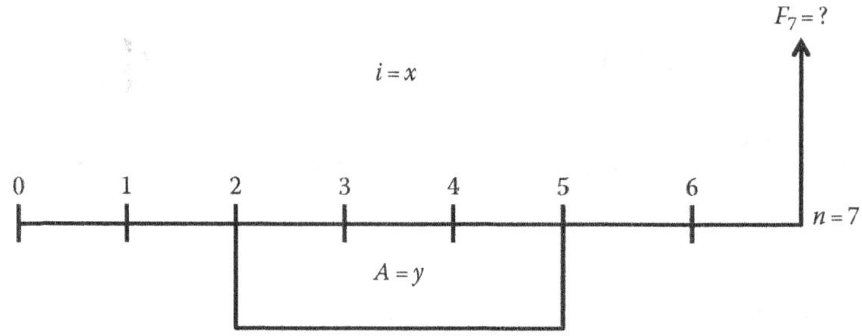

FIGURE 6.3 Sample C—cash flow diagram of problem requiring multiple factors.

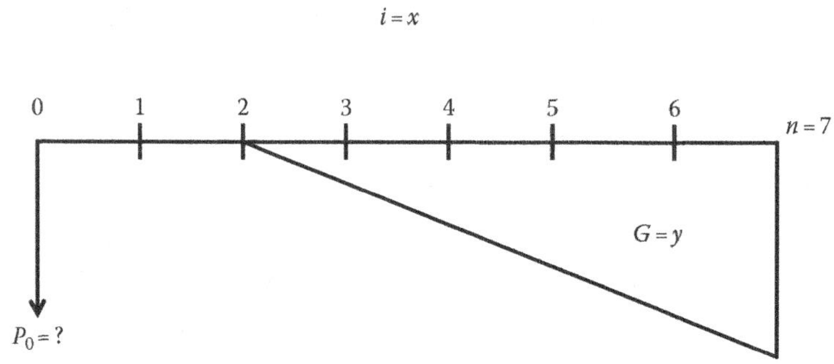

FIGURE 6.4 Sample D—cash flow diagram of problem requiring multiple factors.

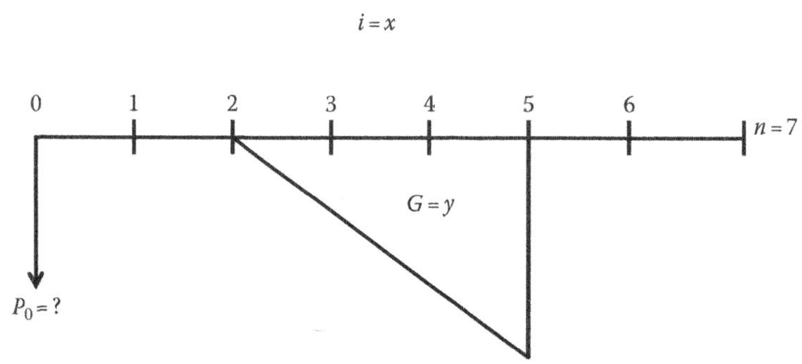

FIGURE 6.5 Sample E—cash flow diagram of problem requiring multiple factors.

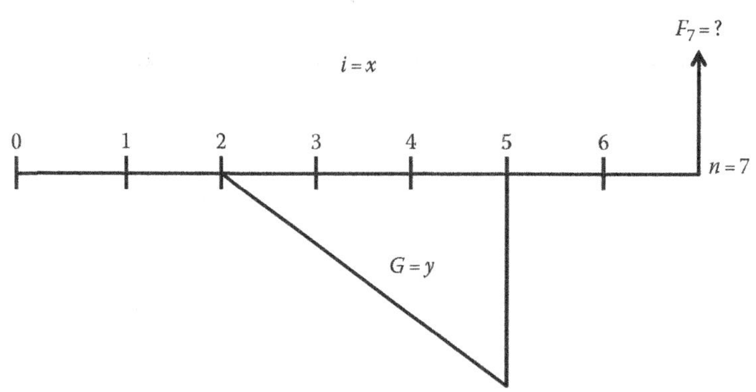

FIGURE 6.6 Sample F—cash flow diagram of problem requiring multiple factors.

In addition to using multiple factors to solve for the present or future worth of the payments and disbursements shown in Figures 6.1 through 6.6, there could be problems combining the different elements of the cash flow diagrams in Figures 6.1 through 6.6. Case Study 6.1 provides solutions to the economic situations represented by the cash flow diagrams in Figures 6.1 through 6.6.

Case Study 6.1 Solutions to the Economic Situations in Figures 6.1 Through 6.6

The following solutions represent the present or the future worth of the payments or disbursements represented in Figures 6.1 through 6.6 with an annuity of $1,000.00 or a gradient of $500.00 and an interest rate of 5% (all of the values are negative values).

Solution for Figure 6.1

ETZ is at year 3 and $n = 7 - 3 = 4$

$$P_3 = A(P/A,i,n) = \$1{,}000.00(P/A,5,4) = \$1{,}000.00(3.5459) = \$3{,}545.90$$

$$P_3 = F_3$$

$$P_0 = F_3(P/F,i,n) = \$3{,}545.90(P/F,5,3) = \$3{,}545.90(0.86384) = \$3{,}063.09$$

or

$$F_7 = A(F/A,i,n) = \$1{,}000.00(F/A,5,4) = \$1{,}000.00(4.3101) = \$4{,}310.00$$

$$P_0 = F_7(P/F,i,n) = \$4{,}310.00(P/F,5,7) = \$4{,}310.00(0.71068) = \$3{,}063.03$$

Solution for Figure 6.2

ETZ is at year 1 and $n = 5 - 1 = 4$

$$P_1 = A(P/A,i,n) = \$1{,}000.00(P/A,5,4) = \$1{,}000.00(3.5459) = \$3{,}545.90$$

$$P_1 = F_1$$

$$P_0 = F_1(P/F,i,n) = \$3{,}545.90(P/F,5,1) = \$3{,}545.90(0.95238) = \$3{,}377.04$$

or

$$F_5 = A(F/A,i,n) = \$1{,}000.00(F/A,5,4) = \$1{,}000.00(4.3101) = \$4{,}310.10$$

$$P_0 = F_5(P/F,i,n) = \$4{,}310.10(P/F,5,5) = \$4{,}310.10(0.78353) = \$3{,}377.09$$

Solution for Figure 6.3

ETZ is at year 1 and $n = 5 - 1 = 4$

$$F_5 = A(F/A,i,n) = \$1{,}000.00(F/A,5,4) = \$1{,}000.00(4.3101) = \$4{,}310.10$$

$$F_5 = P_5$$

$$F_7 = P_5(F/P,i,n) = \$4{,}310.10(F/P,5,2) = \$4{,}310.10(1.1025) = \$4{,}751.89$$

or

$$P_1 = A(P/A,i,n) = \$1{,}000.00(P/A,5,4) = \$1{,}000.00(3.5459) = \$3{,}545.9$$

$$F_7 = P_1(F/P,i,n) = \$3{,}545.90(F/P,5,6) = \$3{,}545.90(1.3401) = \$4{,}751.86$$

Solution for Figure 6.4

ETZ is at year 1 and $n = 7 - 1 = 6$

$$P_1 = G(P/G,i,n) = \$500.00(P/G,5,6) = \$500.00(11.968) = \$5{,}984.00$$

$$P_1 = F_1$$

$$P_0 = F_1(P/F,i,n) = \$5{,}984.00(P/F,5,1) = \$5{,}984.00(0.95238) = \$5{,}699.04$$

or

$$F_7 = G(F/G,i,n) = \$500.00(F/G,5,6) = \$500.00(16.038) = \$8{,}019.00$$

$$P_0 = F_7(P/F,i,n) = \$8{,}019.00(P/F,5,7) = \$8{,}019.00(0.71068) = \$5{,}698.94$$

Solution for Figure 6.5

ETZ is at year 1 and $n = 5 - 1 = 4$

$$P_1 = G(P/G,i,n) = \$500.00(P/G,5,4) = \$500.00(5.1028) = \$2{,}551.40$$

$$P_1 = F_1$$

$$P_0 = F_1(P/F,i,n) = \$2{,}551.40(P/F,5,1) = \$2{,}551.40(0.95238) = \$2{,}429.90$$

or

$$F_5 = G(F/G,i,n) = \$500.00(F/G,5,4) = \$500.00(6.2025) = \$3{,}101.25$$

$$P_0 = F_5(P/F,i,n) = \$3{,}101.25(P/F,5,5) = \$3{,}101.25(0.78353) = \$2{,}429.92$$

Solution for Figure 6.6

ETZ is at year 1 and $n = 5 - 1 = 4$

$$P_1 = G(P/G,i,n) = \$500.00(P/G,5,4) = \$500.00(5.1028) = \$2{,}551.40$$

$$n = 7 - 1 = 6$$

$$F_7 = P_1(F/P,i,n) = \$2{,}551.40(F/P,5,6) = \$2{,}551.40(1.3401) = \$3{,}419.13$$

or

$$F_5 = G(F/G,i,n) = \$500.00(F/G,5,4) = \$500.00(6.2025) = \$3{,}101.25$$

$$F_5 = P_5$$

$$F_7 = P_5(F/P,i,n) = \$3{,}101.25(F/P,5,2) = \$3{,}101.25(1.1025) = \$3{,}419.13$$

Many economic situations require multiple factors to represent combinations of payment and disbursement streams. Situations where there are savings accounts, trust funds, maintenance contracts, and investment instruments, all represent deposits and withdrawals occurring at various times; therefore, they all require multiple factors to solve for either the present or future worth or equivalent uniform annual series.

Example 6.2 provides a problem with a trust fund scenario demonstrating the use of multiple factors.

Example 6.1

As soon as their first child is born, a couple wants to start a trust fund they will pay into on a yearly basis until their child reaches the age of 18. Once their child reaches 18, the child would be able to withdraw $20,000.00 for years 18 through 21. If the interest rate is 4%, what amount does the couple need to invest each year for 18 years to have enough money in the trust fund to meet the four-year payout amounts? Figure 6.7 is the cash flow diagram for the family investment.

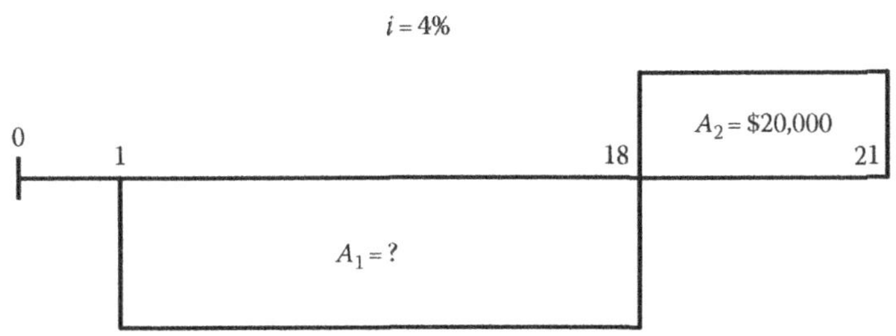

FIGURE 6.7 Cash flow diagram for the family investment in Example 6.1.

Solution

First, calculate the present worth of the three-year annuity of $20,000.00 at equation time zero of year 18 and add this value to the $20,000.00 at year 18:

$$P_{18} = A_2(P/A, i, n) + A_2 = \$20{,}000.00(P/A, 4, 3) + \$20{,}000.00$$
$$= \$20{,}000.00(2.7750) + \$20{,}000.00$$
$$= \$55{,}500.00 + \$20{,}000.00$$
$$= \$75{,}500.00$$

Second, use the present worth at year 18 of $75,500.00 as the future worth of the 18-year deposit annuity and solve for the yearly deposit amount:

$$P_{18} = F_{18}$$
$$A_1 = F_{18}(A/F, i, n) = \$75{,}500.00(A/F, 4, 18)$$
$$= \$75{,}500.00(0.03899)$$
$$= \$2{,}943.74$$

Case Study 6.2 demonstrates the process for using multiple factors to calculate the amount of money a student will owe on their student loan when they graduate from college and the amount they will have to repay each year.

Case Study 6.2 Repaying Student Loans

A student borrows \$20,000.00 per year for the four years he is in college. One year after graduating, the student starts to repay his student loan. The interest rate is 8%. Part I—How much would the student be paying each month for the next 10 years and what is the total amount he will have repaid after 10 years? Figure 6.8 is the cash flow diagram for the student loan in Part I. Part II—What would the student pay per month if he repaid the loan in 20 years instead of 10 years and what is the total amount repaid after 20 years? Figure 6.9 is the cash flow diagram for the student loan in Part II.

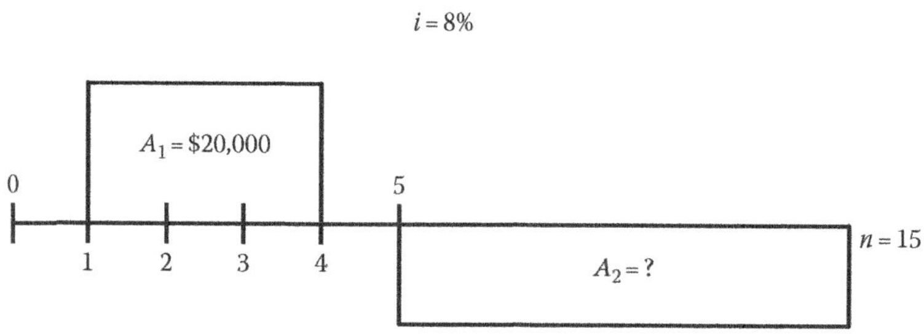

FIGURE 6.8 Cash flow diagram for the 10-year repayment plan in Case Study 6.2 Part A.

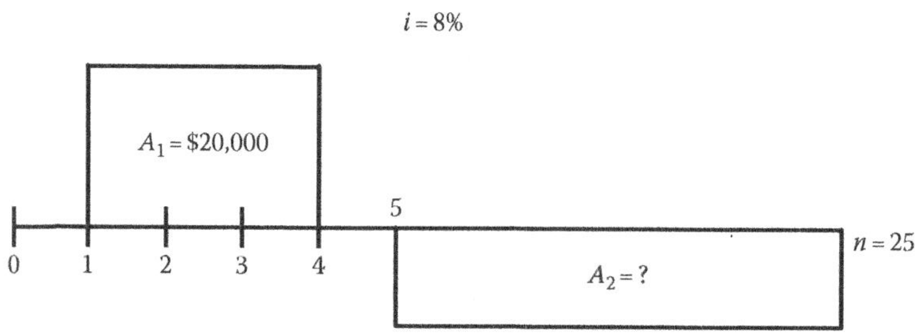

FIGURE 6.9 Cash flow diagram for the 20-year repayment plan in Case Study 6.2 Part B.

Solution to Part I

The student will have borrowed \$80,000.00 by year four and there is no time value of money on the \$80,000.00 since no interest is charged on the loan until a student graduates. Therefore, \$80,000.00 is the amount owed when the student starts to repay the loan in year five.

Procedure

First, calculate the monthly interest rate and the total numbers of months:

$$i = \frac{i_n}{m} = \frac{8\%/\text{Year}}{12\ \text{periods/Year}} = 0.667\%\ \text{per period}$$

$$n = \frac{\text{Number of compounding periods}}{\text{Year}} \times \text{Number of years}$$

$$= \frac{12\ \text{periods}}{\text{Year}} \times 10\ \text{years} = 120\ \text{periods}$$

Second, calculate the uniform monthly series starting at year five until year 15:

$$A_2 = P\left[\frac{i(1+i)^n}{(1+i)^n - 1}\right] = \$80,000.00\left[\frac{0.00667(1+0.00667)^{120}}{(1+0.00667)^{120} - 1}\right]$$

$$= \$80,000.00\left(\frac{0.014811}{1.22052}\right) = \$80,000.00(0.012135)$$

$$= \$970.80 \text{ per month}$$

Third, the amount repaid is the monthly payment times the total number of months:

$$F_{15} = \$970.80 \times 120 \text{ months} = \$116,496.00$$

Therefore, the student paid \$36,400.00 in interest on the \$80,000.00 repaid monthly over 10 years on the student loan.

Fourth, calculate the future worth of the uniform series at year 15:

$$F_{15} = \$970.80\left[\frac{(1+0.00667)^{120} - 1}{0.00667}\right] = \$970.80\left(\frac{1.220522}{0.00667}\right)$$

$$= \$970.80(182.9869)$$

$$= \$177,643.68$$

This is the amount the funds would have been worth at year 15 if the monthly payments were invested at an interest rate of 8% per year compounded monthly instead of repaying the student loan.

Solution to Part II

The student will have borrowed \$80,000.00 by year four and there is no time value of money on the \$80,000.00 since no interest is charged on the loan until a student graduates. Therefore, \$80,000.00 is the amount owed when the student starts to repay the loan at year five.

Procedure

First, solve for the monthly interest rate and the total numbers of months:

$$i = \frac{i_n}{m} = \frac{8\%/\text{Year}}{12 \text{ periods/Year}} = 0.667\% \text{ per period}$$

$$n = \frac{\text{Number of compounding periods}}{\text{Year}} \times \text{Number of years}$$

$$= \frac{12 \text{ periods}}{\text{Year}} \times 20 \text{ years} = 240 \text{ periods}$$

Second, calculate the uniform monthly series starting at year five until year 15:

$$A = P_0\left[\frac{i(1+i)^n}{(1+i)^n-1}\right] = \$80,000.00\left[\frac{0.00667(1+0.00667)^{240}}{(1+0.00667)^{240}-1}\right]$$

$$= \$80,000.00\left(\frac{0.03289}{3.93072}\right) = \$80,000.00(0.008367)$$

$$= \$669.40 \text{ per month}$$

Third, the amount repaid is the monthly payment times the total number of months:

$$F_{15} = \$669.40\times 240 \text{ months} = \$160,656.00$$

Therefore, the student paid \$80,656.00 in interest on the \$80,000.00 repaid monthly over 20 years on the student loan.

Fourth, calculate the future worth of the uniform series at year 25:

$$F_{25} = \$669.40\left[\frac{(1+0.00667)^{240}-1}{0.00667}\right] = \$669.40\left(\frac{3.93072}{0.00667}\right)$$

$$= \$669.40(589.3133)$$

$$= \$394,486.32$$

Thus is the amount the funds would have been worth at year 25 if the monthly payments were invested at an interest rate of 8% per year compounded monthly instead of repaying the student loan.

6.2 TWO SEQUENTIAL SERIES WITH DIFFERENT INTEREST RATES

In some instances, interest rates may change during a sequence of payments or the amount of the payments may increase or decrease. These two situations could occur with home mortgages when the interest rate charged on the home mortgage loan is adjusted periodically. When this occurs, the solution requires multiple factors to compensate for different interest rates or uniform annual series.

Example 6.2 demonstrates calculating solutions when there are different interest rates and uniform annual series with different values.

Example 6.2

A biomedical engineering firm needs to save funds to pay a large tax bill due at the end of 15 years. For the first four years, the firm will be able to save \$10,000.00 per year at an interest rate of 6%. Then starting at year six, the firm will save \$20,000.00 per year until year 15 at an interest rate of 7%. How much will the firm have in 15 years to pay their taxes? Figure 6.10 is the cash flow diagram for the tax bill savings plan.

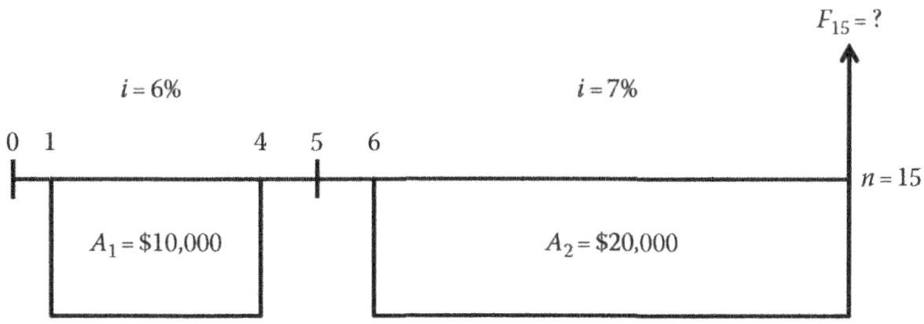

FIGURE 6.10 Cash flow diagram for the biomedical engineering firm tax bill in Example 6.2.

Solution

First, calculate the future worth at year four of the first uniform annual series using the interest rate of 6%:

$$F_4 = A_1(F/A,i,n) = \$10,000.00(F/A,6,4) = \$10,000.00(4.3746)$$
$$= \$43,746.00$$

Second, calculate the future worth of the present value at year four, which is the future worth of the first uniform annual series using the interest rate of 7%:

$$F_4 = P_4$$
$$F_{15\,of\,A_1} = P_4(F/P,i,n) = \$43,746.00(F/P,7,11) = \$43,746.00(2.1048)$$
$$= \$92,076.58$$

Third, calculate the future worth of the second uniform annual series at year 15 using an interest rate of 7%:

$$F_{15\,of\,A_2} = A_2(F/A,i,n) = \$20,000.00(F/A,7,10) = \$20,000.00(13.816)$$
$$= \$276,320.00$$

Fourth, solve for the total future worth at year 15 by summing up the future worth of the first and second uniform annual series:

$$F_{total} = F_{15\,of\,A_1} + F_{15\,of\,A_2} = \$92,076.58 + \$276,320.00$$
$$= \$368,396.58$$

Example 6.3 illustrates calculating the present worth of both uniform annual series and future values.

Example 6.3

An industrial engineer has determined the amount of money required to be withdrawn from an investment to meet periodic expenses for a project the firm is contemplating undertaking soon. She has determined the firm will be required to withdraw $10,000.00 a year starting at year two through year 10. At year 10, $8,000.00 will be deposited into the investment account paying 10% interest. At year 15, the firm will start depositing $3,000.00 a year until year 25. Also, at year 15, $5,000.00 will be withdrawn from the account. In the 20th year $7,000.00 will be withdrawn, and at year 25, $9,000.00 will be withdrawn. What is the present worth of the future payment and disbursement streams? Figure 6.11 is the cash flow diagram of the industrial engineering project.

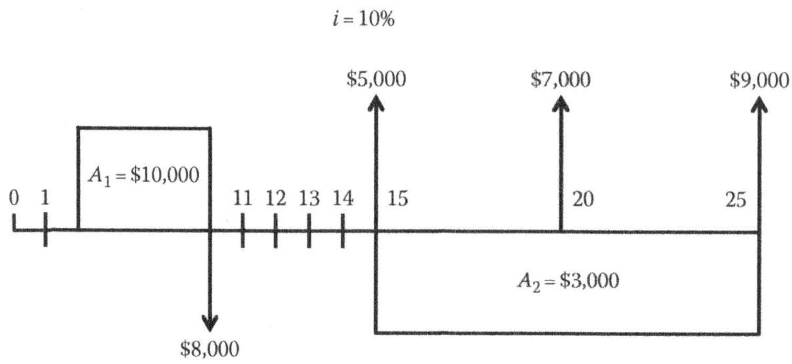

FIGURE 6.11 Cash flow diagram for the industrial engineering project in Example 6.3.

Solution

$$\text{For } A_1 \text{ the ETZ is at year 1 and } n = 10 - 1 = 9 \text{ years}$$

$$\text{For } A_2 \text{ the ETZ is at year 14 and } n = 25 - 14 = 11 \text{ years}$$

First, calculate the present worth of the uniform annual series A_1 at the ETZ at year one:

$$\begin{aligned} P_{1\,\text{of}\,A_1} &= A_1(P/A,i,n) = \$10{,}000.00(P/A,i,n) = \$10{,}000.00(P/A,10,9) \\ &= \$10{,}000.00(5.7590) \\ &= \$57{,}590.00 \end{aligned}$$

Second, calculate the present worth at year zero of the present worth of the uniform annual series A_1 at year one:

$$\begin{aligned} P_1 &= F_1 \\ P_{0\,\text{of}\,A_1} &= F_1(P/F,i,n) = \$57{,}590.00(P/F,10,1) = \$57{,}590.00(0.90909) \\ &= \$52{,}354.49 \end{aligned}$$

Third, calculate the present worth of the uniform annual series A_2 at the ETZ of year 14:

$$\begin{aligned} P_{14\,\text{of}\,A_2} &= A_2(P/A,i,n) = \$3{,}000.00(P/A,10,11) = \$3{,}000.00(6.4950) \\ &= -\$19{,}485.00 \end{aligned}$$

Fourth, calculate the present worth at time zero of the future worth at year 14:

$$\begin{aligned} P_{14} &= F_{14} \\ P_{0\,\text{of}\,A_2} &= F_{14}(P/F,i,n) = -\$19{,}485.00(P/F,10,14) = -\$19{,}485.00(0.26333) \\ &= -\$5{,}130.98 \end{aligned}$$

Fifth, calculate the present worth of the future values at year 10, 15, 20, and 25:

$$
\begin{aligned}
P_0 &= F_{10}(P/F,i,n)+F_{15}(P/F,i,n)+F_{20}(P/F,i,n)+F_{25}(P/F,i,n)\\
&= -\$8,000.00(P/F,10,10)+\$5,000.00(P/F,10,15)+\$7,000.00(P/F,10,20)\\
&\quad +\$9,000.00(P/F,10,25)\\
&= -\$8,000.00(0.38554)+\$5,000.00(0.23939)+\$7,000.00(0.14864)\\
&\quad +\$9,000.00(0.09230)\\
&= -\$3,084.32+\$1,196.50+\$1,040.48+\$830.70\\
&= -\$16.64
\end{aligned}
$$

Sixth, calculate the total present worth by summing up the present worth of the uniform annual series A_1 and A_2 and the future values:

$$
\begin{aligned}
P_{total} &= P_{0\ of\ A_1}+P_{0\ of\ A_2}+P_{0\ of\ future\ values}\\
&= \$52,354.49-\$5,130.48-\$16.64\\
&= \$47,207.37
\end{aligned}
$$

A second method for solving for the present worth is the following:

$$
\begin{aligned}
P_0 &= \$10,000.00(P/A,10,9)(P/F,10,1)-\$8,000.00(P/F,10,10)\\
&\quad +\$5,000.00(P/F,10,15)+\$7,000.00(P/F,10,20)+\$9,000.00(P/F,10,25)\\
&\quad -\$3,000.00(P/A,10,11)(P/F,10,14)\\
&= \$10,000.00(5.7590)(0.90909)-\$8,000.00(0.38554)+\$5,000.00(0.23939)\\
&\quad +\$7,000.00(0.14864)+\$9,000.00(0.09230)-\$3,000.00(6.4950)(0.26333)\\
&= \$52,354.49-\$3,084.32+\$1,196.95+\$1,040.48+\$830.70-\$5,130.99\\
&= \$47,207.31
\end{aligned}
$$

Example 6.4 demonstrates calculating the future worth of a gradient and a uniform annual series using multiples factors.

Example 6.4

A process engineer starts investing his money when he graduates from college. He is able to afford investing \$10,000.00 a year from the time he graduates in four years until the end of eight years. He also plans to invest an additional \$2,500.00 per year increasing by \$2,500.00 per year at the end of the year after he graduates until year eight. How much will the process engineer have saved by the end of year eight and what is its present worth if the interest rate is 10%? Figure 6.12 is the cash flow diagram for the investment plan of the process engineer.

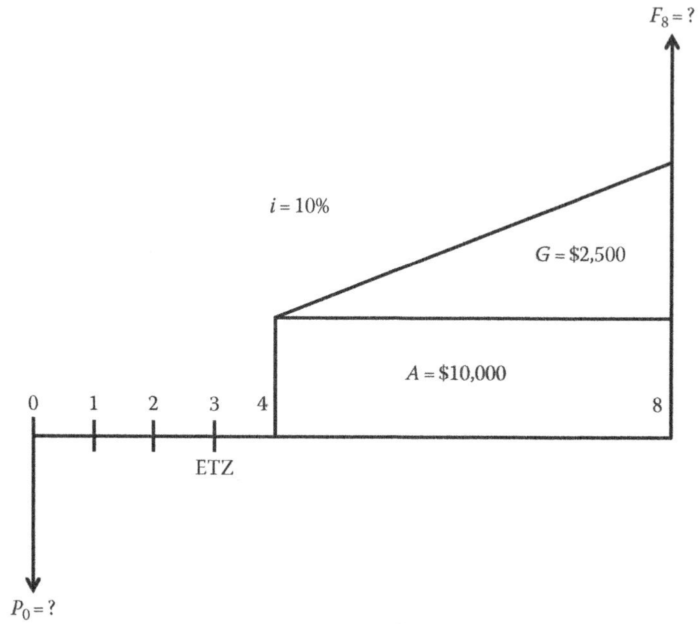

FIGURE 6.12 Cash flow diagram for the investments by the process engineer in Example 6.4.

Solution

The ETZ for the uniform series is at year 3 and $n = 8 - 3 = 5$

The ETZ for the gradient is at year 3 and $n = 8 - 3 = 5$

$$F_{8\,of\,A} = A(F/A,i,n) = \$10,000.00(F/A,10,5) = \$10,000.00(6.1051)$$
$$= \$61,051.00$$

$$F_{8\,of\,C} = G(F/G,i,n) = \$2,500.00(F/G,10,5) = \$2,500.00(11.051)$$
$$= \$27,627.50$$

$$F_{total} = F_{8\,of\,A} + F_{8\,of\,C}$$
$$= \$61,051.00 + \$27,627.50$$
$$= \$88,678.50$$

$$P_0 = F_8\,(P/F,i,n) = \$88,678.50(P/F,10,8) = \$88,678.50(0.46651)$$
$$= \$41,369.41$$

6.3 COMPOUNDING PERIOD NOT EQUAL TO THE PAYMENT PERIOD

When money is deposited into savings or investment accounts in many instances it is not deposited simultaneously with the compounding of the interest unless in situations where it is yearly or monthly. For annuities where the compounding period is more frequent than the payment period, such as semiannual payments at an interest rate compounded quarterly, the interest rate needs to be converted to a rate for the payment period or the payment period is converted to the same period as the interest rate if the payment period is equal or less frequent than the compounding period.

Whenever either of the single payment factors (*P*/*F*) or (*F*/*P*) are used, divide the nominal interest rate by the number of compounding periods per year and multiply the number of years by the number of compounding periods. Equations 6.1 and 6.2 are the factor table equations that compensate for converting the interest rate and compounding periods into equivalent terms:

$$P_0 = F_n\left(P/F, \frac{i}{m}, n \times m\right) \tag{6.1}$$

$$F_n = P_0\left(F/P, \frac{i}{m}, n \times m\right) \tag{6.2}$$

where

n is the number of years
m is the number of compounding periods per year

In most situations, funds are deposited or withdrawn monthly such as automatic deposits for paychecks or withdrawals for bills. In some situations, the interest is compounded daily or continuously, or funds are deposited at the end of the year and the compounding of interest occurs monthly, as is demonstrated by Example 6.5.

Example 6.5

A systems engineer deposits \$12,000.000 into a savings account each year paying 12% per year compounded monthly. How much will be in the account at the end of three years if the interest rate is 12%? Figure 6.13 is the cash flow diagram for the investment plan of the systems engineer.

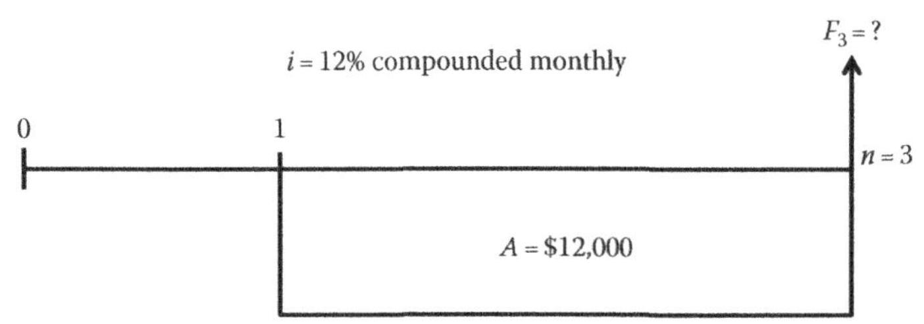

FIGURE 6.13 Cash flow diagram for the investments by the systems engineer in Example 6.5.

Solution

First, determine (i) for the payment period:

$$i = \frac{i_n}{m} = \frac{12\%/\text{Year}}{12 \text{ compounding periods/Year}} = 1\% \text{ per month}$$

Second, calculate the effective interest for one year:

$$i_e = (1+i)^n - 1 = (1+0.01)^{12} - 1 = 0.1268 = 12.68\% \text{ per year}$$

Third, calculate the future worth at the end of three years:

$$F_3 = A\left[\frac{(1+i)^n - 1}{i}\right] = \$5,000.00\left[\frac{(1+0.01268)^3 - 1}{0.01268}\right] = \$5,000.00\left(\frac{0.038524}{0.01268}\right)$$

$$= \$5,000.00(3.0382)$$

$$= \$15,191.00$$

Example 6.6 calculates the interest rate for the payment period when the compounding period occurs at a different time than the payment period.

Example 6.6

An agricultural engineer deposits $5,000.000 into a savings account every month for seven years for his firm. How much would be in the account after the last deposit if the interest rate is 8% compounded quarterly? Figure 6.14 is the cash flow diagram for the savings plan of the agricultural engineer.

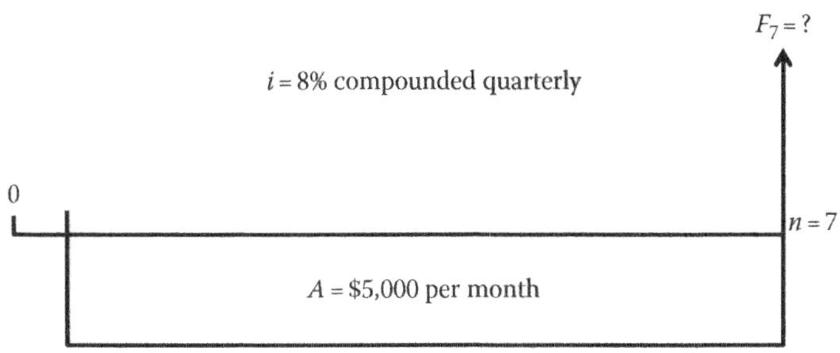

FIGURE 6.14 Cash flow diagram for the investments of the agricultural engineer in Example 6.6.

Solution

First, determine (i) for the payment period:

$$i = \frac{i_n}{m} = \frac{8\%/\text{Year}}{4 \text{ compounding periods/Year}} = 2\% \text{ per quarter and two quarters per 6 months}$$

Second, calculate the effective interest for six months:

$$i_e = (1+i)^m - 1 = (1+0.02)^2 - 1 = 0.0404 = 4.04\% \text{ for 6 months}$$

$$\text{Total number of compounding periods} = \frac{2 \text{ periods}}{\text{Year}} \times 7 \text{ years} = 14 \text{ periods}$$

Third, calculate the future worth at year 14:

$$F_{14} = A\left[\frac{(1+i)^n - 1}{i}\right] = \$5,000.00\left[\frac{(1+0.0404)^{14} - 1}{0.0404}\right] = \$5,000.00\left(\frac{0.741024}{0.0404}\right)$$

$$= \$5,000.00(18.3422)$$

$$= \$91,711.00$$

6.4 SUMMARY

This chapter demonstrated combining multiple factors when calculating the present or future worth of multiple uniform annual series, gradients, and present and future values. The first part of this chapter provided six examples of situations where multiple factors are required to solve problems and included a case study with solutions to the six examples. Another case study was provided where the monthly payments and the total amount paid were calculated for student loans repaid over 10 and 20 years. Additional example problems were provided, which converted various payment and disbursement streams into either present or future worth using multiple factors. An explanation of the process for solving for the present or future worth of two sequential series with different interest rates was included in this chapter along with problems demonstrating the process for solving these types of problems. The last part of this chapter explained the procedures for calculating present or future worth using multiple factors when the compounding period is not equal to the payment period and example problems were provided in this section to illustrate this process.

KEY TERM

Multiple factors

PROBLEMS

6.1 A petroleum engineering firm manager is planning on investing in land so the firm would be able to expand its drilling operations. By investing in the drilling operation, the firm will realize a profit of $20,000,000.00 each year for 20 years. The firm will also sell part of the land at year four for $10,000,000.00 and at year 16 for $15,000,000.00. If the interest rate is 6%, what should the firm pay for the land and setting up the drilling operations to justify the investment?

6.2 If the petroleum engineering firm in Problem 6.1 is able to obtain an interest rate of 15%, what should be paid for the land?

6.3 Calculate the present worth of the oil field investment in Problem 6.1 if the profits do not start until year three.

6.4 What would be the future worth of the oil field data in Problem 6.3?

6.5 An electrical engineer plans to purchase new components for one of the firm's assembly lines. The first purchases will be at years three, four, and five and will cost $100,000.00. The second purchases will occur at years nine through 13 and will cost $150,000.00 each year. Using an interest rate of 15%, calculate the present worth of the two uniform annual series.

6.6 Calculate the future worth of the two uniform annual series in Problem 6.5.

6.7 A biomedical engineer is planning on purchasing a new microscope that will increase yearly profits for her firm by $4,600.00 a year starting at year zero and continuing until year six. The microscope has a salvage value of $50,000.00 at year seven. What is the present worth of this investment if the interest rate is 8%?

6.8 What is the present worth of the investment in Problem 6.7 if the initial cost of the microscope is $75,000.00?

6.9 A manufacturing engineer would be able to increase her firm's yearly profits if it purchases a new computer network control (CNC) machine. If the profits increase by $10,000.00 starting at year four and continuing until year nine and the machine will have a salvage value of $70,000.00 in year nine, how much should the firm pay for the CNC machine? Use an interest rate of 11%.

6.10 If the initial cost of the CNC machine in Problem 6.9 is $150,000.00, what is the net present worth?

6.11 A construction firm is borrowing funds to purchase a scraper. The firm has to start repaying the loan at year two at an interest rate of 12% and a payment of $80,000.00 per year until year six. Starting at year seven, the payment increases to $100,000.00 a year at an interest rate of 15% until year 12. What is the present worth of the payments on the loan for the scraper?

6.12 Calculate the future worth of the two uniform annual series in Problem 6.11.

6.13 The owner of an agricultural engineering firm has leased an office on a 10-year lease. The office space is 10,000 square feet. The rent is paid once a year at a rate of $110.00 per square foot. At the end of four years, the owner decides to relocate the firm to another city. Calculate the amount the owner of the leased office space will have to be compensated to pay off the balance owed on the lease. The interest rate is 4%.

6.14 A nuclear engineer purchased a monitoring system for $130,000,000.00. The maintenance and operating costs are $1,700,000.00 per year. After five years, the engineer purchases a system to upgrade the monitoring system at a cost of $70,000,000.00 and the operating and maintenance costs will be $900,000.00 per year. The machine will be used for 16 years and then sold for $18,000,000.00. Calculate the present worth of the machine using an interest rate of 9%.

6.15 An investor is selling his shares in oil wells in East Texas. The wells produce 6,000 barrels a day and they are projected to keep producing for 10 years. The oil is currently selling for $45.00 a barrel and the price will increase by $5.00 per barrel for the next six years starting at year three. If the interest rate is 8%, what would someone be willing to pay for the oil well shares?

6.16 What would someone be willing to pay for the oil well shares in Problem 6.15 if the wells produce 3,000 barrels a day?

6.17 A process engineer purchases a machine that costs $120,000.00 that will have a salvage value of $20,000.00 at the end of 20 years. The operating and maintenance costs are $8,000.00 per year. Every five years the machine will be overhauled at a cost of $25,000.00. If the interest rate is 5%, what is the net present worth of the machine?

6.18 What is the future worth of the machine in Problem 6.16?

6.19 How much does an engineer need to deposit now into an account earning 3% interest if she plans on withdrawing $4,000.00 per year for 10 years starting in 25 years?

6.20 An engineer deposits $30,000.00 into an account that pays 6% interest for 10 years. How much is the engineer able to withdraw from the account per year for five years starting at year 11?

7 Present Worth Capitalized Cost Analysis

Present Worth Method of Comparing Alternatives

Capitalized cost calculations are required by municipalities, cities, counties, state and federal government agencies, and may be used by private owners whenever there is a project being proposed and an agency or owner needs to determine an equivalent present worth so that the agency or owner is able to determine the amount to finance when obtaining funding for a project. Government agencies fund projects through *appropriations* voted on by the legislature or citizens or they sell bonds to the public that pay interest or are redeemable for a higher value at the end of the term of the bond issue. Private owners secure financing by borrowing from banks or other institutions that finance private projects. If a firm has enough capital, they may fund their own projects, but most managers of firms prefer to use the capital of others rather than risk the capital of their firm. There are many types of financing schemes in the public and private sector for funding projects and some of them were mentioned in Section 1.6.

This chapter introduces a method for comparing two or more alternatives by determining their capitalized cost, which helps decision makers to select the alternative that will either minimize costs or maximize profits or savings, since these are the goals of firms.

7.1 COMPARING ALTERNATIVES ON THE BASIS OF EQUIVALENT PRESENT WORTH

When managers of firms decide which alternative to select when there is more than one alternative, they need a method for comparing the alternatives based on equivalent terms. One method for comparing alternatives is to compare them based on their *equivalent present worth*. In order to be able to compute an equivalent present worth for each alternative, each person performing an engineering economic analysis needs to know the interest rate and the time frame each alternative will be analyzed over. If the alternatives being compared have the same life spans, then they are compared by calculating the net present worth for each alternative to determine which alternative has the highest net present worth. Equation 7.1 is the formula for calculating net present worth.

$$\text{NPW} = \pm P_0 \pm \sum A(P/A,i,n) \pm \sum F(P/F,i,n) \pm \sum G(P/G,i,n) \tag{7.1}$$

If the alternatives have different life spans, then the least common multiple of the life spans of the alternatives is used when comparing alternatives and this method is introduced in Section 7.3.

Analyzing multiple alternatives to determine which one has the highest net present worth or the lowest net present cost is called analyzing mutually exclusive alternatives. Mutually exclusive means one alternative will be selected to the exclusion of all of the other alternatives. When preparing to calculate the net present worth or cost of multiple alternatives, the person performing the analysis needs to separate the tangibles from the intangibles. Tangibles are items expressed in economic terms and intangibles cannot be expressed in economic terms. Only tangible items are included in net present worth calculations.

Table 7.1 provides a synopsis of the types of input or output, the situations, and the criterion to be achieved when solving an engineering economic problem.

TABLE 7.1
Analyzing Alternatives Based on Fixed Input or Output

Type of Input or Output	Situation	Criterion
Fixed input	Amount of money or other resources are fixed.	Maximize the present worth of benefits or other outputs.
Fixed output	There is a fixed task, benefit, or other output to be accomplished.	Minimize the present worth of the costs or other inputs.

Example 7.1 illustrates comparing the net present worth of three alternatives with equal life spans.

Example 7.1

A mechanical engineer is comparing three alternative processes for increasing the efficiency of a wastewater treatment facility. The costs and benefits determined for the three alternatives are listed in Table 7.2. The interest rate is 7% and each of the processes being analyzed has a life of 12 years. Figures 7.1 through 7.3 are the cash flow diagrams for the wastewater treatment process alternatives.

TABLE 7.2
Mechanical Processes for Wastewater Treatment Facility

Alternative	Initial Cost	Yearly Benefit	Resale Value at the End of 12 Years
Process 1	$1,700,000.00	$275,000.00	$325,000.00
Process 2	$1,400,000.00	$150,000.00 increasing by $10,000.00 per year starting at year 2	$210,000.00
Process 3	$1,200,000.00	$110,000.00	$250,000.00

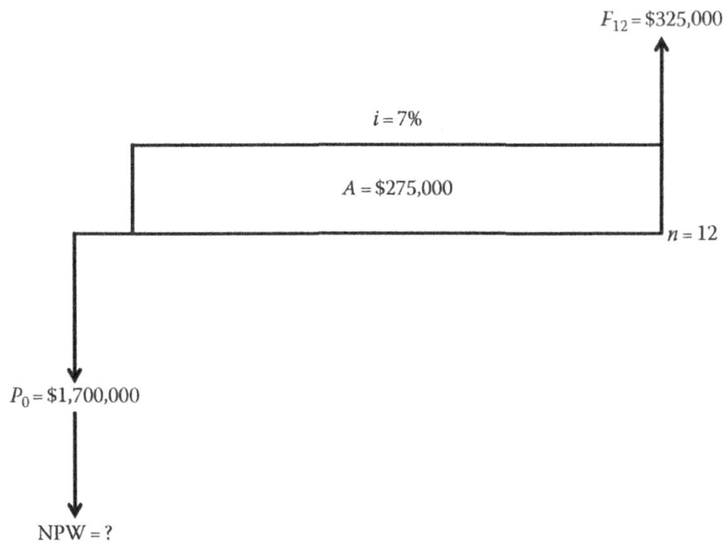

FIGURE 7.1 Cash flow diagram for mechanical process 1 in Example 7.1.

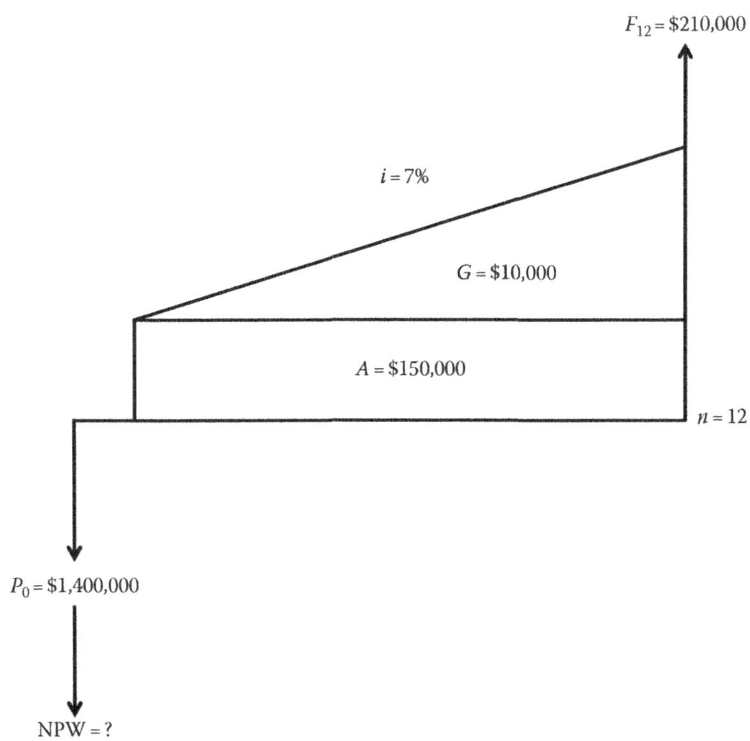

FIGURE 7.2 Cash flow diagram for mechanical process 2 in Example 7.1.

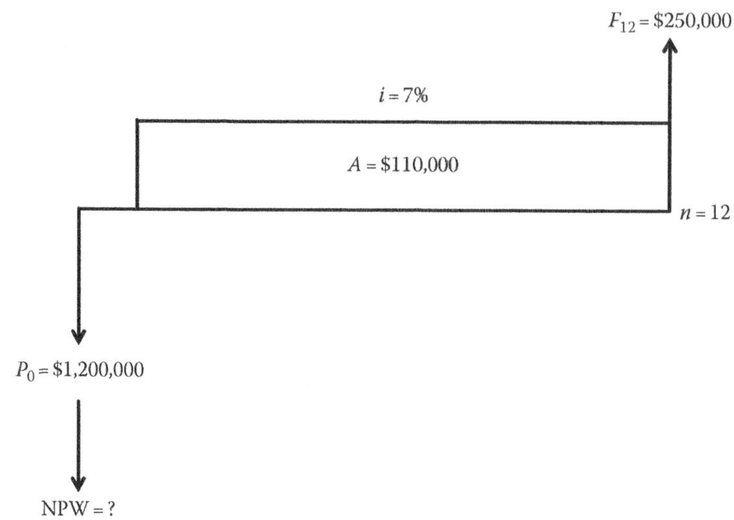

FIGURE 7.3 Cash flow diagram for mechanical process 3 in Example 7.1.

Solution

Process 1

$$\begin{aligned}
\text{NPW}_1 &= P + A(P/A,i,n) + F(P/F,i,n) \\
&= -\$1{,}700{,}000.00 + \$275{,}000.00(P/A,7,12) + \$325{,}000.00(P/F,7,12) \\
&= -\$1{,}700{,}000.00 + \$275{,}000.00(7.9426) + \$325{,}000.00(0.44401) \\
&= -\$1{,}700{,}000.00 + \$2{,}184{,}215.00 + \$144{,}303.25 \\
&= \$628{,}518.25
\end{aligned}$$

Process 2

$$\begin{aligned}
\text{NPW}_2 &= P + A(P/A,i,n) + G(P/G,i,n) + F(P/F,i,n) \\
&= -\$1{,}400{,}000.00 + \$150{,}000.00(P/A,7,12) + \$10{,}000.00(P/G,7,12) \\
&\quad + \$210{,}000.00(P/F,7,12) \\
&= -\$1{,}400{,}000.00 + \$150{,}000.00(7.9426) + \$10{,}000.00(37.350) \\
&\quad + \$210{,}000.00(0.44401) \\
&= -\$1{,}400{,}000.00 + \$1{,}191{,}390.00 + \$373{,}500.00 + \$93{,}242.10 \\
&= \$258{,}132.10
\end{aligned}$$

Process 3

$$
\begin{aligned}
\text{NPW}_3 &= P + A(P/A,i,n) + F(P/F,i,n) \\
&= -\$1,200,000.00 + \$110,000.00(P/A,7,12) + \$250,000.00(P/F,7,12) \\
&= -\$1,200,000.00 + \$110,000.00(7.9426) + \$250,000.00(0.44401) \\
&= -\$1,200,000.00 + \$873,686.00 + \$111,002.50 \\
&= -\$215,311.50
\end{aligned}
$$

Therefore, select process 1 since it has the highest net present worth

7.1.1 Decisions with a Do Nothing Alternative

Some decisions include analyzing whether *do nothing* is a viable alternative and this alternative is also compared to the other available alternatives. If the other alternatives being analyzed all have a negative net present worth, then the do nothing alternative would be selected as the most economical alternative. Example 7.2 is an economic analysis of alternatives that includes the do nothing alternative in the analysis process.

Example 7.2

A chemical engineer is determining the most economical alternative by comparing three alternatives, as well as the do nothing alternative, for a new chemical process. He investigates the costs and the benefits for each of the alternatives and determines the net present worth of each alternative using a minimum attractive rate of return of 10%. All of the alternatives will have a life span of 20 years. The three potential alternatives being analyzed are listed in Table 7.3. Figures 7.4 through 7.6 are the cash flow diagrams for the potential chemical processes.

TABLE 7.3
Chemical Engineering Process Alternatives

Alternative	Total Investment	Uniform Net Annual Benefit	Terminal Value at the End of 20 Years
Chemical process 1	\$500,000.00	\$51,000.00	\$300,000.00
Chemical process 2	\$950,000.00	\$105,000.00	\$300,000.00
Chemical process 3	\$1,500,000.00	\$150,000.00	\$400,000.00
Do nothing	0	0	0

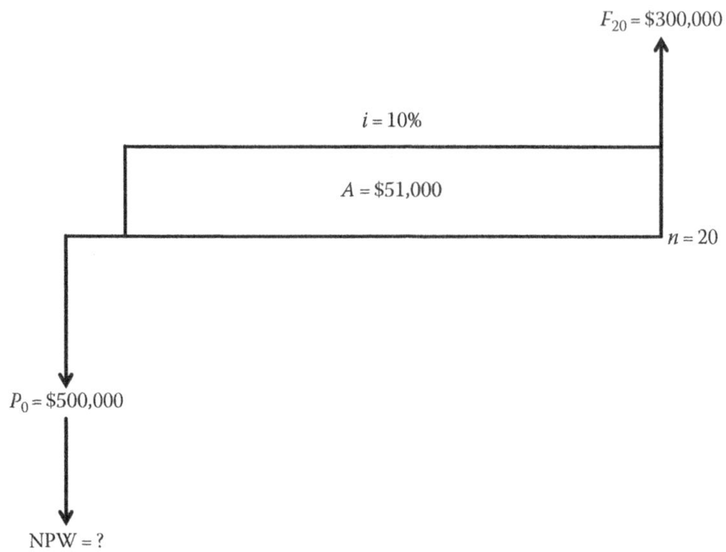

FIGURE 7.4 Cash flow diagram for chemical process 1 in Example 7.2.

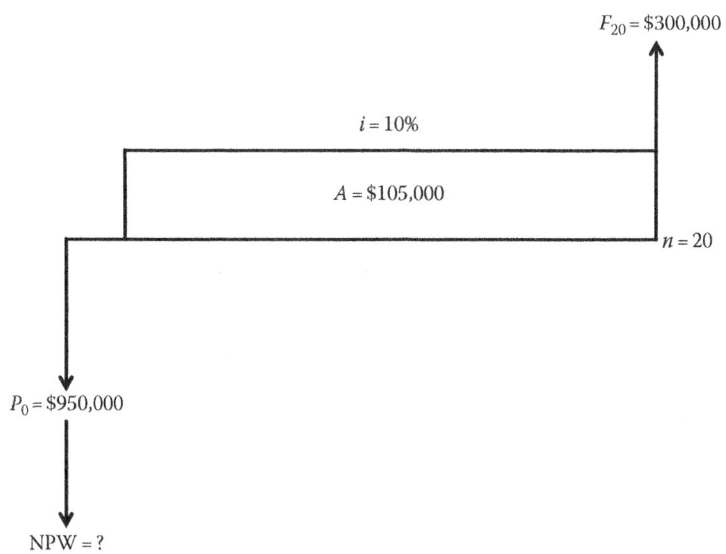

FIGURE 7.5 Cash flow diagram for chemical process 2 in Example 7.2.

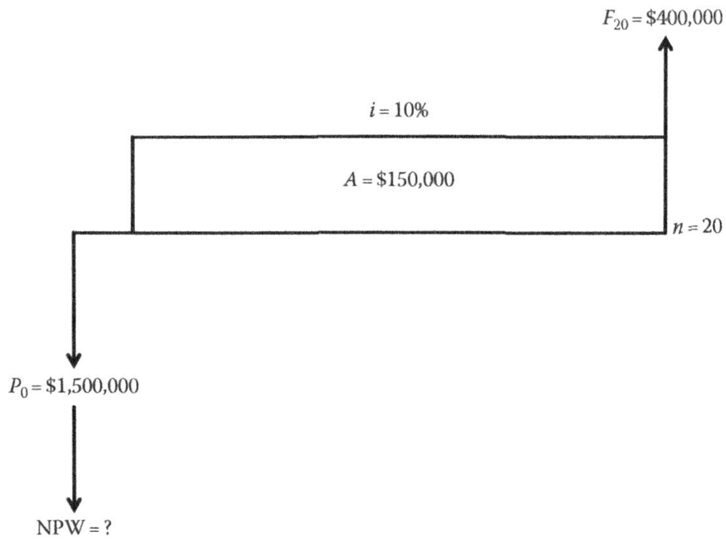

FIGURE 7.6 Cash flow diagram for chemical process 3 in Example 7.2.

Solution

Chemical process 1

$$\begin{aligned}\text{NPW}_1 &= P + A(P/A,i,n) + F(P/F,i,n)\\ &= -\$500{,}000.00 + \$51{,}000.00(P/A,10,20) + \$300{,}000.00(P/F,10,20)\\ &= -\$500{,}000.00 + \$51{,}000.00(8.5136) + \$300{,}000.00(0.14864)\\ &= -\$500{,}000.00 + \$434{,}193.60 + \$44{,}592.00\\ &= -\$21{,}214.40\end{aligned}$$

Chemical process 2

$$\begin{aligned}\text{NPW}_2 &= P + A(P/A,i,n) + F(P/F,i,n)\\ &= -\$950{,}000.00 + \$105{,}000.00(P/A,10,20) + \$300{,}000.00(P/F,10,20)\\ &= -\$950{,}000.00 + \$105{,}000.00(8.5136) + \$300{,}000.00(0.14864)\\ &= -\$950{,}000.00 + \$893{,}928.00 + \$44{,}592.00\\ &= -\$11{,}480.00\end{aligned}$$

Chemical process 3

$$\begin{aligned}\text{NPW}_3 &= P + A(P/A,i,n) + F(P/F,i,n)\\ &= -\$1{,}500{,}000.00 + \$150{,}000.00(P/A,10,20) + \$400{,}000.00(P/F,10,20)\\ &= -\$1{,}500{,}000.00 + \$150{,}000.00(8.5136) + \$400{,}000.00(0.14864)\\ &= -\$1{,}500{,}000.00 + \$1{,}277{,}040.00 + \$59{,}456.00\\ &= -\$163{,}504.00\end{aligned}$$

Do nothing alternative

$$\text{NPW} = 0$$

Therefore, select the do nothing alternative since the other three alternatives all have a negative net present worth.

Appendix C provides spreadsheet formulas for solving for net present worth.

7.2 CAPITALIZED COST CALCULATIONS FOR PERPETUAL LIFE SERIES

When engineers are working in the public sector, or in some private organizations, they may be called upon to economically analyze alternatives projected to have an infinite life (*perpetual life*). Engineers are required to determine the lump sum amount required to fund a project forever ($n = \infty$) including future expenditures for maintenance and other related costs, and this initial lump sum is referred to as the capitalized cost of the project. Examples of perpetual life projects are bridges, civic centers, culverts, dams, right-of-ways, road systems, water supply systems, and water treatment facilities.

The technique for calculating the capitalized cost of projects with an infinite life involves solving for the net present worth of all reoccurring future costs, uniform annual series, and gradients using the perpetual life present worth formulas introduced in Sections 4.5 and 4.6 for uniform annual series and in Section 5.6 for gradients. To determine the capitalized cost of infinite future values, uniform annual series, and gradients calculate the following:

1. Present worth of reoccurring future values: Convert each reoccurring future value into an equivalent uniform annual series using the single payment sinking fund factor (A/F) as if it occurred over the length of time between the reoccurring future values, and divide the resulting uniform annual series by the interest rate to determine the present worth of the perpetual life reoccurring series $\left[A = F_n\left(A/F,i,n\right) \text{ then use the resulting } A \text{ in } P_0 = \frac{A}{i}\right]$.
2. Present worth of infinite series: Divide the uniform annual series by the interest rate $\left(P_0 = \frac{A}{i}\right)$.
3. Present worth of reoccurring gradients: Divide the gradient by the interest rate squared $\left(P_0 = \frac{G}{i^2}\right)$.

When analyzing perpetual life projects using capitalized cost calculations, the first step is to draw a cash flow diagram. The cash flow diagram should use a technique for indicating which costs or disbursements are one-time costs or disbursements and which are reoccurring ones. An arrow pointing to the future could be used to indicate a uniform annual series or a gradient is an infinite series and a line drawn over a future value or annuity could be used to indicate it is a reoccurring value. Only two cycles of reoccurring costs and disbursements need to be shown on cash flow diagrams.

The second step in the analysis process requires calculating the present worth of all nonreoccurring costs and disbursements using the single payment present worth factor (P/F). The third step entails converting reoccurring future values into equivalent uniform annual series occurring over the time period between the reoccurring costs or disbursements.

The fourth step is to add or subtract each of the uniform annual series determined in step three to or from all other reoccurring uniform annual series and then divide the total of all of the uniform annual series by the interest rate. If a uniform annual series starts at an ETZ different from the PTZ, the present worth of the infinite uniform annual series is calculated at the ETZ by dividing the uniform annual series by the interest rate $\left(P_0 = \frac{A}{i}\right)$ and then solving for the present worth of this value at the PTZ of the value using the single payment present worth factor (*P*/*F*).

The fifth step requires converting reoccurring gradients into a present worth by dividing the gradient by the interest rate squared $\left(P_0 = \frac{G}{i^2}\right)$. If any of the gradients start at an ETZ different than the PTZ, the present worth of the infinite gradient is first calculated at the ETZ and then converted into a present worth at the PTZ using the single payment present worth factor (*P*/*F*).

All of the present worth values calculated in the five previous steps are summed up to obtain the capitalized cost of the project. The capitalized cost represents the amount of funds to be deposited at the PTZ at the indicated interest rate in order for the funds to cover the cost of all future cash flows indefinitely.

Case Study 7.1 demonstrates the process for using the previous five steps to calculate the capitalized cost of a public project.

Case Study 7.1 Capitalized Cost of a Public Project

A municipality needs to determine the capitalized cost of a project for rebuilding a small roadway bridge over a culvert. The municipality plans on depositing funds into an interest-bearing account paying 6% interest and the funds in the account will cover all of the current and future costs related to the bridge. The bridge is expected to last indefinitely. Its initial cost is $1,500,000.00. During the first five years, bridge maintenance will be $5,000.00 per year, and then at year six, it will increase to $6,000.00 per year. The bridge will have to be resurfaced every 20 years at a cost of $1,200,000.00. Determine the capitalized cost of the bridge. Figure 7.7 is the cash flow diagram for the capitalized cost for Case Study 7.1.

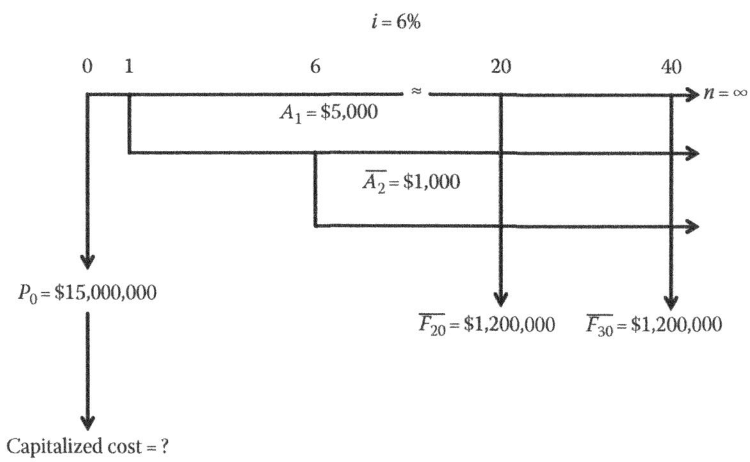

FIGURE 7.7 Cash flow diagram for the capitalized cost of the bridge in Case Study 7.1.

Solution

Part A—Calculate for the present worth of the first infinite uniform annual series of \$5,000.00:

$$P_{A_1} = \frac{A_1}{i} = \frac{-\$5,000.00}{0.06} = -\$83,333.33$$

Part B—Calculate for the present worth of the second annuity of \$1,000.00 at the ETZ of year five:

$$P_{A_2\ at\ year\ 5} = \frac{A_2}{i} = \frac{-\$1,000.00}{0.06} = -\$16,666.67$$

Part C—Calculate for the present worth of the future value at year five $\left(P_{A_2\ at\ year\ 5} = F_5\right)$:

$$P_{0\ of\ A_2} = F_5\left(P/F,i,n\right) = -\$16,666.67\left(P/F,6,5\right) = -\$16,666.67\left(0.74726\right)$$

$$= -\$12,454.33$$

Part D—Convert the reoccurring cost of \$1,200,000.00 every 20 years into an equivalent uniform annual series:

$$A_3 = F_{20}\left(A/F,i,n\right) = -\$1,200,000.00\left(A/F,6,20\right) = -\$1,200,000.00\left(0.02718\right)$$

$$= -\$32,616.00$$

Part E—Calculate the present worth of the infinite uniform annual series A_3:

$$P_{A_3} = \frac{A_3}{i} = \frac{-\$32,616.00}{0.06} = -\$543,600.00$$

Part F—Sum up the present worth of all of the equivalent uniform annual series:

$$P_{0\ for\ annuities} = P_{A_1} + P_{A_2} + P_{A_3} = -\$83,333.33 - \$12,454.33 - \$543,600.00$$

$$= -\$639,387.66$$

Part G—Add the initial cost of \$15,000,000.00 to the present worth for all of the equivalent uniform annual series:

$$P_{total} = -\$15,000,000.00 - \$639,387.66$$

$$= -\$15,639,387.66$$

Therefore, the capitalized cost for the bridge project is \$15,639,387.66.

7.2.1 Comparing Alternatives Using Capitalized Cost Calculations

Capitalized cost calculations are one method for comparing two or more perpetual life alternatives to determine which alternative has the lowest capitalized cost. If there are costs common to all of the alternatives, they could be eliminated from the calculations if the objective is to only determine which alternative has the lowest capitalized cost and not to calculate the actual capitalized cost.

By eliminating common costs from all of the alternatives under consideration, it simplifies the calculations; therefore, only the difference in cash flows becomes part of the analysis.

When comparing alternative projects with perpetual lives through capitalized cost calculations, the same procedures outlined in Section 7.2.1 are followed to solve for the capitalized cost of each alternative, and then the alternatives are compared to determine which alternative has the lowest capitalized cost. Case Study 7.2 demonstrates the process for comparing alternatives based on their capitalized cost.

Case Study 7.2 Comparing Alternatives Based on Capitalized Cost

A public works engineer is determining which alternative for an addition to a wastewater treatment facility would have the lowest capitalized cost if the interest rate is 4%. She has gathered data on two alternatives and determined the costs associated with each alternative and these are shown in Table 7.4. Determine which of the two alternatives has the lowest capitalized cost. Figure 7.8 is the cash flow diagram for the wastewater treatment facility alternative 1 and Figure 7.9 is the cash flow diagram for the wastewater treatment facility alternative 2.

TABLE 7.4
Data for the Wastewater Treatment Facility Addition Alternatives

Costs	Addition Alternative 1	Addition Alternative 2
Initial cost	\$190,000,000.00	\$240,000,000.00
Yearly maintenance	\$15,000,000.00	\$9,000,000.00 increasing by \$10,000.00 every year
Periodic upgrades	\$50,000,000.00 every 10 years	\$40,000,000.00 every 12 years
Overhaul costs	\$90,000,000 at year 50	\$80,000,000.00 at year 60

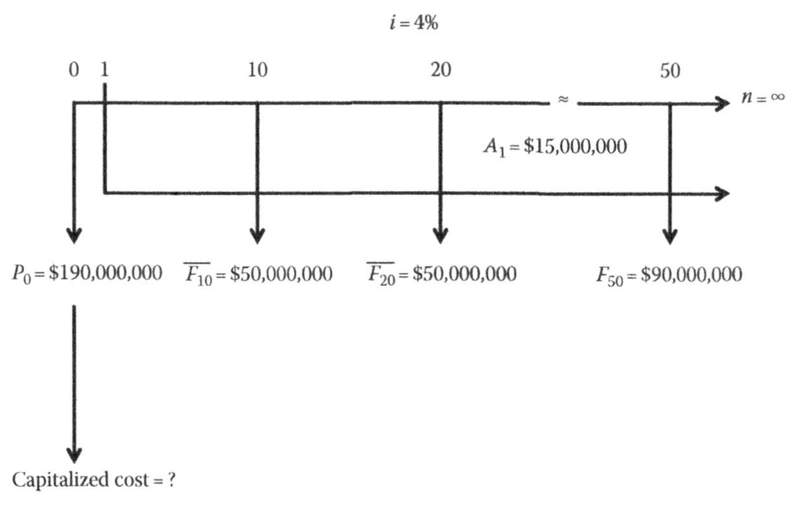

FIGURE 7.8 Cash flow diagram for wastewater treatment facility alternative 1 in Case Study 7.2.

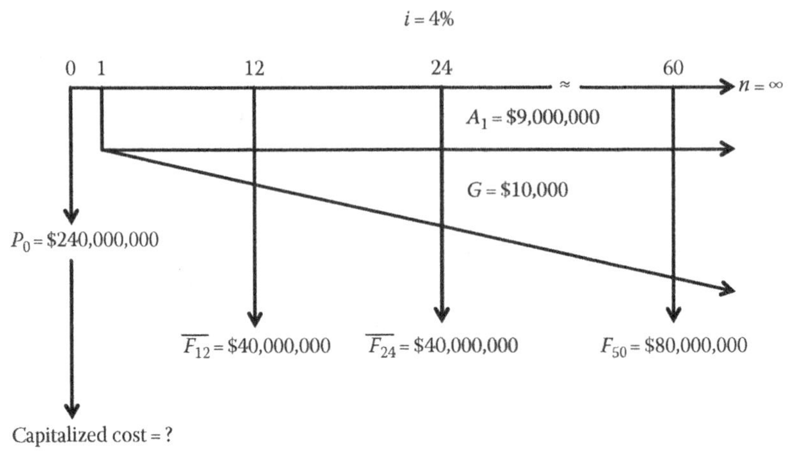

FIGURE 7.9 Cash flow diagram for wastewater treatment facility alternative 2 in Case Study 7.2.

Solution

Wastewater treatment facility alternative 1

Part A—Calculate the present worth of the \$15,000,000.00 infinite uniform annual series A_1:

$$P_{A_1} = \frac{A_1}{i} = \frac{-\$15,000,000.00}{0.04} = -\$375,000,000.00$$

Part B—Calculate the equivalent uniform annual series for the \$50,000,000.00 periodic upgrade by converting it into an equivalent uniform series:

$$A_2 = P_0\left(A/F,i,n\right) = -\$50,000,000.00\left(A/F,4,10\right) = -\$50,000,000.00\left(0.08329\right)$$
$$= -\$4,164,500.00$$

Part C—Convert the infinite uniform annual series A_2 into a present worth:

$$P_{A_2} = \frac{A_2}{i} = \frac{-\$4,164,500.00}{0.04} = -\$104,112,500.00$$

Part D—Calculate the present worth of the one-time overhaul cost of \$90,000,000.00 at year 50:

$$P_0 = F_{50}\left(P/F,i,n\right) = -\$90,000,000.00\left(P/F,4,50\right) = -\$90,000,000.00\left(0.14071\right)$$
$$= -\$12,663,900.00$$

Part E—Calculate the capitalized cost by summing up the initial cost and the present worth of the two equivalent uniform annual series and the future value:

$$P_{total} = P_0 + P_{A_1} + P_{A_2} + P_{F50}$$
$$= -\$190,000,000.00 - \$375,000,000.00 - \$104,112,500.00 - \$12,663,900.00$$
$$= -\$681,776,400.00$$

Wastewater treatment facility alternative 2

Part A—Calculate the present worth of the \$9,000,000.00 infinite uniform annual series A_1:

$$P_{A_1} = \frac{A_1}{i} = \frac{-\$9,000,000.00}{0.04} = -\$225,000,000.00$$

Part B—Calculate the equivalent uniform series for the \$40,000,000.00 periodic upgrade by converting it into an equivalent uniform annual series:

$$A_2 = P_0(A/F,i,n) = -\$40,000,000.00(A/F,4,12) = -\$40,000,000.00(0.06655)$$

$$= -\$2,662,000.00$$

Part C—Convert the infinite uniform annual series A_2 into a present worth:

$$P_{A_2} = \frac{A_2}{i} = \frac{-\$2,662,000.00}{0.04} = -\$66,550,000.00$$

Part D—Calculate the present worth of the one-time overhaul cost of \$80,000,000.00 at year 60:

$$P_0 = F_{50}(P/F,i,n) = -\$80,000,000.00(P/F,4,60) = -\$80,000,000.00(0.09506)$$

$$= -\$7,604,800.00$$

Part E—Calculate the present worth of the \$10,000.00 gradient:

$$P_G = \frac{G}{i^2} = \frac{-\$10,000.00}{0.04^2} = -\$6,250,000.00$$

Part F—Calculate the capitalized cost by adding the initial cost to the present worth of all of the equivalent uniform annual series, the gradient, and the future value:

$$P_{total} = P_0 + P_{A_1} + P_{A_2} + P_{F_{50}} + P_G$$

$$= -\$240,000,000.00 - \$225,000,000.00 - \$66,550,000.00 - \$7,604,800.00$$

$$- \$6,250,000.00$$

$$= -\$545,404,800.00$$

Therefore, select alternative 2 since it has a lower capitalized cost $-\$545,404.800.00 < -\$681,776,400.00$

7.3 PRESENT WORTH USING LEAST COMMON MULTIPLES OF LIFE SPANS

If the alternatives have different life spans, then they are analyzed over the least common multiple of the life spans of all of the alternatives. For example, if the alternatives have a life span of two, three, and four years, then the least common multiple is 12 years. Another example is three alternatives having life spans of two, four, and five years; then the least common multiple would be 20 years. Example 7.3 provides a problem where the present worth of three mutually exclusive alternatives with different life spans are compared to each other.

Example 7.3

A nuclear engineer needs a set of wires installed on a cable tray and she receives bids from three subcontractors for the cost of installing the cable tray and wires. The three bids are listed in Table 7.5. To analyze the three bids, the net present worth of each bid is calculated using the least common multiple of the life spans of the alternatives and an interest rate of 5%. Figures 7.10 through 7.12 are the cash flow diagrams for the three bids.

TABLE 7.5
Nuclear Company Bid Estimates for the Installation of Cable Tray and Wires

Bid Estimate	Company 1	Company 2	Company 3
Initial cost	$75,000.00	$110,000.00	$150,000.00
Resale value	$30,000.00	$50,000.00	$70,000.00
Operating and maintenance costs	$8,000.00 per year	$6,000.00 per year	$7,000.00 the first year increasing by $250.00 per year
License fee every time reinstalled	$400.00	$750.00	$1,000.00
Life in years	4	6	6

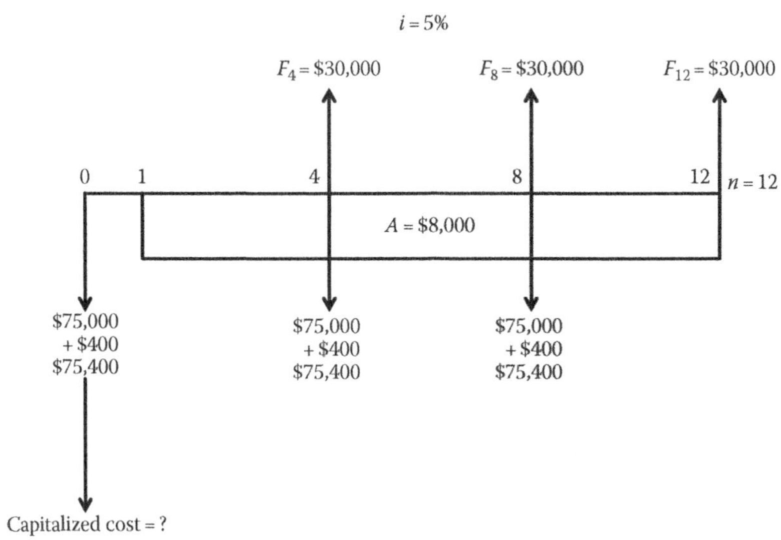

FIGURE 7.10 Cash flow diagram for installation of the cable tray for company 1 in Example 7.3.

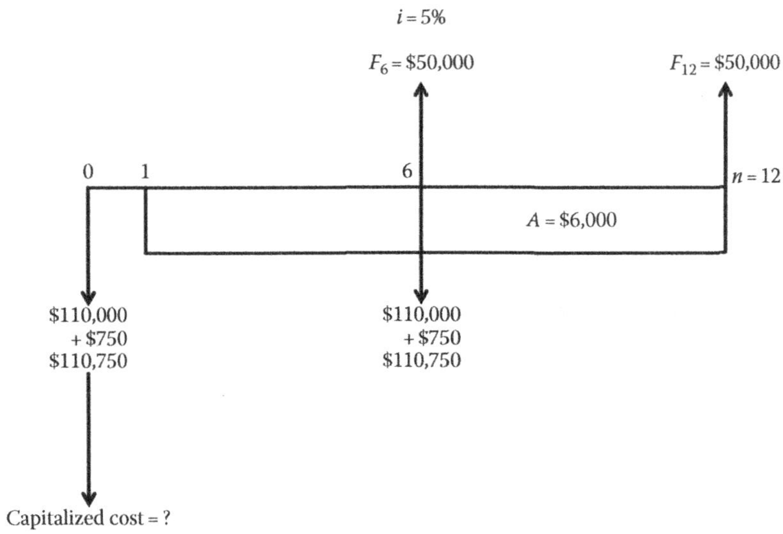

FIGURE 7.11 Cash flow diagram for installation of the cable tray for company 2 in Example 7.3.

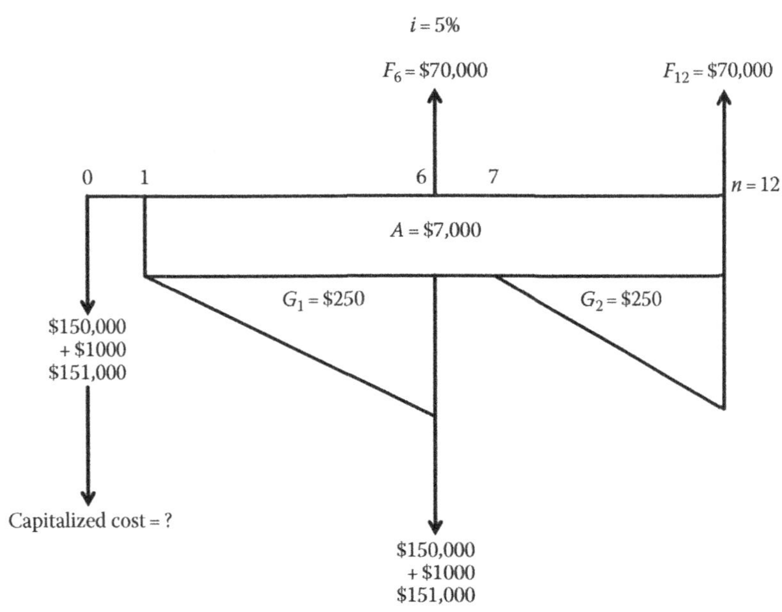

FIGURE 7.12 Cash flow diagram for installation of the cable tray for company 3 in Example 7.3.

Solution

The solution requires all three of the alternatives to be analyzed over the least common multiple, which in this case would be 12 years.

Company 1

$$\begin{aligned}\text{NPW}_1 &= P_0 + A(P/A,i,n) + F_4(P/F,i,n) + F_8(P/F,i,n) + F_{12}(P/F,i,n)\\ &= -\$75,400.00 - \$8,000.00(P/A,5,12) + (-\$75,400.00 + \$30,000.00)(P/F,5,4)\\ &\quad + (-\$75,400.00 + \$30,000.00)(P/F,5,8) + \$30,000.00(P/F,5,12)\\ &= -\$75,400.00 - \$8,000.00(8.8632) - \$45,400.00(0.82270)\\ &\quad - \$45,400.00(0.67684) + \$30,000.00(0.55684)\\ &= -\$75,400.00 - \$70,905.60 - \$37,350.00 - \$30,728.54 + \$16,705.20\\ &= -\$197,678.94\end{aligned}$$

Company 2

$$\begin{aligned}\text{NPW}_2 &= P_0 + A(P/A,i,n) + F_6(P/F,i,n) + F_{12}(P/F,i,n)\\ &= -\$110,750.00 - \$6,000.00(P/A,5,12)\\ &\quad + (-\$110,750.00 + \$50,000.00)(P/F,5,6) + \$50,000.00(P/F,5,12)\\ &= -\$110,750.00 - \$6,000.00(8.8632) - \$60,750.00(0.74622)\\ &\quad + \$50,000.00(0.55684)\\ &= -\$110,750.00 - \$53,179.20 - \$45,332.87 + \$27,842.00\\ &= -\$181,420.07\end{aligned}$$

Company 3

$$\begin{aligned}\text{NPW}_3 &= P_0 + A(P/A,i,n) + F_6(P/F,i,n) + F_{12}(P/F,i,n) + G_1(P/G,i,n)\\ &\quad + G_2(P/G,i,n)(P/F,i,n)\\ &= -\$151,000.00 - \$7,000.00(P/A,5,12)\\ &\quad + (-\$151,000.00 + \$70,000.00)(P/F,5,6) + \$70,000.00(P/F,5,12)\\ &\quad - \$250.00(P/G,5,6) - \$250.00(P/G,5,6)(P/F,5,6)\\ &= -\$151,000.00 - \$7,000.00(8.8632) - \$81,000.00(0.74622)\\ &\quad + \$70,000.00(0.55684) - \$250.00(11.968) - \$250.00(11.968)(0.74622)\\ &= -\$151,000.00 - \$62,042.40 - \$60,443.82 + \$38,978.80\\ &\quad - \$2,992.00 - \$2,232.69\\ &= -\$239,732.11\end{aligned}$$

Therefore, select the bid from company 2 since it has the lowest net present cost.

7.4 SUMMARY

This chapter introduced capitalized cost calculations and explained the process public agencies perform when calculating the capitalized cost of a project to determine the amount of the initial investment to cover the costs associated with the project for its infinite life. A table was provided containing the situations and criteria considered when analyzing alternatives based on whether they have fixed input or fixed output. Several example problems illustrated the process for selecting between project alternatives by calculating the capitalized cost for each of the project alternatives.

The second section of this chapter provided steps for calculating the present worth of infinite future values, uniform annual series, and gradients when determining the capitalized cost of projects. A case study demonstrated the procedures for determining the capitalized cost of a public project. A method for comparing alternatives based on the their capitalized cost was introduced and a second case study was included that covered calculating capitalized cost and comparing alternatives based on their capitalized cost. The last section of this chapter addressed calculating the capitalized cost of alternatives using the least common multiple of the life spans of the alternatives under consideration, and a detailed example was provided to demonstrate this process.

KEY TERMS

Appropriations
Capitalized cost
Do nothing alternative
Equivalent present worth
Perpetual life series

PROBLEMS

7.1 An engineer working at a process plant is analyzing two machines to determine which machine to purchase for his firm. The data for the two machines are listed in Table 7.6. If the interest rate is 15%, which machine should be selected for the processing plant based on net present worth analysis?

TABLE 7.6
Data for Process Plant Machine Alternatives

Costs or Disbursements	Processing Plant Machine 1	Processing Plant Machine 2
Initial cost	$110,000.00	$180,000.00
Operating and maintenance costs	$35,000.00	$31,000.00
Salvage value	$10,000.00	$20,000.00
Life in years	6	9

7.2 If the interest rate is 4%, which machine should be selected for the processing plant in Problem 7.1?

7.3 Two machines are being considered as replacements in a chemical processing plant. The data for the potential replacement machines are listed in Table 7.7. Using an interest rate of 10%, which machine should be selected by the chemical processing company based on net present worth analysis?

TABLE 7.7
Data for Chemical Processing Plant Machine Alternatives

Costs or Disbursements	Chemical Plant Machine 1	Chemical Plant Machine 2
Initial cost	$250,000.00	$350,000.00
Operating and maintenance costs	$90,000.00	$70,000.00
Salvage value	$20,000.00	$35,000.00
Life in years	5	5

7.4 Using an interest rate of 30%, determine which machine should be selected for the chemical processing plant in Problem 7.3.

7.5 A mining company is considering two alternatives for hauling debris out of one of the mines. The data for the two alternatives are provided in Table 7.8. If the company uses an interest rate of 15%, which alternative should be selected by the mining company based on net present worth analysis?

TABLE 7.8
Data for Mining Debris Removal Alternatives

Costs or Disbursements	Alternative 1 Tracks	Alternative 1 Carts	Alternative 2 Conveying System
Initial cost	$450,000.00	$280,000.00	$1,750,000.00
Operating and maintenance costs	$60,000.00	$3,000.00	$35,000.00
Salvage value	$50,000.00	$20,000.00	$100,000.00
Life in years	8	12	24

7.6 Determine which alternative should be selected in Problem 7.5 if the initial cost of the tracks is $950,000.00 instead of $450,000.00.

7.7 A municipal engineer is evaluating two alternatives for adding capacity to the water supply system for the city. The first alternative is a dam with a cost of $80,000,000.00 that will require operating and maintenances costs of $250,000.00 per year. The dam will have an infinite life. The second alternative under consideration is a pipeline to transport water from a nearby lake. The pipeline costs $4,500,000.00, it will last five years, and it has operating and maintenance costs of $500,000.00 per year. Using an interest rate of 5%, which alternative should be selected by the city based on capitalized cost?

7.8 Using an initial cost for the first dam alternative in Problem 7.7 of $15,000,000.00, determine which dam alternative should be selected by the municipal engineer.

7.9 An aerospace engineer is analyzing two alternatives for a mechanical arm for the space station. The first alternative costs $22,000,000.00 and will have a salvage value of $2,000,000.00 in 10 years. The operating and maintenance costs are $500,000.00 per year increasing by $100,000.00 per year starting at year two. The second alternative costs $1,000,000.00 per year and the cost will increase by $300,000.00 per year starting at year two until year 10. Determine which alternative to select using net present worth analysis and an interest rate of 6%.

7.10 Use an interest rate of 25% to determine which mechanical arm alternative in Problem 7.9 should be selected by the aerospace engineer.

7.11 Compare the two alternatives in Problem 7.10 if the life span is six years instead of 10 years.

7.12 A county is considering paving a road with either concrete or asphalt. Concrete will last 10 years and an asphalt roadway will last five years. The data for each alternative are listed in Table 7.9. The county uses an interest rate of 6%. Determine which road surface the country should select based on net present worth analysis.

TABLE 7.9
Data for Roadway Paving Process Alternatives

Costs or Disbursements	Concrete Paving Process	Asphalt Paving Process
Initial cost	$20,000,000.00	$5,000,000.00
Operating and maintenance costs	$1,000,000.00	$2,000,000.00
Salvage value	$2,500,000.00	—
Life in years	10	5

7.13 Determine which road surface to select in Problem 7.12 if the concrete paving process has a life of 13 years instead of 10 years.

7.14 A county is securing the right-of-way for a roadway and it will cost $1,000,000.00 per year increasing by $200,000.00 per year starting at year two. If the interest rate is 8% and the county will have the right-of-way forever, what is the capitalized cost of the right-of-way?

7.15 A city planning commission is analyzing a project to determine its capitalized cost. The project is a water park that has an initial cost of $11,000,000.00 and operating and maintenance costs of $100,000.00 in the first year increasing by $10,000.00 a year for five years, and then the operating and maintenance costs return to $100,000.00 per year for the infinite life. The water park will generate profits of $1,300,000.00 per year. The interest rate is 13%. Determine the capitalized cost of the water park using net present worth analysis.

7.16 If the water park in Problem 7.15 generates profits of $1,900,000.00 per year, calculate its capitalized cost.

7.17 A city planning commission is evaluating two proposals for a new civic auditorium. The first proposal has an initial cost of $200,000,000.00. The operating and maintenance costs would be $2,500,000.00 and the yearly income would be $19,000,000.00 the first year increasing by $100,000.00 per year. The second proposal costs $210,000,000.00 and has operating and maintenance costs of $3,000,000.00 per year. The projected income of the second proposal is $21,000,000.00 per year increasing by $80,000.00 per year starting at year two. Using an interest rate of 9%, determine the capitalized cost for each alternative.

7.18 Determine the capitalized cost for each alternative in Problem 7.17 if the interest rate is 15%.

7.19 An environmental engineer is analyzing two proposals for mitigating hazardous waste. The first mitigation process costs $1,350,000.00, it will require annual operating and maintenance costs of $122,500.00, and the salvage value is $109,000.00. The second mitigation process has an initial cost of $1,125,000.00, the annual operating and maintenance costs will be $92,500.00, and the salvage value is $83,000.00. The interest rate is 15%. If the mitigation process will be used for eight years, which process should be selected based on net present worth analysis?

7.20 If the first mitigation process in Problem 7.19 has a life of 12 years instead of eight years, determine which mitigation process should be selected by the environmental engineer.

8 Equivalent Uniform Annual Worth Comparison Method

Along with comparing alternatives on the basis of their net present worth, members of firms and individuals are able to compare alternatives based on their *equivalent uniform annual worth* (EUAW). This chapter introduces procedures for calculating the EUAW of alternatives; and using EUAW calculations to compare alternatives; and incorporating salvage values, trade-in values, and sunk costs into EUAW calculations. This chapter also covers calculating the equivalent uniform annual cost of perpetual life alternatives and provides example problems comparing alternatives on the basis of their EUAW.

8.1 EQUIVALENT UNIFORM ANNUAL WORTH OF ALTERNATIVES

The equivalent uniform annual worth comparison method is only used to evaluate multiple alternatives in terms of their equivalent yearly value since it does not determine actual costs. It is only a technique for comparing the cost and benefits of alternatives converted into an equivalent form. The EUAW method solves the problem of having to use least common multiples of life spans when comparing alternatives since all of the income and disbursements are converted into a yearly basis. This method converts all values into yearly equivalent costs and disbursements and then these values are summed up to determine the EUAW of each alternative. The EUAW for each alternative is then compared to the EUAW of the other alternatives to determine which alternative has either the lowest equivalent uniform annual cost or the highest equivalent uniform annual net worth or benefit.

The EUAW method of analyzing alternatives is a faster technique for comparing alternatives than solving for the net present worth of alternatives since it does not involve using the least common multiples of life spans. The EUAW method is also used to reduce the list of alternatives down to two or three alternatives and then the net present worth method of comparing alternatives is used to calculate the actual net present worth of the remaining alternatives. The resulting net present worth for each of the remaining alternatives is then compared to determine which alternative has the lowest cost or provides the highest net worth. In some situations, when the perpetual life of a uniform series needs to be determined, it is calculated using the perpetual life formula $\left(A = \frac{P_0}{i}\right)$ derived from the present worth perpetual life formula $\left(P_0 = \frac{A}{i}\right)$. If the only cost of an investment is the amount of interest paid per year, then the annual cost of the investment is merely the interest payments.

One difficulty in implementing the EUAW method is in ensuring the equivalent uniform annual costs or disbursements represent the annual value over the entire life of a project. If an annuity starts in the future, it must be converted into an equivalent uniform annual series over the entire life of the project, not just the number of years the annuity occurs over. An annuity is converted into a uniform series over the life of a project by calculating its present or future worth, then converting this value into a present worth at the PTZ or a future worth at the end of the project, and then annualizing either of these values over the entire life of the project. The same method is used for calculating the EUAW of future values occurring at any time except for at the end of a project.

The following cash flow diagrams and related formulas illustrate the process for converting different payment or disbursement streams or future values into equivalent uniform annual series over the life of a project:

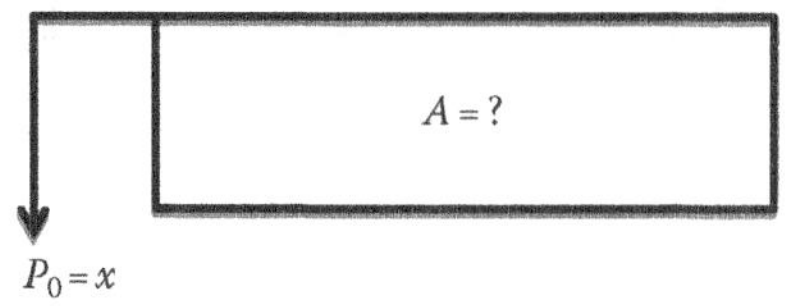

$A = P_0\,(A/P, i, n)$

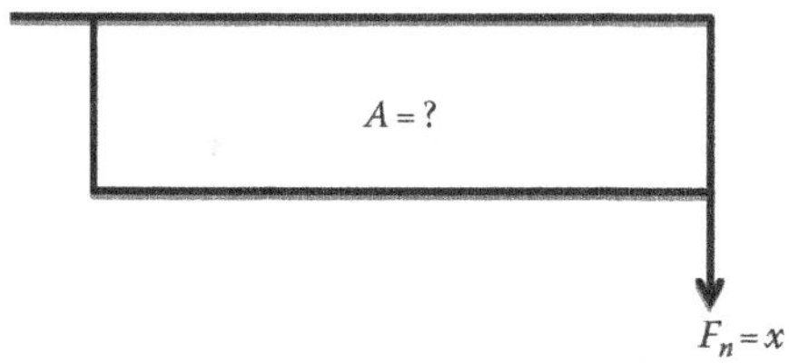

$A = F_n\,(A/F, i, n)$

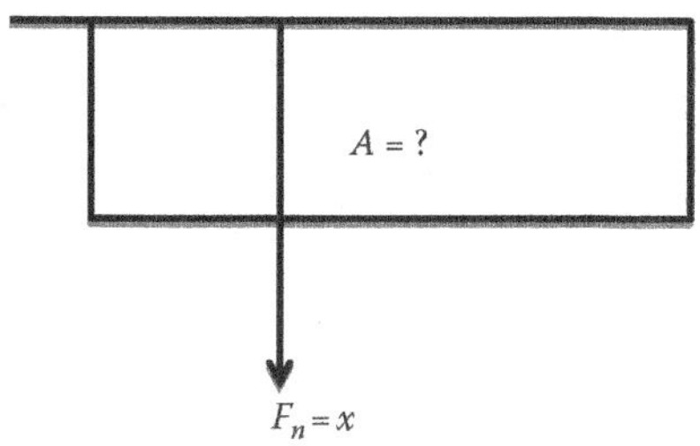

$P_0 = F_n\,(P/F, i, n)$

then

$A = P_0\,(A/P, i, n)$

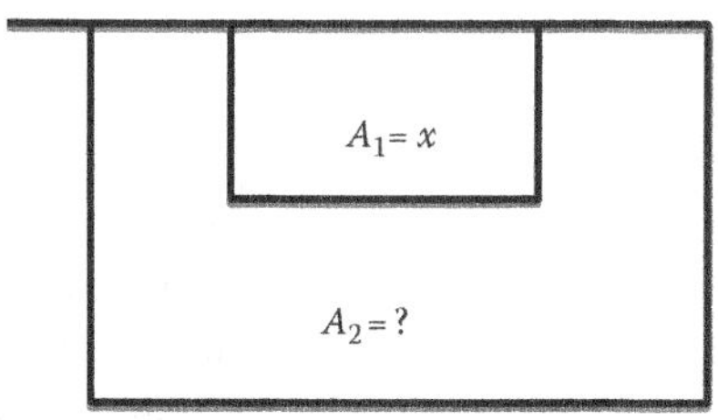

$P_{ETZ} = A_1\,(P/A, i, n)$

$P_{ETZ} = F_{ETZ}$

then

$P_0 = F_{ETZ}\,(P/F, i, n)$

then

$A_2 = P_0\,(A/P, i, n)$

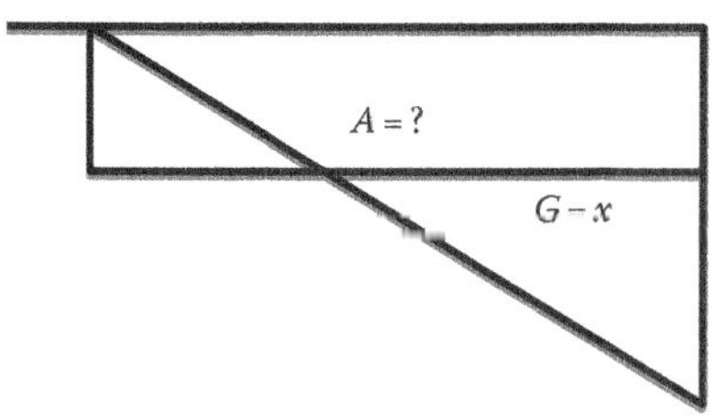

$P_0 = G\,(P/G, i, n)$

then

$A = P_0\,(A/P, i, n)$

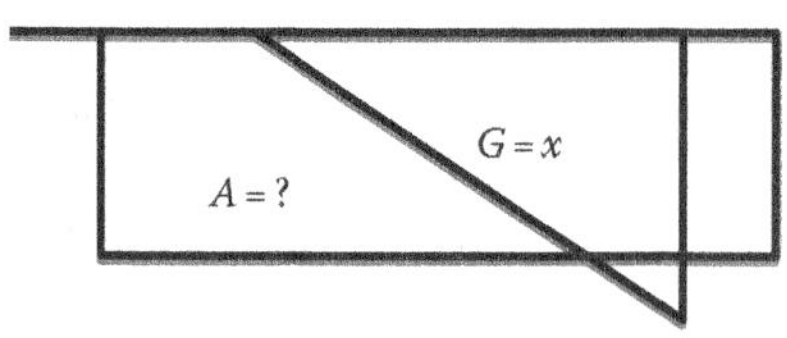

$F_G = G\,(F/G, i, n)$

$F_G = P_G$

then

$F_n = P_G\,(F/G, i, n)$

then

$A = F_n\,(A/F, i, n)$

Equivalent uniform annual worth is the total of all payment and disbursement streams converted to yearly amounts and summed up to give an equivalent yearly cost or worth. Equation 8.1 is the formula for calculating EUAW:

$$\text{EUAW} = \pm P_0\left(A/P,i,n\right) \pm \sum A \pm \sum G\left(A/G,i,n\right) \pm \sum F\left(A/F,i,n\right) \tag{8.1}$$

8.2 SALVAGE VALUE, TRADE-IN VALUE, AND SUNK COSTS

This section explains incorporating salvage values, trade-in values, and sunk costs into EUAW analysis.

8.2.1 Salvage Value

Engineering economic analysis techniques are used to analyze facilities, structures, equipment, or products and each of these may have a *salvage value* they could be sold for at the end of their useful life. Therefore, the salvage value is included in the analysis as a positive future value.

8.2.2 Salvage Sinking Fund Method

The *salvage sinking fund method* allows for the inclusion of salvage values in engineering economic analysis. This method involves converting the initial cost into a uniform annual series using the uniform series capital recovery factor (A/P) and the salvage value into uniform annual series using the single payment sinking fund factor (A/F) and then adding the salvage value equivalent uniform annual series to the initial cost uniform annual series, as shown in Equation 8.2.

$$\text{EUAW} = -P_0\left(A/P,i,n\right) + \text{SV}\left(A/F,i,n\right) \tag{8.2}$$

8.2.3 Salvage Present Worth Method

In addition to the salvage sinking fund method, there is also a *salvage present worth method* for addressing the incorporation of salvage values. This method uses Equation 8.3 to calculate the EUAW of the present value and a future salvage value:

$$\text{EUAW} = \left[-P_0 + \text{SV}\left(P/F,i,n\right)\right]\left(A/P,i,n\right) \tag{8.3}$$

8.2.4 Capital-Recovery-Plus Interest Method

Another method for including the salvage value in calculations is the *capital-recovery-plus interest method*. This method uses Equation 8.4 to calculate the EUAW:

$$\text{EUAW} = \left(P_0 - \text{SV}\right)\left(A/P,i,n\right) + \text{SV}\left(i\right) \tag{8.4}$$

The capital-recovery-plus interest method acknowledges that the salvage value is included in the analysis, but the salvage value will not be obtained until the end of the project; therefore, this must be accounted for by including the value of the interest lost during the project [SV(i)]. Without including [SV(i)], the formula would calculate the salvage value as if it was obtained at time zero.

8.2.5 Trade-In Value Method

If one of the alternatives being compared during an engineering economic analysis includes a situation where the current facility, equipment, or product will continue to be used, then the salvage value, or *trade-in value*, remains invested in it and the salvage value is charged as a cost of keeping the facility, equipment, or product.

The value not obtained when a facility, equipment, or product is not traded-in is considered to be an expenditure of assets since assets will be spent to cover the amount that would have been obtained if the facility, equipment, or product was traded-in and a salvage value was obtained from the trade-in.

To obtain the EUAW in this method, the salvage value is subtracted from the initial cost, then the resulting value is multiplied by the uniform series payment capital recovery factor (*A/P*) to convert the present value into an equivalent uniform series, and then the salvage value is multiplied by the interest rate and added to the uniform annual series.

8.2.6 Sunk Costs

Sometimes when performing engineering economic analysis, it is difficult to determine what are *sunk costs*. A sunk cost is a nonrecoverable cost because it has already been paid, and invested, or the transaction has been completed; therefore, the cost is not retrievable. In many instances, a person conducting an engineering economic analysis is not able to discern which costs are sunk costs and they include them in the analysis. Any reduction in the cost of an asset to the current market value is a sunk cost and it cannot be regained; therefore, it is not included in the engineering economic analysis.

8.3 EQUIVALENT UNIFORM ANNUAL WORTH OF PERPETUAL LIFE ALTERNATIVES

In order to calculate the perpetual life equivalent uniform annual series of an initial cost or disbursement, the cost or disbursement is multiplied by the interest rate $[A = P_0(i)]$. When comparing alternatives on the basis of their EUAW, there may be one or more alternatives with a perpetual life, meaning their time horizon is infinity. These alternatives are addressed in the same manner as other alternatives and an EUAW is calculated using infinity as the number of periods. To illustrate this concept if there is a choice between having \$10,000.00 now or receiving \$600.00 per year forever with an interest rate of 6%, both alternatives would be equivalent since:

$$A = P_0(i) = \$10{,}000.00(0.06) = \$600.00$$

8.4 SOLVED EXAMPLE PROBLEMS

This section provides example problems to illustrate the process for calculating the EUAW of alternatives.

Example 8.1

Two different alternatives are being analyzed by an electrical engineer for upgrading a testing machine. Table 8.1 contains the values obtained by the engineer for the costs and disbursements associated with the two machines. Using the data in Table 8.1 and an interest rate of 3%, calculate the equivalent uniform annual worth for each alternative to determine which alternative should be selected by the engineer. Figures 8.1 and 8.2 are the cash flow diagrams for the two testing machine upgrades.

TABLE 8.1
Electrical Testing Machine Alternatives

Cost or Disbursements	Testing Machine 1	Testing Machine 2
Initial cost	\$16,000.00	\$24,000.00
Annual operating cost	\$600.00	\$400.00
Annual labor cost	\$8,000.00	\$6,000.00
Annual maintenance cost	\$2,000.00	0
Resale value	\$3,000.00 at year 6	\$4,000.00 at year 10
Life in years	6	10

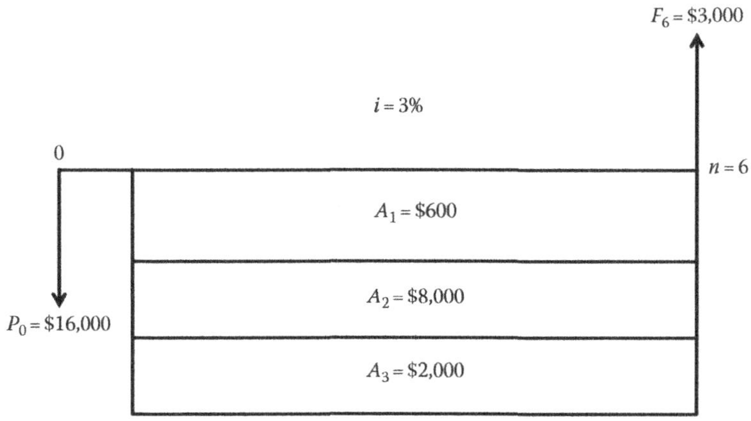

FIGURE 8.1 Cash flow diagram for testing machine 1 for Example 8.1.

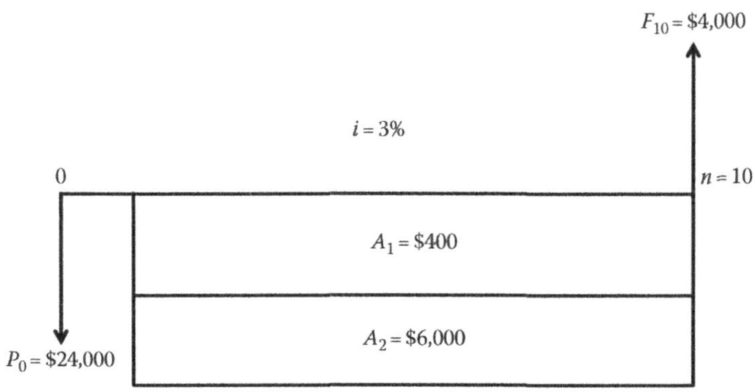

FIGURE 8.2 Cash flow diagram for testing machine 2 for Example 8.1.

Solution

Testing machine 1

$$A_1 = -\$600.00$$

$$A_2 = -\$8,000.00$$

$$A_3 = -\$2,000.00$$

First, calculate the equivalent uniform annual series of the initial cost (A_4):

$$A_4 = P_0(A/P,i,n) = -\$16,000.00(A/P,3,6) = -\$16,000.00(0.18460)$$
$$= -\$2,953.60$$

Second, calculate the equivalent uniform annual series for the future value at year six (A_5):

$$A_5 = F_6(A/F,i,n) = \$3,000.00(A/F,3,6) = \$3,000.00(0.15460)$$
$$= \$463.80$$

Third, solve for the EUAW by summing up all of the uniform annual series:

$$\text{EUAW} = A_1 + A_2 + A_3 + A_4 + A_5$$
$$= -\$600.00 - \$8,000.00 - \$2,000.00 - \$2,953.60 + \$464.80$$
$$= -\$13,088.80$$

Testing machine 2

$$A_1 = -\$400.00$$

$$A_2 = -\$6,000.00$$

First, calculate the equivalent uniform annual series of the initial cost (A_3):

$$A_3 = P_0(A/P,i,n) = -\$24,000.00(A/P,3,10) = -\$24,000.00(0.11723)$$
$$= -\$2,813.52$$

Second, calculate the equivalent uniform annual series for the future value at year six (A_4):

$$A_4 = F_{10}(A/F,i,n) = \$4,000.00(A/F,3,10) = \$4,000.00(0.08723)$$
$$= \$348.92$$

Third, solve for the EUAW by summing up all of the uniform annual series:

$$\text{EUAW} = A_1 + A_2 + A_3 + A_4$$
$$= -\$4,00.00 - \$6,000.00 - \$2,813.52 + \$348.92$$
$$= -\$8,864.60$$

Therefore, select alternative 2 since it has a lower equivalent uniform annual cost $-\$8,664.60 < -\$13,088.80$

Example 8.2

A mechanical engineer is investigating two systems for regulating the flow into a wastewater treatment plant. He has collected data for the potential costs and disbursements for each alternative and the data are shown in Table 8.2. Using equivalent uniform annual worth analysis techniques and an interest rate of 5%, determine which alternative should be selected based on the lowest equivalent uniform annual cost. Figures 8.3 and 8.4 are the cash flow diagrams for the two flow regulators.

TABLE 8.2
Wastewater Flow Regulator Alternatives

Cost or Disbursements	Flow Regulator 1	Flow Regulator 2
Initial cost	\$6,500,000.00	\$6,500,000.00
Annual maintenance cost	\$220,000.00	\$10,000.00
Annual labor cost	\$120,000.00	—
Repairs	—	\$100,000.00 every 5 years
Salvage value	\$700,000.00	—
Life in years	10	∞

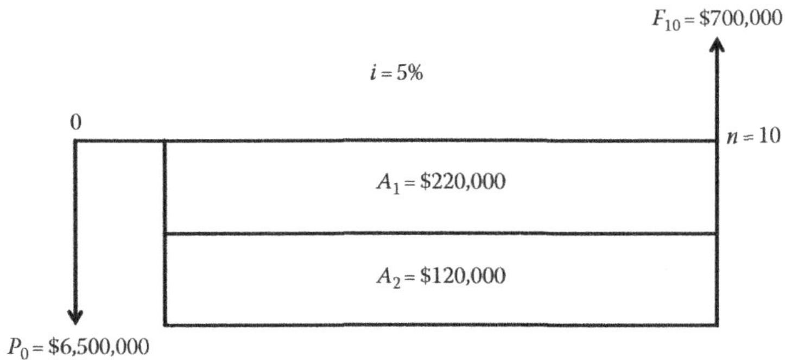

FIGURE 8.3 Cash flow diagram for flow regulator 1 for Example 8.2.

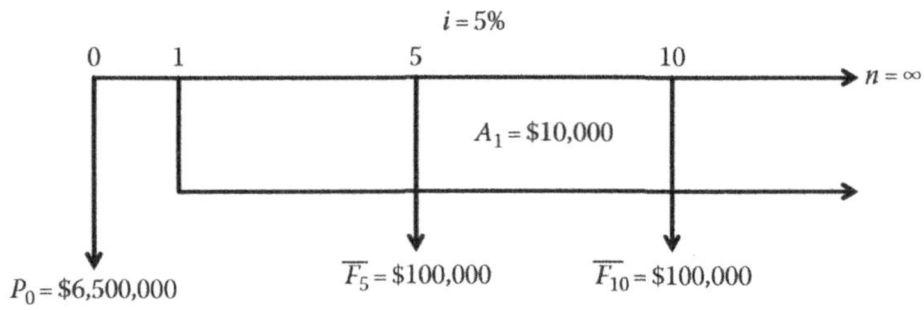

FIGURE 8.4 Cash flow diagram for flow regulator 2 for Example 8.2.

Solution

Flow regulator 1

$$\begin{aligned}
\text{EUAW}_1 &= P(A/P,i,n) + A_1 + A_2 + F_{10}(A/F,i,n) \\
&= -\$6,500,000.00(A/P,5,10) - \$220,000.00 - \$120,000.00 \\
&\quad + \$700,000.00(A/F,5,10) \\
&= -\$6,500,000.00(0.12950) - \$220,000.0 - \$120,000.00 \\
&\quad + \$700,000.00(0.07950) \\
&= -\$841,750.00 - \$220,000.00 - \$120,000.00 + \$55,650.00 \\
&= -\$1,126,100.00
\end{aligned}$$

Flow regulator 2

$$\begin{aligned}\text{EUAW}_2 &= P_0(i) + A_1 + F_5(A/F,i,n)\\ &= -\$650{,}000.00(0.05) - \$10{,}000.00 - \$100{,}000.00(A/F,5,5)\\ &= -\$6{,}500{,}000.00(0.05) - \$10{,}000.00 - \$100{,}000.00(0.18097)\\ &= -\$325{,}000.00 - \$10{,}000.00 - \$18{,}097.00\\ &= -\$353{,}097.00\end{aligned}$$

Therefore, select flow regulator 2 since it has a lower equivalent uniform annual cost than regulator 1 –\$353,097.00 < –\$1,126,100.00

Example 8.3

A civil engineer is comparing two alternatives for a building using equivalent uniform annual worth analysis techniques. The first building has a current salvage value of \$50,000,000.00 and it would cost \$20,000,000.00 to remodel the building. The second building would be a new building costing \$65,000,000.00. Use an interest rate of 9% and a life of the buildings of 50 years.

Solution

If the first building were remodeled, then it would forego the salvage value of \$50,000,000.00 and this would be part of the cost of this building; therefore, the present worth of the first building would be the following:

$$P_0 = -\$50{,}000{,}000.00 - \$20{,}000{,}000.00 = -\$70{,}000{,}000.00$$

The equivalent uniform annual worth of the first building is the following:

$$\begin{aligned}\text{EUAW}_1 &= P_0(A/P,i,n)\\ &= -\$70{,}000{,}000.00(A/P,9,50)\\ &= -\$70{,}000{,}000.00(0.09123)\\ &= -\$6{,}386{,}100.00\end{aligned}$$

The second building costs \$65,000,000.00 and its equivalent uniform annual worth is the following:

$$\begin{aligned}\text{EUAW}_2 &= P_0(A/P,i,n)\\ &= -\$65{,}000{,}000.00(A/P,9,50)\\ &= -\$65{,}000{,}000.00(0.09123)\\ &= -\$5{,}929{,}950.00\end{aligned}$$

Therefore, select building 2 since it has a lower equivalent uniform annual cost

$-\$5{,}929{,}950.00 < -\$6{,}386{,}100.00$

Example 8.4

An industrial engineer is comparing two machines his firm is considering installing in their manufacturing plant, one is a new machine and the other is a refurbished used machine. Table 8.3 includes the costs and disbursements associated with each machine. The interest rate is 5%. Determine which machine the industrial engineer should select using equivalent uniform annual worth analysis methods. Figures 8.5 and 8.6 are the cash flow diagrams for the new and used machines.

TABLE 8.3
Data for Industrial Machine Alternatives

Cost or Disbursements	New Machine	Refurbished Machine
Initial cost	\$440,000.00	\$230,000.00
Annual operating cost	\$70,000.00 increasing by \$4,000.00 per year starting in year 5	\$90,000.00 increasing by \$10,000.00 per year starting in year 7
Annual repairs	\$4,000.00	\$2,000.00
Salvage value	\$40,000.00	\$60,000.00
Life in years	12	15

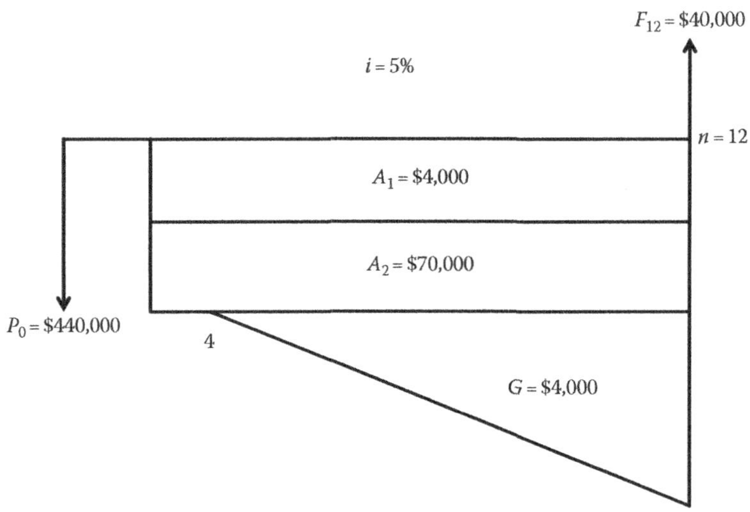

FIGURE 8.5 Cash flow diagram for the new machine in Example 8.4.

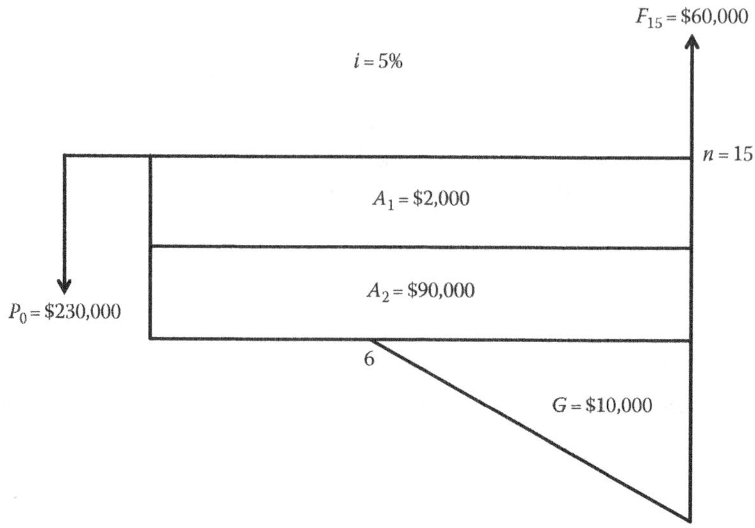

FIGURE 8.6 Cash flow diagram for the refurbished machine in Example 8.4.

Solution

New machine

$$
\begin{aligned}
\text{EUAW}_N &= P_0(A/P,i,n) + A_1 + A_2 + G(F/G,i,n)(A/F,i,n) + F(A/F,i,n) \\
&= -\$440{,}000.00(A/P,5,12) - \$4{,}000.00 - \$70{,}000.00 \\
&\quad -\$4{,}000.00(F/G,5,9)(A/F,5,12) + \$40{,}000.00(A/F,5,12) \\
&= -\$440{,}000.00(0.11283) - \$74{,}000.00 \\
&\quad -\$4{,}000.00(40.531)(0.06283) + \$40{,}000.00(0.06283) \\
&= -\$49{,}645.20 - \$74{,}000.00 - \$10{,}186.25 + \$2{,}513.20 \\
&= -\$131{,}318.25
\end{aligned}
$$

Refurbished machine

$$
\begin{aligned}
\text{EUAW}_R &= P_0(A/P,i,n) + A_1 + A_2 + G(F/G,i,n)(A/F,i,n) + F(A/F,i,n) \\
&= -\$230{,}000.00(A/P,5,15) - \$2{,}000.00 - \$90{,}000.00 \\
&\quad -\$10{,}000.00(F/G,5,9)(A/F,5,15) + \$60{,}000.00(A/F,5,15) \\
&= -\$230{,}000.00(0.09634) - \$2{,}000.00 - \$90{,}000.00 \\
&\quad -\$10{,}000.00(40.531)(0.04634) + \$60{,}000.00(0.04634) \\
&= -\$22{,}158.20 - \$2{,}000.00 - \$90{,}000.00 - \$18{,}782.07 + \$2{,}780.40 \\
&= -\$130{,}159.87
\end{aligned}
$$

Therefore, select the used machine since it has a lower equivalent uniform annual cost than the refurbished machine $-\$130{,}159.87 < -\$131{,}318.25$

Example 8.5

Three different types of four-inch piping are under consideration by a petroleum engineer for use in a refinery. The three types of pipe are copper, thermoplastic, and polyvinyl chloride (PVC). Table 8.4 lists the costs and disbursements associated with installing and using these three types of pipe for this particular application. Determine which type of pipe would have the lowest equivalent uniform annual cost using an interest rate of 6%. Figure 8.7 through 8.9 are the cash flow diagrams for the three types of pipes.

TABLE 8.4
Data for Piping Alternatives

Cost or Disbursements	Copper Pipe	Thermoplastic Pipe	PVC Pipe
Initial cost of piping and fitting joints	$203.00 per foot	$160.00 per foot	$171.00 per foot
Installation cost	$240.00 per foot	$180.00 per foot	$162.00 per foot
Operating and maintenance cost	$40,000.00 per year	$90,000.00 per year	$70,000 per year
Repairs	$100,000.00 every 2 years	$250,000.00 every 15 years	$300,000.00 every 10 years
Salvage value	$90.00 per foot	$50.00 per foot	$60.00 per foot
Number of feet required	4,000 ft	4,000 ft	4,000 ft
Life in years	100	60	70

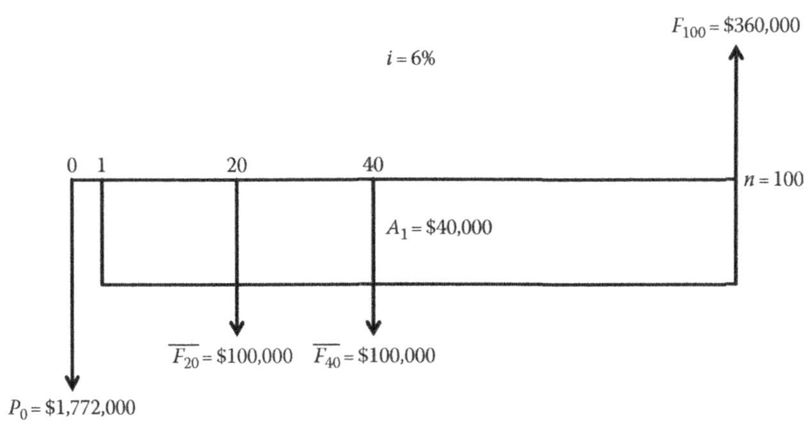

FIGURE 8.7 Cash flow diagram for copper piping in Example 8.5.

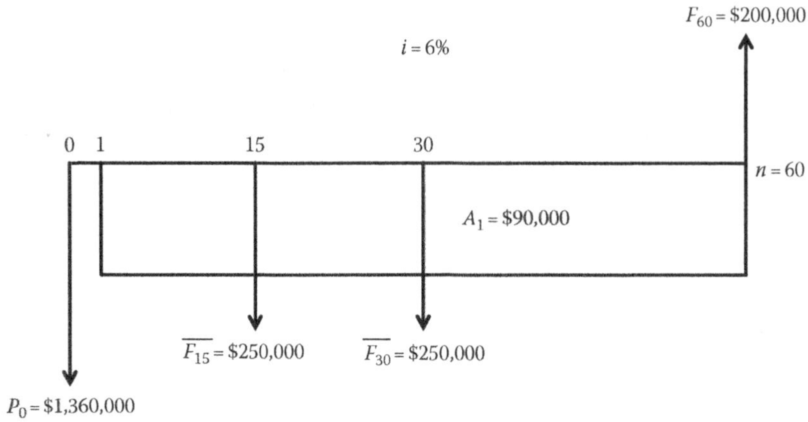

FIGURE 8.8 Cash flow diagram for the thermoplastic piping in Example 8.5.

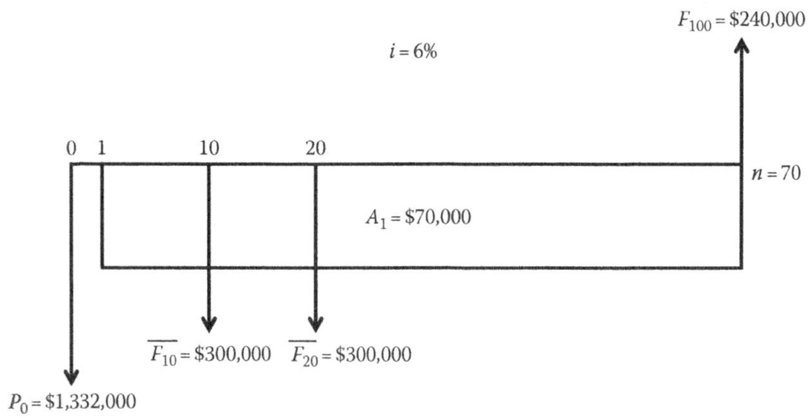

FIGURE 8.9 Cash flow diagram for the polyvinyl chloride piping in Example 8.5.

Solution

Copper pipe

$$\text{Initial cost of pipe} = (\text{Cost of pipe and fitting joints per foot} + \text{Installation cost per foot}) \times \text{Length of pipe}$$

$$\text{Initial cost of pipe} = \left(\frac{\$203.00}{\text{foot}} + \frac{\$240.00}{\text{foot}}\right) \times 4{,}000 \text{ ft} = \$1{,}772{,}000.00$$

$$\text{Salvage value of pipe} = \text{Salvage value per foot} \times \text{Length of the pipe}$$

$$\text{Salvage value of pipe} = \frac{\$90.00}{\text{foot}} \times 4{,}000 \text{ ft} = \$360{,}000.00$$

$$\begin{aligned}
\text{EUAW}_C &= P_0(A/P,i,n) + A_1 + F_{20}(A/F,i,n) + F_{100}(A/F,i,n) \\
&= -\$1{,}772{,}000.00(A/P,6,100) - \$40{,}000.00 - \$100{,}00.00(A/F,6,20) \\
&\quad + \$360{,}000.00(A/F,6,100) \\
&= -\$1{,}772{,}000.00(0.06018) - \$40{,}000.00 - \$100{,}00.00(0.02718) \\
&\quad + \$360{,}000.00(0.00018) \\
&= -\$106{,}638.96 - \$40{,}000.00 - \$2{,}718.00 + \$64.80 \\
&= -\$149{,}292.16
\end{aligned}$$

Thermoplastic pipe

Initial cost of pipe = (Cost of pipe and fitting joints per foot + Installation cost per foot) × Length of pipe

$$\text{Initial cost of pipe} = \left(\frac{\$160.00}{\text{foot}} + \frac{\$180.00}{\text{foot}}\right) \times 4{,}000 \text{ ft} = \$1{,}360{,}000.00$$

Salvage value of pipe = Salvage value per foot × Length of the pipe

$$\text{Salvage value of pipe} = \frac{\$50.00}{\text{foot}} \times 4{,}000 \text{ ft} = \$200{,}000.00$$

$$\begin{aligned}
\text{EUAW}_T &= P_0(A/P,i,n) + A_1 + F_{15}(A/F,i,n) + F_{60}(A/F,i,n) \\
&= -\$1{,}360{,}000.00(A/P,6,60) - \$90{,}000.00 - \$250{,}000.00(A/F,6,15) \\
&\quad + \$200{,}000.00(A/F,6,60) \\
&= -\$1{,}360{,}000.00(0.06188) - \$90{,}000.00 - \$250{,}000.00(0.04296) \\
&\quad + \$200{,}000.00(0.00188) \\
&= -\$84{,}156.80 - \$90{,}000.00 - \$10{,}740.00 + \$376.00 \\
&= -\$184{,}520.80
\end{aligned}$$

PVC pipe

Initial cost of pipe = (Cost of pipe and fitting joints per foot + Installation cost per foot) × Length of pipe

$$\text{Initial cost of pipe} = \left(\frac{\$171.00}{\text{foot}} + \frac{\$162.00}{\text{foot}}\right) \times 4{,}000 \text{ ft} = \$1{,}332{,}000.00$$

Salvage value of pipe = Salvage value per foot × Length of the pipe

$$\text{Salvage value of pipe} = \frac{\$60.00}{\text{foot}} \times 4{,}000 \text{ ft} = \$240{,}000.00$$

$$\text{EUAC}_P = P_0(A/P,i,n) + A_1 + F_{10}(A/F,i,n) + F_{70}(A/F,i,n)$$

$$= -\$1,332,000.00(A/P,6,70) - \$70,000.00 - \$300,00.00(A/F,6,10)$$

$$+ \$240,000.00(A/F,6,70)$$

$$= -\$1,332,000.00(0.06103) - \$70,000.00 - \$300,00.00(0.07587)$$

$$+ \$240,000.00(0.00103)$$

$$= -\$81,291.96 - \$70,000.00 - \$22,761.00 + \$247.20$$

$$= -\$173,805.76$$

Therefore, select the copper pipe since it has the lowest equivalent uniform annual cost of the three types of pipes.

Example 8.6

A civil engineer is asked to select a new scarper for his construction firm. He collects data from two equipment manufacturers and the data are shown in Table 8.5. Determine which scraper he should recommend based on equivalent uniform annual worth analysis using an interest rate of 9%. Figures 8.10 and 8.11 are the cash flow diagrams for the two scarpers.

TABLE 8.5
Data for Scraper Alternatives

Cost or Disbursements	Scraper 1	Scraper 2
Initial cost	$1,600,000.00	$1,200,000.00
Cost to operate per hour	$120.00 per hour	$150.00 per hour
Number of hours used per year	2,000 hours	2,000 hours
Increase in operating cost	$1,000.00 per year starting in year 7 until year 12	$1,500.00 per year starting in year 7 until year 10
Repairs	$15,000.00 years 6 through 9	$20,000.00 years 6 through 9
Major overhaul	$200,000.00 at year 5	$270,000.00 at year 5
Salvage value	$260,000.00	$200,000.00
Life in years	12	10

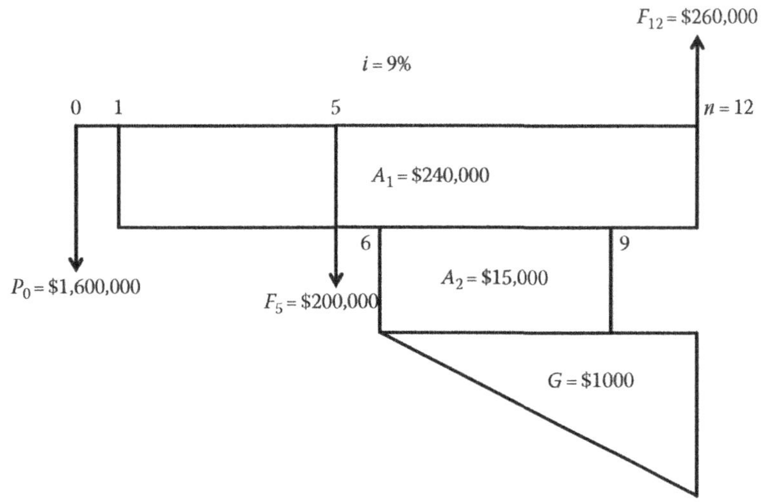

FIGURE 8.10 Cash flow diagram for scraper 1 in Example 8.6.

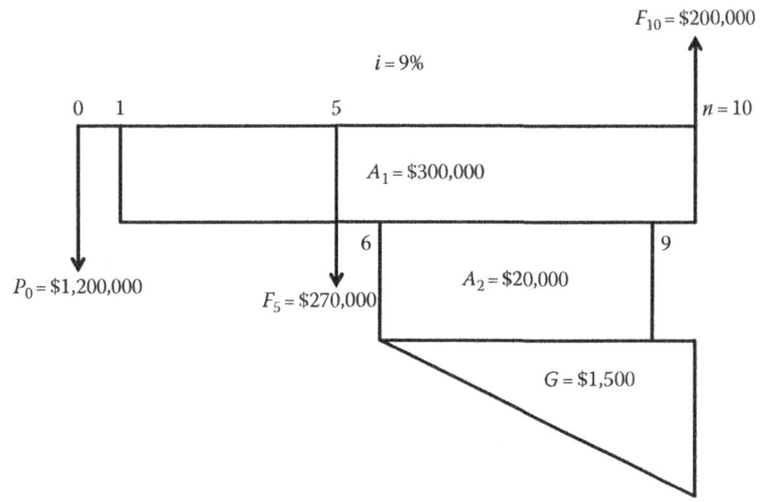

FIGURE 8.11 Cash flow diagram for scraper 2 in Example 8.6.

Solution

Scraper 1

The yearly operating cost is the cost per hour times the number of hours per year the scraper is operated during the year:

$$\text{Operating cost} = \frac{\$120.00}{\text{hour}} \times \frac{2{,}000 \text{ hours}}{\text{year}} = \$240{,}000.00 \text{ per year}$$

$$\begin{aligned}
\text{EUAW}_1 &= P_0(A/P,i,n) + A_1 + A_2(P/A,i,n)(P/F,i,n)(A/P,i,n) \\
&\quad + G(F/G,i,n)(A/F,i,n) + F_5(P/F,i,n)(A/P,i,n) + F_{12}(A/F,i,n) \\
&= -\$1,160,000.00(A/P,9,12) - \$240,000.00 \\
&\quad - \$15,000.00(P/A,9,4)(P/F,9,5)(A/P,9,12) \\
&\quad -\$1,000.00(F/G,9,7)(A/F,9,12) - \$200,000.00(P/F,9,5)(A/P,9,12) \\
&\quad + \$260,000.00(A/F,9,12) \\
&= -\$1,160,000.00(0.13965) - \$240,000.00 \\
&\quad - \$15,000.00(3.2397)(0.64993)(0.13965) \\
&\quad - \$1,000.00(24.449)(0.04965) - \$200,000.00(0.64993)(0.13965) \\
&\quad + \$260,000.00(0.04965) \\
&= -161,994.00 - \$240,000.00 - \$4,410.66 - \$1,213.89 - \$18,152.55 \\
&\quad + \$12,909.00 \\
&= -\$412,862.10
\end{aligned}$$

Scraper 2

The yearly operating cost is the cost per hour times the number of hours per year the scraper is operated during the year:

$$\text{Operating cost} = \frac{\$150.00}{\text{hour}} \times \frac{2,000 \text{ hours}}{\text{year}} = \$300,000.00 \text{ per year}$$

$$\begin{aligned}
\text{EUAW}_2 &= P_0(A/P,i,n) + A_1 + A_2(P/A,i,n)(P/F,i,n)(A/P,i,n) \\
&\quad +G(F/G,i,n)(A/F,i,n) + F_5(P/F,i,n)(A/P,i,n) + F_{10}(A/F,i,n) \\
&\quad -\$1,200,000.00(A/P,9,10) - \$3,000,000.00 \\
&\quad -\$20,000.00(P/A,9,4)(P/F,9,5)(A/P,9,10) \\
&\quad -\$1,500.00(F/G,9,5)(A/F,9,10) - \$270,000.00(P/F,9,5)(A/P,9,10) \\
&\quad +200,000.00(A/F,9,10) \\
&= -\$1,200,000.00(0.15582) - \$300,000.00 \\
&\quad -\$20,000.00(3.2397)(0.64993)(0.15582) \\
&\quad -1,500.00(10.941)(0.06582) - \$270,000.00(0.64993)(0.15882) \\
&\quad + \$200,000.00(0.06582) \\
&= -\$186,984.00 - \$300,000.00 - \$6,561.82 - \$1,080.20 - \$27,869.91 \\
&\quad +13,164.00 \\
&= -\$509,331.93
\end{aligned}$$

Therefore, select scraper 1 since it has a lower equivalent uniform annual cost

$-\$412,862.10 < -\$509,331.93$

8.5 SUMMARY

This chapter discussed comparing alternatives based on their equivalent uniform annual worth. Procedures were introduced for calculating the equivalent uniform annual worth of alternatives along with the process for using this method to compare alternatives. Salvage values, trade-in values, and sunk costs were defined along with an explanation of their relationship to equivalent uniform annual worth calculations. The process for calculating equivalent uniform annual cost of perpetual life alternatives was covered and detailed examples were provided demonstrating comparing alternatives by calculating their equivalent uniform annual worth.

KEY TERMS

Capital-recovery-plus interest method
Equivalent uniform annual worth
Salvage present worth method
Salvage sinking fund method
Salvage value
Sunk costs
Trade-in value

PROBLEMS

8.1 An agricultural engineer is comparing two potato-peeling machines. The data for the two machines are listed in Table 8.6. Using an interest rate of 15%, equivalent uniform annual worth analysis, and the salvage sinking fund method, which peeling machine should be selected by the firm?

TABLE 8.6
Data for Potato-Peeling Machine Alternatives

Costs or Disbursements	Potato-Peeling Machine 1	Potato-Peeling Machine 2
Initial cost	$260,000.00	$360,000.00
Operating and maintenance costs	$8,000.00	$3,000.00
Labor cost	$110,000.00	$70,000.00
Salvage value	$10,000.00	$20,000.00
Repair cost per year	—	$26,000.00
Life in years	6	10

8.2 Calculate the equivalent uniform annual cost of the potato-peeling machines in Problem 8.1 if the interest rate is 4%.

8.3 A biomedical engineer is purchasing a fleet of five electric vehicles. The initial cost is $46,000.00 per vehicle and the vehicles will each have a salvage value of $3,000.00 after five years. The cost of operating all of the vehicles is $6,500.00 in the first year increasing by $500.00 per year starting in the second year. The electric vehicles will save $12,000.00 per vehicle per year in gasoline costs. Using an interest rate of 10% and equivalent uniform annual worth analysis, determine if the vehicles should be purchased by the firm.

8.4 For the data provided in Problem 8.3, determine the equivalent uniform annual worth if the interest rate is 3%.

8.5 A process engineer has invested \$10,000.00 into an interest-bearing account. She will be investing another \$30,000.00 in three years and \$6,000.00 per year for five years starting at year four. The process engineer needs to determine the amount of money she may withdraw forever starting at year 12 if the interest rate is 8%.

8.6 Determine the amount of money that could be withdrawn forever in Problem 8.5 if the amount initially invested is \$20,000.00.

8.7 A mechanical engineer deposits \$100,000.00 into an interest-bearing account with an interest rate of 7%. How long do the funds need to be invested for the engineer to be able to withdraw \$14,000.00 per year forever?

8.8 How long do the funds in Problem 8.7 need to be invested for the mechanical engineer to be able to withdraw \$20,000.00 forever?

8.9 A scraper is due for a major overhaul at year three costing \$100,000.0. If the scraper has an economic life of seven years, what is the equivalent uniform annual cost of the overhaul over seven years if the interest rate is 10%?

8.10 What is the equivalent uniform annual cost of the overhaul over seven years for the scraper in Problem 8.9 if the interest rate is 20%?

8.11 An industrial engineer has estimated that a processing machine will need repairs in years two through five that will cost \$50,000.00 per year, but no additional repairs will be required after year five. If the machine is used for 12 years and the interest rate is 7%, what is the equivalent uniform annual cost of the repairs?

8.12 What would be the equivalent uniform annual cost of the processing machine repairs in Problem 8.11 if the machine requires repairs from year four until year 10?

8.13 A roadway will not have any maintenance costs until year six when they will be \$10,000.00 increasing by \$10,000.00 each year until year 15. If the interest rate is 7%, what is the equivalent uniform annual cost of the maintenance costs?

8.14 If the roadway in Problem 8.13 requires maintenance starting at year four of \$12,000.00 per year increasing by \$12,000.00 each year until year 12, what is the equivalent uniform annual cost of the maintenance over 15 years?

8.15 If the roadway in Problem 8.13 has an initial cost of \$15,000,000.00, what would be the equivalent uniform annual cost?

8.16 A city will be constructing a college gym that will be used for 40 years. The gym will cost \$70,000,000.00 of which \$60,000,000.00 is for gym construction and \$10,000,000.00 is the cost of the land. The maintenance costs include \$100,000.00 for painting and staining every four years, \$5,000,000.00 for air conditioning unit replacements every 10 years, and \$15,000,000.00 to replace the wood floors every 20 years. At the end of 40 years, the salvage value will be \$5,000,000.00. The gym will be paid for through the sale of bonds that pay 9% interest. What would be the yearly revenue required to repay the financing on this project?

8.17 What would be the yearly revenue required to repay the financing for the gym in Problem 8.16 if the interest rate is 20%?

8.18 Table 8.7 lists the costs and disbursements for an excavator owned by a heavy construction company. If the interest rate is 20%, what amount does the construction company need to earn each year to cover the costs associated with the excavator?

TABLE 8.7
Data for Excavator

Costs or Disbursements	Excavator Data
Initial cost	$600,000.00
Operating and maintenance costs	$70,000.00 increasing by $10,000.00 per year
Overhaul	$200,000.00 every 5 years
Repairs from years 7 to 10	$15,000.00
Salvage value	$100,000.00
Life in years	10

8.19 If the excavator in Problem 8.18 has operating and maintenance costs of $100,000.00 per year increasing by $20,000.00 per year, what amount does the construction company need to earn per year to cover the costs associated with the excavator?

8.20 What amount does the construction company in Problem 8.18 need to earn per year to cover the excavator expenses if the interest rate is 8%?

9 Rate of Return Method for Comparing Alternatives

Chapters 7 and 8 introduced the net present worth and the equivalent uniform annual worth methods for comparing alternatives, and this chapter introduces another method for comparing alternatives, the *rate of return (ROR) method.* This chapter demonstrates developing ROR equations using net present worth and equivalent uniform annual worth methods and explains the process for calculating rates of return using interpolation. The last section explains the process for evaluating alternatives using incremental investment analysis—incremental rate of return (IROR) analysis.

Once a ROR is calculated, it is normally compared to the ROR a firm would be able to obtain through a secure investment such as a certificate of deposit or a savings account to determine whether the return on the proposed investment is higher than the secure investment. The ROR a company compares the proposed project to is the *minimum attractive rate of return* (MARR). If proposed investments or projects are projected to have a higher rate of return than the MARR, then a firm would be willing to invest in the proposed investment or project. Firms also evaluate the risks associated with proposed projects in terms of the calculated ROR for the project to determine if the ROR justifies the level of risk for the project. The higher the potential risk involved in a project, the higher the expected ROR should be to compensate for the risk. In some instances, the ROR could be higher than 50% if a project has elements considered to be risky because either they have never been built before, they use untested technology, there are unforeseen sight conditions, there are only a few vendors who are able to fabricate the elements, or the project is being built in a country with a high level of political risk, which could cause the project to be nationalized by the native government.

The MARR used by firms varies from company to company since some companies are willing to build projects for lower rates of return than other companies for a variety of reasons. The MARR also varies based on current interest rates in the economy. If interest rates in the economy are high, then a firm would use a MARR higher than current interest rates, and if interest rates are low, the MARR may be lower. To summarize, if the ROR for a project is higher than the MARR (ROR > MARR), a firm will invest in the project. If the ROR is less than the MARR (ROR < MARR), a firm will not invest in the project unless there are extenuating circumstances. Rate of return was defined in Equation 2.9 as:

$$\text{Rate of return (percent)} = \frac{\text{Total amount of money received} - \text{Original investment}}{\text{Original investment}} \times 100\%$$

Rate of return was also defined in Equation 2.10 as:

$$\text{Rate of return (percent)} = \frac{\text{Profit}}{\text{Original investment}} \times 100\%$$

9.1 SOLVING FOR RATES OF RETURN

A ROR represents the percentage return on an investment but it also represents the interest rate where the present worth of initial investments and all future values are equivalent.

9.1.1 Solving for Rates of Return Using Net Present Worth

A *net present worth equation* is used to solve for a ROR by setting the net present worth equation equal to zero. The interest rate where the net present worth equation equals zero is the ROR. Rate of return equations equate a present value to the present worth of all future sums as shown in Equation 9.1:

$$0 = \pm P_0 \pm \sum F(P/F,i,n) \pm \sum A(P/A,i,n) \pm \sum G(P/G,i,n) \quad (9.1)$$

Not all equations will have every one of the factors listed in Equation 9.1; therefore, only the relevant factors are used where appropriate.

Example 9.1 demonstrates the process for developing ROR equations.

Example 9.1

A company invests \$10,000.00 now and is able to withdraw \$5,000.00 at the end of three years and \$15,000.00 at the end of five years. What is the net present worth equation for solving for the rate of return? Figure 9.1 is the cash flow diagram for the investment.

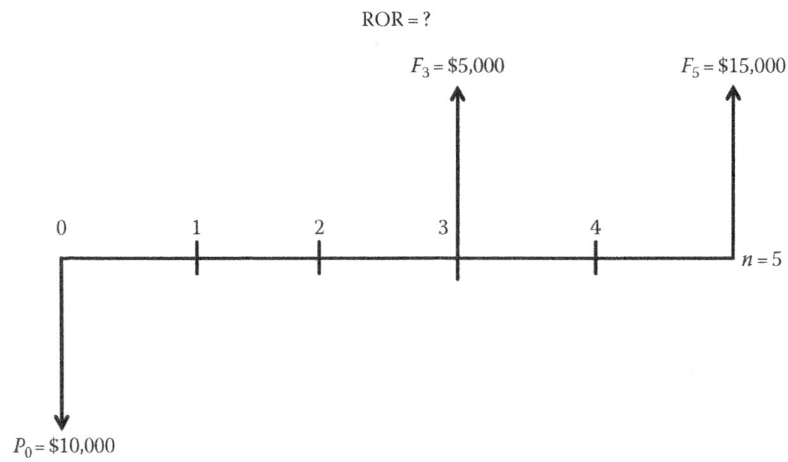

FIGURE 9.1 Cash flow diagram for the investment in Example 9.1.

Solution

$$0 = -P_0 + F_3(P/F,i,n) + F_5(P/F,i,n)$$

$$= -\$10,000.00 + \$5,000.00(P/F,i,3) + \$15,000.00(P/F,i,5)$$

9.2 SOLVING FOR UNKNOWN INTEREST RATES USING INTERPOLATION

The method for solving for a ROR requires developing a net present worth equation in the format of Equation 9.1 and then testing different interest rates until a positive answer is obtained that approaches zero and then testing interest rates again until a negative answer is obtained that approaches zero. Next, the actual interest rate is calculated by interpolating between the positive and negative answers. The formula for interpolation, Equation 3.14, and the table used to develop interpolation problems, Table 3.1, were introduced in Section 3.5.2. The interpolation formula Equation 3.14 is the following:

$$\text{ROR} = c + \left(\frac{a}{b}\right)d$$

The format for interpolation shown in Table 3.1 is modified to solve for net present worth and it is shown in Table 9.1.

TABLE 9.1
Table for Developing Interpolation Problems for Unknown Rate of Return

	ROR	Net Present Worth	
	i_1	A	a (A to 0)
d	Unknown ROR	0	b (A to C)
	i_2	C	

Note: $a = A - B$, $b = A - C$, $d = i_1 - i_2$.

The following demonstrates the process for calculating the rate of return in Example 9.1:

$$\begin{aligned} 0 &= -P_0 + F_3(P/F,i,n) + P_5(P/A,i,n) \\ &= -\$10{,}000.00 + \$5{,}000.00(P/F,i,3) + \$15{,}000.00(P/F,i,5) \end{aligned}$$

Try 14%

$$\begin{aligned} 0 &= -\$10{,}000.00 + \$5{,}000.00(0.67497) + \$15{,}000.00(0.51937) \\ &= -\$10{,}000.00 + \$3{,}374.85 + \$7{,}790.55 \\ &= \$1{,}165.40 \end{aligned}$$

Try 20%

$$\begin{aligned} 0 &= -\$10{,}000.00 + \$5{,}000.00(0.57870) + \$15{,}000.00(0.40188) \\ &= -\$10{,}000.00 + \$2{,}893.50 + \$6{,}028.20 \\ &= -\$1{,}078.30 \end{aligned}$$

The interest rate is between 14% and 20%; therefore, interpolation and Table 9.2 are used to determine the actual interest rate.

TABLE 9.2
Table for Developing Interpolation Problem for Unknown Rate of Return with n = 5 Years for Example 9.1

	ROR	Net Present Worth		
d	14%	1165.40	a	b
	Unknown ROR	0		
	20%	−1078.30		

$$\text{ROR} = c + \left(\frac{a}{b}\right)d$$

$$\text{ROR} = 14 + \left(\frac{1,165.40 - 0}{1,165.40 - (-1,078.30)}\right) \times (20 - 14)$$

$$= 14 + \left(\frac{1,165.40}{2,243.70}\right) \times 6$$

$$= 14 + (0.51941 \times 6)$$

$$= 14 + 3.1164$$

$$= 17.12\%$$

Another example demonstrating the process for calculating rates of return using the net present worth method is Example 9.2.

Example 9.2

An aerospace engineer has suggested his firm introduce a new type of rotor into the design of a plane. The rotor costs $250,000.00 per engine and the plane has two engines. The rotor will save the company $50,000.00 per year in operating costs. The salvage value of the rotors at the end of 10 years is $70,000.00 per rotor. Determine the rate of return if the rotors are used for 10 years using net present worth analysis techniques. Figure 9.2 is the cash flow diagram for the rotors.

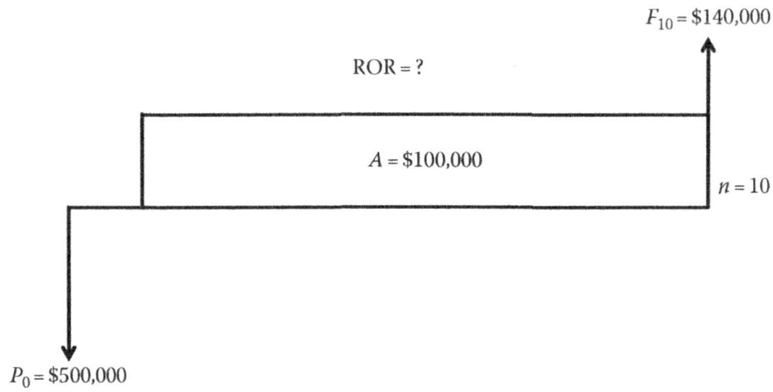

FIGURE 9.2 Cash flow diagram for the rotors in Example 9.2.

Solution

Since there are two rotors per plane, the initial costs, the yearly benefits, and the salvage value are all multiplied by two.

$$\text{Initial cost} = \$250,000 \times 2 = \$500,000.00$$

$$\text{Yearly benefits} = \$50,000.00 \times 2 = \$100,000.00$$

$$\text{Salvage value} = \$70,000.00 \times 2 = \$140,000.00$$

Use Equation 9.1 to develop the net present worth equation for this problem:

$$0 = -P_0 + F_{10}(P/F,i,10) + A(P/A,i,10)$$

Try 10%

$$= -\$500,000.00 + \$140,000.00(P/F,i,10) + \$100,000.00(P/A,i,10)$$

$$= -\$500,000.00 + \$140,000.00(0.38554) + \$100,000.00(6.1445)$$

$$= -\$500,000.00 + \$53,975.60 + \$614,450.00$$

$$= \$168,425.60$$

Try 20%

$$= -\$500,000.00 + \$140,000.00(P/F,i,10) + \$100,000.00(P/A,i,10)$$

$$= -\$500,000.00 + \$140,000.00(0.16151) + \$100,000.00(4.1924)$$

$$= -\$500,000.00 + \$22,611.40 + \$419,120.00$$

$$= -\$58,268.60$$

Use interpolation and Table 9.3 to solve for the rate of return:

TABLE 9.3
Table for Developing Interpolation Problem for Unknown Rate of Return with n = 10 Years for Example 9.2

	ROR	Net Present Worth	
d	10%	168,425.60	a
	Unknown ROR	0	b
	20%	−58,268.60	

$$\text{ROR} = c + \left(\frac{a}{b}\right)d$$

$$\text{ROR} = 10 + \left(\frac{168{,}425.60 - 0}{168{,}425.60 - (-58{,}268.60)}\right) \times (20 - 10)$$

$$= 10 + \left(\frac{168{,}425.60}{226{,}694.20}\right) \times 10$$

$$= 10 + (0.742964 \times 10)$$

$$= 10 + 7.429639$$

$$= 17.43\%$$

Example 9.3 calculates a ROR using net present worth analysis.

Example 9.3

An industrial design firm has an opportunity to invest $100,000.00 in a project resulting in revenue of $30,000.00 per year for seven years. The cost to maintain the project will be $8,000.00 per year and the salvage value at the end of seven years will be $32,000.00. The industrial design firm will invest in the project if it will have a rate of return above 15% (MARR) per year. Calculate whether this project will earn a rate of return above the acceptable MARR for this firm using net present worth analysis techniques. Figure 9.3 is the cash flow diagram for the industrial design firm investments.

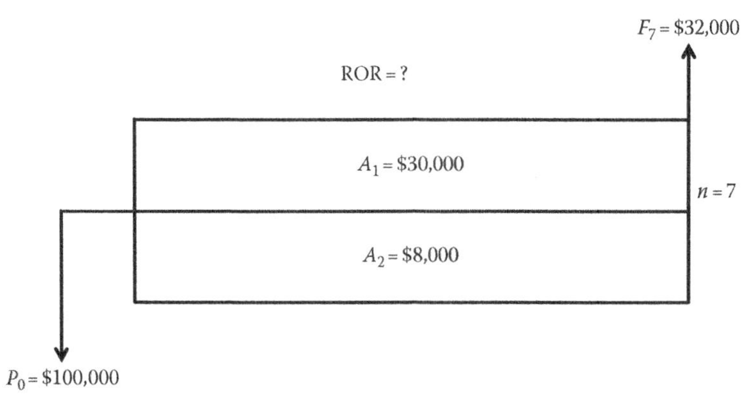

FIGURE 9.3 Cash flow diagram for the industrial design firm project in Example 9.3.

Solution

$$0 = -P_0 + F_7(P/F,i,n) + \sum A(P/A,i,n)$$

Try 10%

$$= -\$100,000.00 + \$32,000.00(P/F,i,7) + (\$30,000.00 - \$8,000.00)(P/A,i,7)$$

$$= -\$100,000.00 + \$32,000.00(0.51316) + \$22,000.00(4.8684)$$

$$= -\$100,000.00 + \$16,421.12 + \$107,104.00$$

$$= \$23,525.12$$

Try 25%

$$= -\$100,000.00 + \$32,000.00(P/F,i,7) + \$22,000.00(P/A,i,7)$$

$$= -\$100,000.00 + \$32,000.00(0.20972) + \$22,000.00(3.1611)$$

$$= -\$100,000.00 + \$6,711.04 + \$69,544.20$$

$$= -\$23,744.76$$

Use interpolation and Table 9.4 to solve for the rate of return:

TABLE 9.4
Table for Developing Interpolation Problem for Unknown Rate of Return with n = 7 Years for Example 9.3

	ROR	Net Present Worth	
	10%	23,525.12	a
d	Unknown ROR	0	b
	25%	−23,744.76	

$$\text{ROR} = c + \left(\frac{a}{b}\right)d$$

$$\text{ROR} = 10 + \left(\frac{23,525.12 - 0}{23,525.12 - (-23,744.76)}\right) \times (25 - 10)$$

$$= 10 + \left(\frac{23,525.12}{47,269.88}\right) \times 15$$

$$= 10 + (0.497677 \times 15)$$

$$= 10 + 7.465151$$

$$= 17.46\%$$

The rate of return is greater than the MARR, 17.45% > 15%; therefore, the firm should invest in this project.

9.3 SOLVING FOR RATES OF RETURN USING EQUIVALENT UNIFORM ANNUAL WORTH

Another analysis method for solving for rates of return requires developing equations using the equivalent uniform annual worth analysis method based on Equation 9.2:

$$0 = \pm P_0\left(A/P,i,n\right) \pm A \pm \sum G\left(A/G,i,n\right) \pm \sum F\left(A/F,i,n\right) \tag{9.2}$$

Not all problems will include all of the factors listed in Equation 9.2; therefore, each equation for solving for the EUAW should only include the relevant factors.

The rate of return in EUAW analysis is calculated using the same steps for net present worth analysis. The trial and error method for calculating the EUAW at different interest rates is employed, and after determining both a positive and a negative answer, an interpolation table is developed to calculate the interest rate. Example 9.4 calculates the ROR for the values provided in Example 9.3 using EUAW analysis methods instead of the net present worth analysis methods used in Example 9.3.

Example 9.4

Solve for the rate of return in Example 9.3 using equivalent uniform annual worth analysis methods. Figure 9.3 in Example 9.3 is also the cash flow diagram for Example 9.4.

Solution

Solve for the equivalent uniform annual cost using Equation 9.2:

$$0 = -P_0(A/P,i,n) + A + F_7(A/F,i,n)$$

$$= -\$100{,}000.00(A/P,i,7) + \$22{,}000.00 + \$32{,}000.00(A,P,i,7)$$

Try 15%

$$= -\$100{,}000.00(0.24036) + \$22{,}000.00 + \$32{,}000.0(0.09036)$$

$$= -\$24{,}036.00 + \$22{,}000.00 + \$2{,}891.52$$

$$= \$855.52$$

Try 20%

$$= -\$100{,}000.00(0.27742) + \$22{,}000.00 + \$32{,}000.00(0.07742)$$

$$= -\$27{,}742.00 + \$22{,}000.00 + \$2{,}477.44$$

$$= -\$3{,}264.56$$

Use interpolation and Table 9.5 to calculate the rate of return:

TABLE 9.5
Table for Developing Interpolation Problem for Unknown Rate of Return with n = 7 Years for Example 9.4

	ROR	EUAW		
d	15%	855.52	a	b
	Unknown ROR	0		
	20%	−3264.56		

$$\text{ROR} = c + \left(\frac{a}{b}\right)d$$

$$\text{ROR} = 15 + \left(\frac{855.52 - 0}{855.52 - (-3264.56)}\right) \times (20 - 15)$$

$$= 15 + \left(\frac{855.52}{4{,}120.08}\right) \times 5$$

$$= 15 + (0.207646 \times 5)$$

$$= 15 + 1.03823$$

$$= 16.04\%$$

The rate of return is greater than the MARR, 16.04% >15%; therefore, the firm should invest in this project.

Note: Since the interpolation was calculated between 15% and 20% the resulting rate of return, it is more accurate than the interpolation between 10% and 25% in Example 9.3.

Example 9.4 illustrates that if the ROR is interpolated between a smaller spread of interest rates, the ROR is more accurate than a larger spread of interest rates. Therefore, if the interest rate is being calculated for a real-world project, the trial and error method is used to narrow down the ROR and then a second interpolation should be conducted to calculate the ROR between a spread of only one or two interest rates rather than a larger spread of interest rates. Example 9.5 demonstrates accomplishing this using equivalent uniform annual worth analysis.

Example 9.5

A process engineer plans on investing $5,000.00 now so he may withdraw $500.00 a year for 12 years and still have $1,000.00 left in the account at the end of 12 years. Part I—calculate the rate of return. Part II—narrow down the spread between the interest rates before calculating the rate of return. Calculate the rate of return on this investment using equivalent uniform annual worth analysis. Figure 9.4 is the cash flow diagram for the investment by the process engineer.

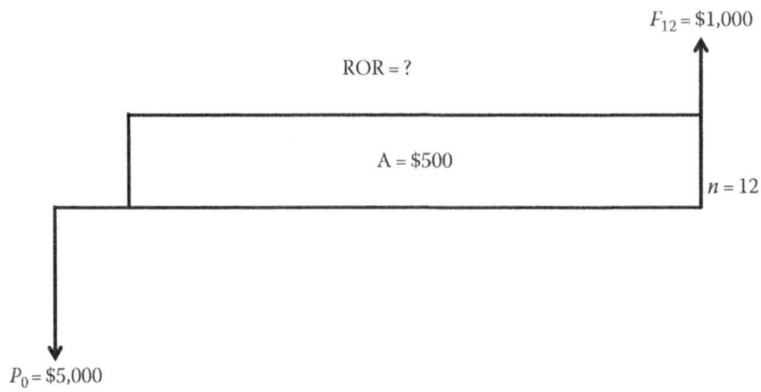

FIGURE 9.4 Cash flow diagram for the investment by the process engineer in Example 9.5.

Solution

Part I

Solve for the rate of return using Equation 9.2:

$$0 = -P_0(A/P,i,n) + A + F_{12}(A/F,i,n)$$

$$= -\$5{,}000.00(A/P,i,12) + \$500.00 + \$1{,}000.00(A/F,i,12)$$

Try 10%

$$= -\$5{,}000.00(0.14676) + \$500.00 + \$1{,}000.00(0.04676)$$

$$= -\$733.80 + \$500.00 + \$46.76$$

$$= -\$187.04$$

Try 5%

$$= -\$5{,}000.00(0.11283) + \$500.00 + \$1{,}000.00(0.06283)$$

$$= -\$584.15 + \$500.00 + \$62.83$$

$$= -\$21.32$$

Try 1%

$$= -\$5{,}000.00(0.08885) + \$500.00 + \$1{,}000.00(0.07885)$$

$$= -\$444.25 + \$500.00 + \$78.85$$

$$= \$134.60$$

Use interpolation and Table 9.6 to solve for the rate of return:

TABLE 9.6
Table for Developing Interpolation Problem for Unknown Rate of Return with n = 12 Years for Example 9.5 Part I

	ROR	EUAW	
d	1%	134.60	a
	Unknown ROR	0	b
	5%	−21.32	

$$\text{ROR} = c + \left(\frac{a}{b}\right)d$$

$$\text{ROR} = 1 + \left(\frac{134.60 - 0}{134.60 - (-21.32)}\right) \times (5 - 1)$$

$$= 1 + \left(\frac{134.60}{155.92}\right) \times 4$$

$$= 1 + (0.863263 \times 4)$$

$$= 1 + 3.453053$$

$$= 4.45\%$$

Solution
Part II

In order to narrow down the interest rate even further, try 4% and then interpolate between 4% and 5%.

Try 4%

$$= -\$5,000.00(0.10655) + \$500.00 + \$1,000.00(0.06655)$$

$$= -\$532.75 + \$500.00 + \$66.55$$

$$= \$33.75$$

At 5%, the equivalent uniform annual worth was −\$21.32.

Interpolate between 4% and 5% and use Table 9.7 to determine the actual rate of return:

TABLE 9.7
Table for Developing Interpolation Problem for Unknown Rate of Return with n = 12 Years for Example 9.5 Part II

	ROR	EUAW	
d [	4%	33.75	] a] b
	Unknown ROR	0	
	5%	−21.32	

$$\text{ROR} = c + \left(\frac{a}{b}\right)d$$

$$\text{ROR} = 4 + \left(\frac{33.75-0}{33.75-(-21.32)}\right) \times (5-4)$$

$$= 4 + \left(\frac{33.75}{55.07}\right) \times 1$$

$$= 4 + (0.612856 \times 1)$$

$$= 4 + 0.612856$$

$$= 4.61\%$$

Therefore, this is the actual rate of return for Example 9.4.

9.4 EVALUATION BY INCREMENTAL INVESTMENT ANALYSIS: INCREMENTAL RATE OF RETURN

Incremental investment analysis, also known as *incremental rate of return* (*IROR*) *analysis*, is another method for evaluating multiple alternatives. The IROR method allows for the evaluation of each increment of increased initial cost for alternatives that could be added to an initial investment. In this analysis process, the lowest initial cost alternative is compared to the second lowest initial cost alternative using ROR analysis techniques. If the added increment of investment for the second alternative compared to the first investment alternative has a ROR higher than the MARR, the higher cost alternative is then compared to the third lowest cost alternative, and if the added increment has a ROR higher than the MARR, the higher cost alternative is selected and so forth until all of the alternatives have had their incremental investment ROR calculated and compared to the MARR.

If at any point when the alternatives are being compared using ROR analysis techniques, the IROR calculated is less than the MARR, the higher cost alternative is eliminated from consideration, and the remaining alternative is compared to the next highest cost alternative.

If there are sufficient funds to pay for the highest initial cost alternative and the highest initial cost alternative is not selected due to its IROR being less than the MARR, the excess funds over and above the cost of the alternative selected are available to be invested somewhere else or in a different project.

If the IROR for each alternative being compared is higher than the MARR, it indicates that each added increment of investment is justified and will provide a return higher than the MARR of the firm considering the alternatives. This analysis technique provides another method for evaluating potential investments in terms of their economic viability.

The process involved in determining incremental rates of return requires several steps that should be followed in their precise order. If the alternatives are evaluated in a different order than the one described in the following steps, the values calculated will not represent the IROR of the incremental investments.

The following are the steps required for calculating the IROR for alternatives:

1. Organize the alternatives from the lowest initial cost alternative to the highest initial cost alternative.
2. Draw a table and fill in the cash flows of the alternatives listing the lowest initial cost alternative on the left and the higher initial cost alternative in the middle of the table. To the right of the higher cost alternative list the difference in the cash flows of the two alternatives—this is the *net cash flow*—the cash flow of the difference between the alternatives.
3. Draw a cash flow diagram for the lowest cost alternative.
4. Draw a cash flow diagram for the higher cost alternative.
5. Draw a cash flow diagram of the net cash flow, which is the higher initial cost alternative minus the lower initial cost alternative. If the two alternatives being compared have different life spans, then the net cash flow has to be a cash flow representing the least common multiple of both alternatives.
6. The IROR for the incremental investment is the ROR of the net cash flow; therefore, either a net present worth or an equivalent uniform annual worth equation is developed using the net cash flow and the IROR is calculated using the trial and error method introduced in Section 9.2.
7. Once the IROR (the ROR of the net cash flow) has been calculated, it is compared to the MARR. If
   a. IROR > MARR, the higher initial cost alternative is selected.
   b. IROR < MARR, the higher cost alternative of the incremental investment is not justified; therefore, the lower cost alternative is selected.
8. Steps 1 through 7 are repeated with the alternative selected in step 6 being compared to the next higher initial cost alternative until only one alternative remains.

If the IROR analysis being performed includes *mutually exclusive alternatives* (one will be selected to the exclusion of the other alternatives), then an alternative being evaluated using this analysis technique may not be compared with another alternative where the additional incremental investment is not greater than the MARR. If the IROR of an alternative is less than the MARR, the higher initial cost alternative is removed from the evaluation process.

When there are mutually exclusive alternatives, the following steps are used to compare the alternatives using IROR analysis:

1. All of the alternatives are first ranked in terms of the lowest initial cost to the highest initial cost and they are evaluated in the order they are ranked during this step.
2. Before comparing the lowest initial cost alternative with the second lowest initial cost alternative, the lowest initial cost alternative is compared to the do nothing alternative. This entails calculating the ROR of the lowest initial cost alternative using net present worth or equivalent uniform annual worth analysis.
3. The ROR of the lowest initial cost alternative calculated in step 2 is compared to the MARR, and if it is less than the MARR (ROR < MARR), the lowest cost alternative is removed from further analysis.

4. If the lowest cost alternative was eliminated during the calculations in step 3, then the ROR of the second lowest initial cost alternative is compared to the do nothing alternative.
5. Step 4 is repeated until one of the alternatives has a ROR higher than the MARR (ROR > MARR). Once one of the alternatives has a ROR greater than the MARR, then this alternative is compared to the next highest initial cost.
6. The net cash flow is calculated for the alternative with the ROR higher than the MARR and the next highest initial cost alternative. Use the least common multiple of lives for the net cash flow if using net present worth analysis techniques.
7. Solve for the IROR of the net cash flow using either net present worth or equivalent uniform annual worth analysis.
8. Compare the IROR obtained in step 7 to the MARR and if the
   a. IROR > MARR, retain the higher initial cost alternative.
   b. IROR < MARR, retain the lower initial cost alternative.
9. Repeat the previous eight steps until only one alternative is remaining from the analysis process.

Example 9.6 illustrates calculating the incremental rate of return for mutually exclusive alternatives.

Example 9.6

A construction field engineer is considering recommending the purchase a new scraper. The firm has enough funds to afford either of the two alternatives under consideration, but the funds could also be invested in other opportunities if the company does not purchase a scraper. The field engineer solves for the rate of return using net present worth analysis of the lower cost scraper compared to the do nothing alternative. If the incremental investment has a rate of return higher than the company MARR of 10%, then the engineer compares the lower cost alternative to the higher cost alternative to determine the incremental rate of return and then compares the incremental rate of return to the company MARR. If the incremental rate of return is higher than the MARR, the company will purchase the higher cost scraper. Table 9.8 contains the data for each of the potential scrapers. Figure 9.5 is the cash flow diagram for scraper 1 compared to the do nothing alternative, Figure 9.6 is the cash flow diagram for scraper 2, and Figure 9.7 is the net cash flow of scraper 2 minus scraper 1.

TABLE 9.8
Construction Scraper Comparison Data

Cost or Disbursements	Scraper 1	Scraper 2	Net Cash Flow of Scraper 2 Minus Scraper 1
Purchase price	$1,400,000.00	$1,900,000.00	$500,000.00
Yearly expenses	$240,000.00	$200,000.00	$40,000.00
Yearly income	$480,000.00	$500,000.00	$20,000.00
Salvage value	$1,250,000.00	$1,550,000.00	$300,000.00
Life in years	10	10	10

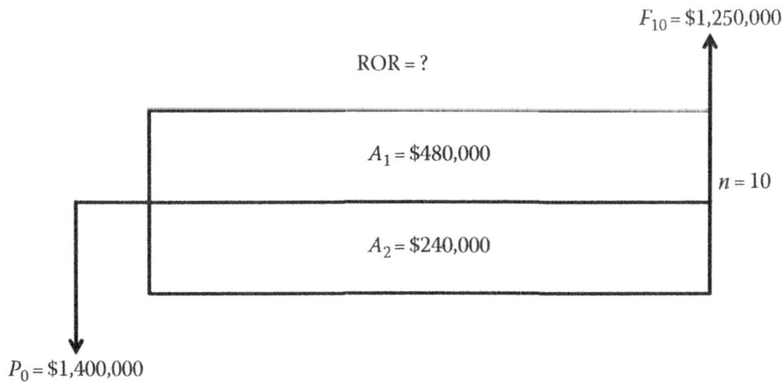

FIGURE 9.5 Cash flow diagram for scraper 1 in Example 9.6.

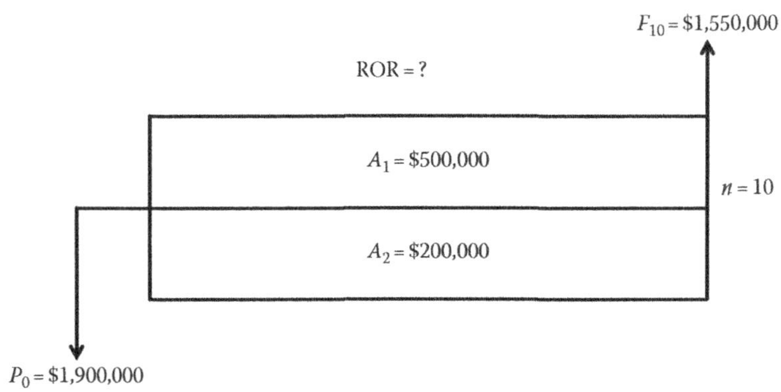

FIGURE 9.6 Cash flow diagram for scraper 2 in Example 9.6.

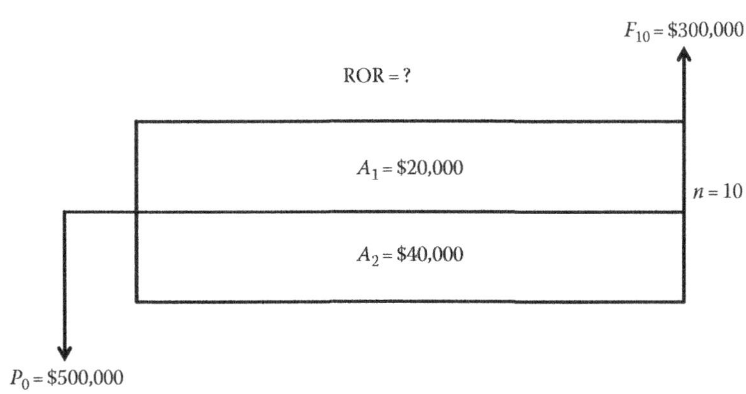

FIGURE 9.7 Net cash flow diagram for scraper 2 minus scraper 1 in Example 9.6.

Solution

First, calculate the incremental rate of return for scraper 1 minus the do nothing alternative using net present worth analysis:

$$0 = -P_0 + F_{10}(P/F,i,n) \pm \sum A(P/A,i,n)$$

Try 15%

$$= -\$1,400,000.00 + \$1,250,000.00(P/F,i,10) + (\$480,000.000 - \$240,000.00)(P/A,i,10)$$

$$= -\$1,400,000.00 + \$1,250,000.00(0.24718) + \$240,000.00(5.0187)$$

$$= -\$1,400,000.00 + \$308,975.00 + \$1,204,488.00$$

$$= \$113,463.00$$

Try 20%

$$= -\$1,400,000.00 + \$1,250,000.00(0.16151) + \$240,000.00(4.1924)$$

$$= -\$1,400,000.00 + \$201,887.50 + \$1,006,176.00$$

$$= -\$191,936.50$$

Use interpolation and Table 9.9 to calculate the incremental rate of return for scraper 1 minus the do nothing alternative:

TABLE 9.9
Table for Developing the Interpolation Problem for Unknown Rate of Return with n = 10 Years for Example 9.6

	IROR	Net Present Worth	
	15%	113,463.00	a
d	Unknown IROR	0	b
	20%	−191,936.50	

$$\text{IROR} = c + \left(\frac{a}{b}\right)d$$

$$\text{IROR} = 15 + \left(\frac{113,463.00 - 0}{113,463.00 - (-191,936.50)}\right) \times (20 - 15)$$

$$= 15 + \left(\frac{113,463.00}{305,399.50}\right) \times 5$$

$$= 15 + (0.371523 \times 5)$$

$$= 15 + 1.857616$$

$$= 16.86\%$$

16.86% > 10% the company MARR; therefore, the incremental rate of return of scraper 1 would justify its purchase.

Second, calculate the incremental rate of return for scraper 2 minus scraper 1 using net present worth analysis:

$$\text{Net income scraper } 2-1 = (\$500{,}000.00 - \$200{,}000.00)$$
$$-(\$480{,}000.00 - \$240{,}000.00) = \$60{,}000.00$$

$$0 = -P_0 + F_{10}\left(P/F,i,n\right) + A\left(P/A,i,n\right)$$

Try 15%

$$= -\$500{,}000.00 + \$300{,}000.00(P/F,i,10) + \$60{,}000.00(P/A,i,10)$$
$$= -\$500{,}000.00 + \$300{,}000.00(0.24718) + \$60{,}000.00(5.0187)$$
$$= -\$500{,}000.00 + \$74{,}154.00 + \$301{,}122.00$$
$$= -\$124{,}724.00$$

Try 10%

$$= -\$500{,}000.00 + \$300{,}000.00(0.38554) + \$60{,}000.00(6.1445)$$
$$= -\$500{,}000.00 + \$115{,}662.00 + \$368{,}670.00$$
$$= -\$15{,}668.00$$

Try 5%

$$= -\$500{,}000.00 + \$300{,}000.00(0.61391) + \$60{,}000.00(7.7217)$$
$$= -\$500{,}000.00 + \$184{,}173.00 + \$463{,}302.00$$
$$= \$147{,}475.00$$

TABLE 9.10
Table for Developing Interpolation Problem for Unknown Rate of Return with n = 10 Years for Example 9.6

	IROR	Net Present Worth		
	5%	147,475.00	*a*	
d	Unknown IROR	0		*b*
	10%	−15,668.00		

Use interpolation and Table 9.10 to calculate the incremental rate of return for scraper 2 minus scraper 1:

$$\text{IROR} = c + \left(\frac{a}{b}\right)d$$

$$\text{IROR} = 5 + \left(\frac{147{,}475.00 - 0}{147{,}475.00 - (-15{,}668.00)}\right) \times (10 - 5)$$

$$= 5 + \left(\frac{147{,}475.00}{163{,}143.00}\right) \times 5$$

$$= 5 + (0.903962 \times 5)$$

$$= 5 + 4.519808$$

$$= 9.52\%$$

9.52% < 10% the company MARR; therefore, scraper 2 is eliminated and scraper 1 is selected for purchase.

The firm is able to invest the incremental cost of $500,000.00 in another investment.

Note: Since the IROR for scraper 2 minus scraper 1 was less than 10%, the actual IROR was not calculated by interpolating between 9% and 10%.

Example 9.7 demonstrates the use of incremental rate of return analysis when there are three alternatives under consideration.

Example 9.7

An agricultural engineer is trying to decide which of three machines she should recommend the company purchase to improve their yearly threshing process. Table 9.11 includes the data for the three machines under consideration with the threshing machines listed in order of the least initial cost to the highest initial cost. For this example, the incremental rate of return for each incremental increase in initial cost using net present worth analysis is calculated including the do nothing alternative. The company uses a MARR of 7% for large equipment. Figures 9.8 through 9.10 are the net cash flow diagrams for the three threshing machines.

TABLE 9.11
Threshing Machine Data

Cost or Disbursements	Threshing Machine 1	Threshing Machine 2	Threshing Machine 3
Initial cost	$140,000.00	$160,000.00	$200,000.00
Yearly revenue	$70,000.00	$87,500.00	$130,000.00
Yearly expenses	$20,000.00	$30,000.00	$50,000.00
Life in years	4	4	4

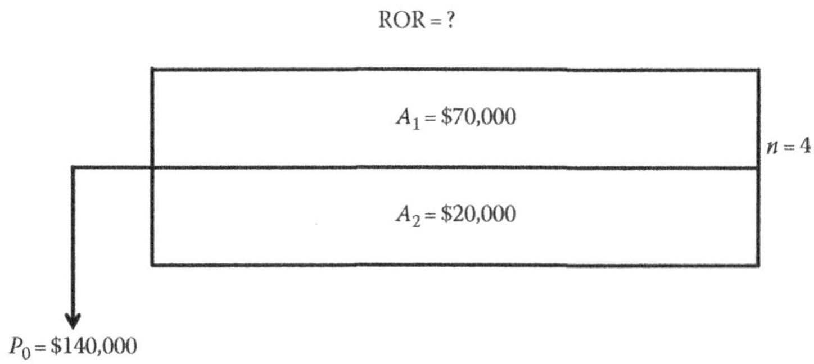

FIGURE 9.8 Net cash flow diagram machine 1 minus the do nothing alternative in Example 9.7.

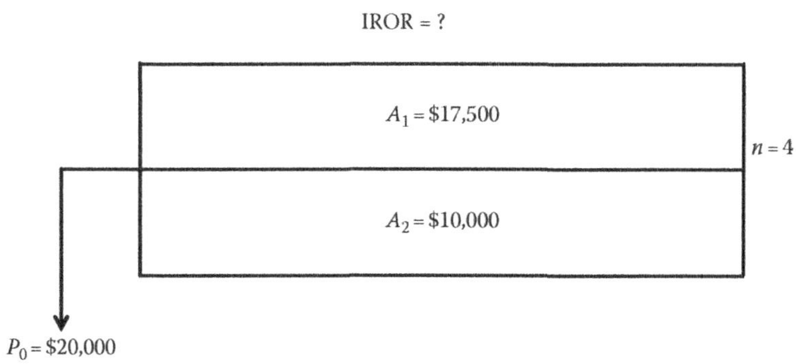

FIGURE 9.9 Net cash flow diagram machine 2 minus machine 1 in Example 9.7.

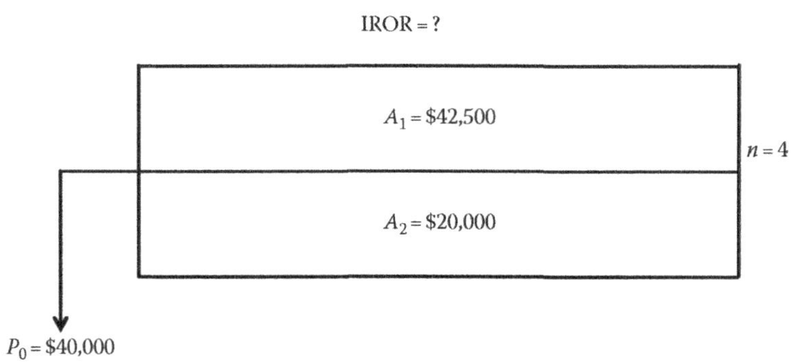

FIGURE 9.10 Net cash flow diagram for machine 3 minus machine 2 in Example 9.7.

Solution

First, calculate the rate of return of the net cash flow between machine 1 and the do nothing alternative using net present worth analysis:

$$0 = -P_0 + (A_1 + A_2)(P/A, i, n)$$

$$= -\$140,000.00 + (\$70,000.00 - \$20,000.00)(P/A, i, 4)$$

Try 20%

$$= -\$140,000.00 + \$50,000.00(2.5887)$$

$$= -\$140,000.00 + \$129,435.00$$

$$= -\$10,565.00$$

Try 15%

$$= -\$140,000.00 + \$50,000.00(2.8549)$$

$$= -\$140,000.00 + \$142,745.00$$

$$= \$2,745.00$$

Use interpolation and Table 9.12 to calculate the rate of return:

TABLE 9.12
Table for Developing the Interpolation Problem for Unknown Rate of Return with n = 4 Years for Example 9.7

	ROR	Net Present Worth		
d	15%	2,745.00	a	b
	Unknown ROR	0		
	20%	−10,565.00		

$$\text{IROR} = c + \left(\frac{a}{b}\right)d$$

$$\text{IROR} = 15 + \left(\frac{2,745.00 - 0}{2,745.00 - (-10,565.00)}\right) \times (20 - 15)$$

$$= 15 + \left(\frac{2,745.00}{13,310.00}\right) \times 5$$

$$= 15 + (0.206236 \times 5)$$

$$= 15 + 1.03118$$

$$= 16.03\%$$

Therefore, since the ROR > MARR, 16.03% > 7%, select machine 1 and compare it to machine 2.

Second, calculate the incremental rate of return of machine 2 minus machine 1 using net present worth analysis:

$$0 = P_0 + (A_1 + A_2)(P/A,i,n)$$
$$= -\$20,000.00 + (\$17,500.00 - \$10,000.00)(P/A,i,4)$$

Try 15%

$$= -\$20,000.00 + \$7,500.00(2.8549)$$
$$= -\$20,000.00 + \$21,411.75$$
$$= \$1,411.75$$

Try 20%

$$= -\$20,000.00 + \$7,500.00(2.5887)$$
$$= -\$20,000.00 + \$19,414.25$$
$$= -\$585.75$$

Use interpolation and Table 9.13 to solve for the incremental rate of return:

TABLE 9.13
Table for Developing the Interpolation Problem for Unknown Incremental Rate of Return with n = 4 Years for Example 9.7

	IROR	Net Present Worth	
d	15%	1,411.75	a
	Unknown IROR	0	b
	20%	−585.75	

$$\text{IROR} = c + \left(\frac{a}{b}\right)d$$

$$\text{IROR} = 15 + \left(\frac{1,411.75 - 0}{1,411.75 - (-585.75)}\right) \times (20 - 15)$$

$$= 15 + \left(\frac{1,411.75}{1,997.50}\right) \times 5$$

$$= 15 + (0.706758 \times 5)$$

$$= 15 + 3.533792$$

$$= 18.53\%$$

Therefore, since the IROR > MARR, 18.52% > 7%, select machine 2 and compare it to machine 3.

Third, calculate the incremental rate of return for machine 3 minus machine 2 using net present worth analysis:

$$0 = P_0 + (A_1 + A_2)(P/A, i, n)$$

$$= -\$40,000.00 + (\$42,500.00 - \$20,000.00)(P/A, i, 4)$$

Try 40%

$$= -\$40,000.00 + \$22,500.00(1.8492)$$

$$= -\$40,000.00 + \$41,607.00$$

$$= \$1,607.00$$

Try 50%

$$= -\$40,000.00 + \$22,500.00(1.6049)$$

$$= -\$40,000.00 + \$36,110.25$$

$$= -\$3,889.75$$

Use interpolation and Table 9.14 to calculate the incremental rate of return:

TABLE 9.14
Table for Developing the Interpolation Problem for Unknown Incremental Rate of Return with n = 4 Years for Example 9.7

	IROR	Net Present Worth	
d	40%	1607.00	a, b
	Unknown IROR	0	
	50%	−3889.75	

$$\text{IROR} = c + \left(\frac{a}{b}\right)d$$

$$\text{IROR} = 40 + \left(\frac{1,607.00 - 0}{1,607.00 - (-3,889.75)}\right) \times (50 - 40)$$

$$= 40 + \left(\frac{1607.00}{5496.75}\right) \times 10$$

$$= 40 + (0.292355 \times 10)$$

$$= 40 + 2.923546$$

$$= 42.92\%$$

Therefore, since the IROR > MARR, 42.92% > 7%, select machine 3.

Example 9.8 demonstrates calculating the incremental rate of return using net present worth techniques where the evaluation requires using of the least common multiples of years.

Example 9.8

A chemical engineer is considering the purchase of a new type of processing machine. There are two companies selling the machine, and the project costs and revenues for the two machines are listed in Table 9.15. The firm uses a MARR of 10%. Solve for the incremental rate of return for the comparison of the two machines using net present worth analysis. Figure 9.11 is the net cash flow diagram for the two processing machines.

TABLE 9.15
Processing Machine Data

Year	Processing Machine 1	Processing Machine 2	Net Cash Flow
0	−$80,000.00	−$130,000.00	−$50,000.00
Years 1–5	−$35,000	−$16,000.00	$19,000.00
Year 5	—	−$110,000.00 (−$130,000.00 + $20,000.00) (Cost plus salvage value)	−$110,000.00
Years 6–10	−$35,000	−$16,000.00	$19,000.00
Year 10	—	$20,000.00	$20,000.00
Life in years	10	10	

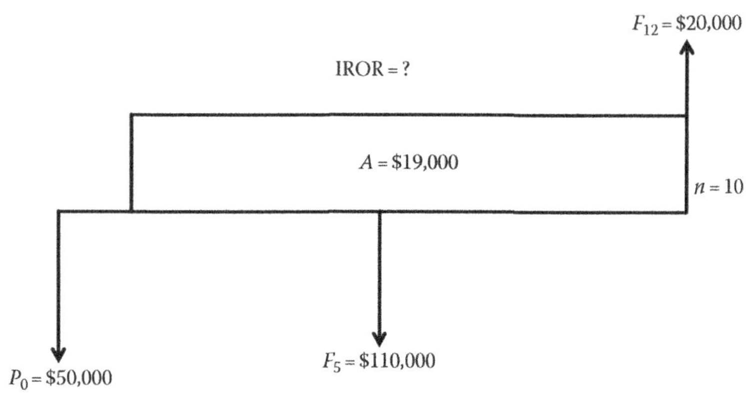

FIGURE 9.11 Net cash flow diagram for processing machines in Example 9.8.

Solution

Calculate the incremental rate of return for machine 2 minus machine 1 using the net cash flow and net present worth analysis:

$$0 = P_0 + A(P/A,i,10) + F_5(P/F,i,5) + F_{10}(P/F,i,10)$$

$$= -\$50{,}000.00 + \$19{,}000.00(P/A,i,10) - \$110{,}000.00(P/F,i,5)$$

$$+ \$20{,}000.00(P/F,i,10)$$

Try 10%

$$= -\$50{,}000.00 + \$19{,}000.00(6.1445) - \$110{,}000.00(0.62092)$$
$$+ \$20{,}000.00(0.38554)$$
$$= -\$50{,}000.00 + \$116{,}745.50 - \$68{,}301.20 + \$7{,}710.80$$
$$= \$6{,}115.10$$

Try 13%

$$= -\$50{,}000.00 + \$19{,}000.00(5.4262) - \$110{,}000.00(0.54276)$$
$$+ \$20{,}000.00(0.29459)$$
$$= -\$50{,}000.00 + \$103{,}097.80 - \$59{,}703.60 + \$5{,}891.80$$
$$= -\$714.00$$

Use interpolation and Table 9.16 to solve for the incremental rate of return:

TABLE 9.16
Table for Developing the Interpolation Problem for Unknown Incremental Rate of Return with n = 10 Years for Example 9.8

	IROR	Net Present Worth	
d [	10%	6,155.10	] *a*] *b*
	Unknown IROR	0	
	13%	−714.00	

$$\text{IROR} = c + \left(\frac{a}{b}\right)d$$

$$\text{IROR} = 10 + \left(\frac{6{,}155.10 - 0}{6{,}155.10 - (-714.00)}\right) \times (13 - 10)$$
$$= 10 + \left(\frac{1{,}607.00}{6{,}829.10}\right) \times 3$$
$$= 10 + (0.235317 \times 3)$$
$$= 10 + 0.70595$$
$$= 10.71\%$$

Therefore, since the IROR > MARR, 10.71% > 10%, select machine 2.

The next example, Example 9.9, provides calculations for determining the incremental rate of return of the two alternatives in Example 9.8 using equivalent uniform annual worth analysis.

Example 9.9

Solve for the incremental rate of return of the net cash flow using equivalent uniform annual worth analysis for the machines in Example 9.8.

Solution

$$0 = -P_0(A/P,i,n) + A + F_5(P/F,i,n)(A/P,i,n) + F_{10}(A/F,i,n)$$

$$= -\$50,000.00(A/P,i,10) + \$19,000.00$$

$$-\$110,000.00(P/F,i,5)(A/P,i,10) + \$20,000.00(A/F,i,10)$$

Try 10%

$$= -\$50,000.00(0.16275) + \$19,000.00$$

$$-\$110,000.00(0.62092)(0.16275) + \$20,000.00(0.06275)$$

$$= -\$8,137.50 + \$19,000.00 - \$11,116.02 + \$1,255.00$$

$$= \$1,001.48$$

Try 13%

$$= -\$50,000.00(0.18429) + \$19,000.00$$

$$-\$110,000.00(0.54276)(0.18429) + \$20,000.00(0.05429)$$

$$= -\$9,214.50 + \$19,000.00 - \$11,002.77 + \$1,085.80$$

$$= -\$131.47$$

Use interpolation and Table 9.17 to calculate the incremental rate of return:

TABLE 9.17
Table for Developing the Interpolation Problem for Unknown Incremental Rate of Return with n = 10 Years for Example 9.9

	IROR	EUAW	
d	10%	1,001.48	*a*
	Unknown IROR	0	*b*
	13%	−131.47	

$$\text{IROR} = c + \left(\frac{a}{b}\right)d$$

$$\text{IROR} = 10 + \left(\frac{1{,}001.48 - 0}{1{,}001.48 - (-131.47)}\right) \times (13 - 10)$$

$$= 10 + \left(\frac{1{,}001.48}{1{,}132.95}\right) \times 3$$

$$= 10 + (0.883958 \times 3)$$

$$= 10 + 2.651873$$

$$= 12.65\%$$

Therefore, since the IROR > MARR, 12.65% > 10%, select machine 2.

Appendix C includes spreadsheet formulas for calculating unknown rates of return.

9.5 SUMMARY

This chapter introduced the rate of return method and the incremental rate of return method for comparing alternative projects. The process for developing the net present worth equations used to solve for rates of return was explained along with the procedure for using interpolation to solve for unknown rates of return. Steps for calculating rates of return using equivalent uniform annual worth techniques were provided and methods for evaluating alternatives using incremental investment analysis, which is referred to as the process for determining incremental rates of return, were covered in the last section of this chapter.

KEY TERMS

Equivalent uniform annual worth analysis
Incremental investment analysis
Incremental rate of return
Minimum attractive rate of return
Mutually exclusive alternatives
Net cash flow
Net present worth equation
Rate of return method

PROBLEMS

9.1 An electrical engineer invests $50,000.00 in bonds paying dividends of $1,000.00 per year for 10 years. At 10 years the engineer sells the bonds for $70,000.00. Calculate the rate of return on the investment using net present worth analysis.

9.2 Calculate the rate of return in Problem 9.1 using equivalent uniform annual worth analysis.

9.3 A petroleum engineer is purchasing new valves and is considering two alternative types of valves. The data for the valves are listed in Table 9.18. The firm uses a MARR of 15%. Calculate the incremental rate of return using net present worth analysis and determine whether the firm should purchase the valves based on the company MARR.

TABLE 9.18
Data for Value Alternatives

Costs or Disbursements	Valve Alternative 1	Valve Alternative 2
Initial cost	$80,000.00	$130,000.00
Operating and maintenance costs	$35,000.00	$16,000.00
Salvage value	—	$20,000.00
Life in years	10	5

9.4 Solve for the incremental rate of return in Problem 9.3 using equivalent uniform annual worth analysis.

9.5 An industrial engineer invests $100,000.00 now, and in three years he adds $3,000.00 to the investment. He is able to withdraw $5,000.00 the first year increasing by $1,000.00 each year for 10 years. He also will be withdrawing $50,000.00 in five years and $20,000.00 in 10 years. Using net present worth analysis, determine the rate of return on the investment.

9.6 Using equivalent uniform annual worth analysis, determine the rate of return for Problem 9.5.

9.7 Two soil testing machines are being evaluated by a geotechnical engineer to determine which machine his company should purchase. The data for the testing machines are listed in Table 9.19. Using net present worth analysis, determine the incremental rate of return and compare it to the company MARR of 10% to determine which machine should be purchased by the company.

TABLE 9.19
Data for Soil Testing Machine Alternatives

Costs or Disbursements	Soil Testing Machine 1	Soil Testing Machine 2
Initial cost	$20,000.00	$4,000.00
Operating and maintenance costs	$7,500.00	$12,000.00
Salvage value	$6,000.00	$12,000.00
Life in years	6	12

9.8 Use equivalent uniform annual worth analysis to determine the incremental rate of return in Problem 9.7.

9.9 A petroleum engineering firm purchased land for drilling for oil for $600,000.00 and sold the property 17 years later for $2,100,000.00. Property taxes on the land were $8,000.00 the first year and they increased by $1,000.00 a year for the rest of the 17 years the land was owned by the firm. Calculate the rate of return on the land investment using net present worth analysis.

9.10 Calculate the rate of return for Problem 9.9 using equivalent uniform annual worth analysis.

9.11 A heavy construction company purchased an excavator for $540,000.00. Operating and maintenance costs will be $36,000.00 for the first year and increase by $30,000.00 each year. The first year the firm will have an income of $660,000.00 and the income will decrease by $5,000.00 each year. The salvage value of the excavator will be $150,000.00 in year 10. Calculate the rate of return for the excavator using net present worth analysis.

9.12 Calculate the rate of return in Problem 9.11 using equivalent uniform annual worth analysis.

9.13 A civil engineering firm purchased a building for $13,000,000.00 and the firm was able to rent out part of the office space for $200,000.00 per month. Property taxes cost $250,000.00 per year. The building was sold at the end of 12 years for $18,000,000.00. Determine the rate of return on the building investment using net present worth analysis.

9.14 Determine the rate of return in Problem 9.13 using equivalent uniform annual worth analysis.

9.15 A mechanical engineer is comparing two parts for incorporation into the design of a truck. The data for the parts are listed in Table 9.20. Calculate the incremental rate of return using equivalent uniform annual worth analysis of the net cash flow between the two potential parts.

TABLE 9.20
Data for Truck Part Alternatives

Costs or Disbursements	Part Alternative 1	Part Alternative 2
Initial cost	$9,000.00	$10,000.00
Operating and maintenance costs	$500.00	$300.00
Labor costs	$5,000.00	$5,000.00
Salvage value	$1,000.00	$1,000.00
Life in years	6	4

9.16 Calculate the rate of return in Problem 9.15 using net present worth analysis.

9.17 An aerospace engineer is analyzing three potential alternatives for a new structural analysis software program. The data for the three software programs are listed in Table 9.21. Using net present worth analysis, calculate the rate of return for all three alternatives, and using a MARR of 12%, determine which alternative the firm should select.

TABLE 9.21
Data for Structural Analysis Software Program Alternatives

Costs or Disbursements	Software Alternative 1	Software Alternative 2	Software Alternative 3
Initial cost	$18,000.00	$22,500.00	$25,000.00
Net annual income	$3,800.00	$4,890.00	$5,000.00
Life in years	8	8	8

9.18 Using net present worth analysis, calculate the incremental rate of return between the alternatives listed in Problem 9.17. Using a MARR of 12%, determine which alternative should be selected by the firm.

9.19 Using equivalent uniform annual worth analysis, calculate the rate of return for each of the alternatives in Problem 9.17.

9.20 Using equivalent uniform annual worth analysis, calculate the incremental rate of return for alternative 2 minus alternative 1 in Problem 9.17.

10 Replacement Analysis

This chapter introduces *replacement analysis*, an engineering economic analysis technique for determining when to replace assets. Definitions for terms related to replacement analysis are discussed including replacements, augmentation, retirement, challengers, defenders, physical life, accounting life, service period, sunk cost, block replacement, reduced performance, alternative requirements, and obsolescence. Methods are presented for determining when to replace equipment or assets, and part of the chapter covers analyzing mutually exclusive alternatives.

One of the challenges of operating a firm that owns equipment or other assets is to determine when to replace them with newer equipment or assets. Before the techniques for determining when to replace equipment or assets are discussed, Section 10.1 introduces replacement analysis terms.

10.1 DEFINITIONS USED IN REPLACEMENT ANALYSIS

This section provides definitions for terms used in engineering economic analysis related to the replacement of equipment and assets.

10.1.1 Replacements

The term *replacement* refers to new equipment or assets purchased and put into service in place of existing equipment or assets. The equipment or assets removed from service are used for other purposes, sold for a salvage value, or disposed of permanently.

10.1.2 Augmentation

Augmentation is the term for equipment or assets purchased and installed to increase the capacity, or to alter the capabilities of, existing equipment. In order to augment existing equipment or other assets, the existing equipment or assets are kept in service.

10.1.3 Retirement

Retirement of equipment or assets occurs when they are removed from service and either repurposed to perform other operations, left idle and only used if other similar equipment or assets are temporarily removed from service for repairs or replacement, or disposed of without a new piece of equipment or other asset being purchased to replace them.

10.1.4 Challengers

When an engineering economic analysis is being performed to evaluate the potential replacements of equipment or other assets, the label of *challenger* is applied to potential new equipment or assets when they are being considered as an alternative.

10.1.5 Defenders

In engineering economic analysis, when existing equipment or assets are being considered for replacement, they are *defenders*.

10.1.6 Economic Life

The *economic life* of equipment or an asset is the time period used for the evaluation process. The economic life is usually different from the physical life of equipment or assets since it is only used for engineering economic evaluations.

10.1.7 Physical Life

The *physical life* of equipment or assets is the amount of time that transpires from when the equipment or asset is created until it is either disposed of or repurposed and used in another application.

10.1.8 Accounting Life

In addition to the economic and physical life of equipment and assets, there is also an *accounting life*. The accounting life is based on the length of time equipment or assets are depreciated for tax purposes. Depreciation is covered in Chapter 13.

10.1.9 Ownership Life

The *ownership life* of equipment and assets includes the length of time from when the equipment or asset is purchased until it is sold.

10.1.10 Service Period

The *service period* of equipment and assets is the time the equipment or asset is available for use within a company.

10.1.11 Sunk Costs

Sunk costs, as described in Sections 2.2.4 and 10.1.11, are expenses already incurred or spent and it is not possible to retrieve them or to be reimbursed for them.

10.1.12 Block Replacements

A *block replacement* occurs when all of the same types of units of equipment are replaced at the same time regardless of whether all of the units are no longer operational.

10.1.13 Reduced Performance

Equipment or assets physically deteriorated, and the deterioration impairs the functioning of the equipment or assets, are considered to be experiencing *reduced performance*.

10.1.14 Alternative Requirements

In some cases, equipment or assets are replaced because of new requirements for items such as speed or accuracy of the equipment or assets, and these are *alternative requirements*.

10.1.15 Obsolescence

Obsolescence of equipment or assets occurs when changing technology creates new equipment or assets that perform more efficiently than existing equipment or assets. An example of obsolescence is when computers with faster processors are developed and being sold in the marketplace.

The existing computers still perform their function but the new ones with faster processors are desired to help increase productivity.

10.2 DETERMINING WHEN TO REPLACE EQUIPMENT OR ASSETS

The process for determining when to replace equipment or assets is unique to each firm. Some firms replace their equipment or assets based on obsolescence, especially if they operate in a cutting-edge profession. Other firms may only keep their equipment or assets as long as they are able to depreciate it and deduct the depreciation from their taxable income. Automobile fleets and office furniture are two examples of assets usually sold at the end of their depreciable life so replacement assets may be purchased to restart the cycle of depreciation. Some construction firms may keep their heavy construction equipment for decades, as long as it is still functioning efficiently, even though the firm is no longer able to depreciate the equipment.

The time when equipment or assets should be replaced occurs when new assets will generate a higher net present worth or equivalent uniform annual worth than the existing equipment or assets or the net present cost or equivalent uniform annual cost of the proposed replacement is less than the existing facility. Engineering economic analysis techniques are used to calculate the point in time where this occurs, which indicates the equipment or assets should be replaced with new ones. In some firms, there may not be any employees who are capable of performing engineering economic analysis and if this is the case, managers may use other criteria to determine when they will replace their equipment or assets.

In order to determine when to replace equipment or assets, net present worth or equivalent uniform annual worth analysis techniques are used to calculate the point in time where a new replacement will generate more income than the existing equipment or asset. The next section provides techniques for analyzing when to replace equipment or assets.

10.3 ANALYZING MUTUALLY EXCLUSIVE ALTERNATIVES FOR REPLACEMENT: SOLVED EXAMPLE PROBLEMS

This section provides example problems that evaluate equipment and assets to determine whether they should be replaced and when they should be replaced with new equipment or assets.

Example 10.1 is a problem addressing the replacement of an existing piece of equipment using equivalent uniform annual worth analysis to determine when the most cost effective time is to replace the equipment.

Example 10.1

A mechanical engineering firm is determining when to replace a piece of equipment bought and put into service four years ago. The original cost of the equipment was $26,000.00. The equipment has a current salvage value of $13,000.00, and the salvage value will decline each year to $10,000.00 at the end of the first year, $8,125.00 at the end of the second year, $7,000.00 at the end of the third year, and $6,250.00 at the end of the fourth year. The operating and maintenance costs are $3,000.00 the first year and increase by $1,000.00 each year until the end of the four years. The interest rate for this analysis is 10%. Out of all of the challengers, the one with the lowest equivalent uniform annual cost is $7,100.00. Calculate when the existing equipment should be replaced by the challenger using equivalent uniform annual cost analysis.

Solution

The salvage value of $13,000.00 is considered to be the first cost for the replacement system and it declines over each of the four years by the salvage values listed in the problem statement. The operating and maintenance costs are $3,000.00 per year increasing by a gradient of $1,000.00

each year. Therefore, the formula for calculating the equivalent uniform annual cost for each year is the following:

$$\begin{aligned}\text{EUAC} &= \text{salvage value}(A/P,i,n) + \text{salvage value at time}(t)(A/F,i,t) + A \\ &\quad + G(A/G,i,n) \\ &= -\$13{,}000(A/P,i,n) + \text{salvage value at time}(t)(A/F,i,t) \\ &\quad + \$3{,}000.00 + \$1{,}000.00(A/G,i,n)\end{aligned}$$

Using this formula, the equivalent uniform annual cost is calculated for years one through four and compared to the equivalent uniform annual cost of the challenger of $7,100.00 to determine which year the equipment should be replaced with the new equipment. Table 10.1 provides the values used in the formula for each year.

TABLE 10.1
Data for Replacement Analysis Calculations

Year	Salvage Value	Salvage Value at Time (*t*)	Annuity	Gradient
1	−$13,000.00	$10,000.00	$3,000.00	—
2	−$13,000.00	$8,125.00	$3,000.00	$1,000.00
3	−$13,000.00	$7,000.00	$3,000.00	$1,000.00
4	−$13,000.00	$6,125.00	$3,000.00	$1,000.00

Calculate the equivalent uniform annual cost for each of the four years in the analysis period.

Year 1

$$\begin{aligned}\text{EUAC}_1 &= -\$13{,}000.00(A/P,i,n) + \text{salvage value at time}(t)(A/F,i,t) \\ &\quad - \$3{,}000.00 \\ &= -\$13{,}000.00(A/P,10,1) + \$10{,}000.00(A/F,10,1) - \$3{,}000.00 \\ &= -\$13{,}000.00(1.1000) + \$10{,}000.00(1.0000) - \$3{,}000.00 \\ &= -\$14{,}300.00 + \$10{,}000.00 - \$3{,}000.00 \\ &= -\$7{,}300.00\end{aligned}$$

Year 2

$$\begin{aligned}\text{EUAC}_2 &= -\$13{,}000.00(A/P,i,n) + \text{salvage value at time}(t)(A/F,i,t) \\ &\quad - \$3{,}000.00 - \$1{,}000.00(A/G,i,n) \\ &= -\$13{,}000.00(A/P,10,2) + \$8{,}125.00(A/F,10,2) - \$3{,}000.00 \\ &\quad - \$1{,}000.00(A/G,10,2) \\ &= -\$13{,}000.00(0.57619) + \$8{,}125.00(0.47619) - \$3{,}000.00 \\ &\quad - \$1{,}000.00(0.4761) \\ &= -\$7{,}490.47 + \$3{,}869.05 - \$3{,}000.00 - \$476.10 \\ &= -\$7{,}097.52\end{aligned}$$

Year 3

$$\begin{aligned}\text{EUAC}_3 &= -\$13{,}000.00(A/P,i,n) + \text{salvage value at time}(t)(A/F,i,t)\\ &\quad - \$3{,}000.00 - \$1{,}000.00(A/G,i,n)\\ &= -\$13{,}000.00(A/P,10,3) + 7{,}000.00(A/F,10,3) - \$3{,}000.00\\ &\quad - \$1{,}000.00(A/G,10,3)\\ &= -\$13{,}000.00(0.40211) + \$7{,}000.00(0.30211) - \$3{,}000.00\\ &\quad - \$1{,}000.00(0.9365)\\ &= -\$5{,}227.43 + \$2{,}114.77 - \$3{,}000.00 - \$963.50\\ &= -\$7{,}076.16\end{aligned}$$

Year 4

$$\begin{aligned}\text{EUAC}_4 &= -\$13{,}000.00(A/P,i,n) + \text{salvage value at time}(t)(A/F,i,t)\\ &\quad - \$3{,}000.00 - \$1{,}000.00(A/G,i,n)\\ &= -\$13{,}000.00(A/P,10,4) + \$6{,}125.00(A/F,10,4) - \$3{,}000.00\\ &\quad - \$1{,}000.00(A/G,10,4)\\ &= -\$13{,}000.00(0.31547) + \$6{,}125.00(0.21547) - \$3{,}000.00\\ &\quad - \$1{,}000.00(1.3811)\\ &= -\$4{,}101.11 + \$1{,}319.75 - \$3{,}000.00 - \$1{,}381.10\\ &= -\$7{,}162.46\end{aligned}$$

Therefore, the yearly equivalent uniform annual cost is more than the challenger after the first year so the challenger should replace the existing piece of equipment after the first year.

Example 10.1 demonstrates determining when to replace an existing piece of equipment by calculating the equivalent uniform annual cost for each of the four years over which the equipment could be replaced at the end of each year. But in some circumstances, the equipment may be replaced earlier than what is indicated by the engineering economic analysis because the new equipment could be faster or more efficient.

The next example, Example 10.2, provides a problem demonstrating determining whether to replace existing vehicles with leased vehicles.

Example 10.2

In the past, an executive from a nuclear power company has purchased vehicles for all of the executives. The executive has been approached by an automobile firm about leasing new vehicles to replace the existing fleet of vehicles. Table 10.2 includes the cost and salvage values associated with the options of purchasing vehicles versus leasing new ones. Using equivalent uniform annual cost analysis, an interest rate of 12%, and a life of 10 years determine whether the firm should replace their existing fleet of vehicles with purchased or leased vehicles. There are eight vehicles in the existing fleet of vehicles. Figures 10.1 and 10.2 are the cash flow diagrams for the purchased and leased vehicles.

TABLE 10.2
Data for Purchasing Vehicles versus Leasing Vehicles

Costs and Disbursements	Purchasing Vehicles (Defender)	Leasing Vehicles (Challenger)
Initial cost	$21,000.00	—
Yearly lease cost	—	$4,500.00 per year
Annual operating cost	$6,000.00	$7,000.00
Salvage value	$4,000.00	—

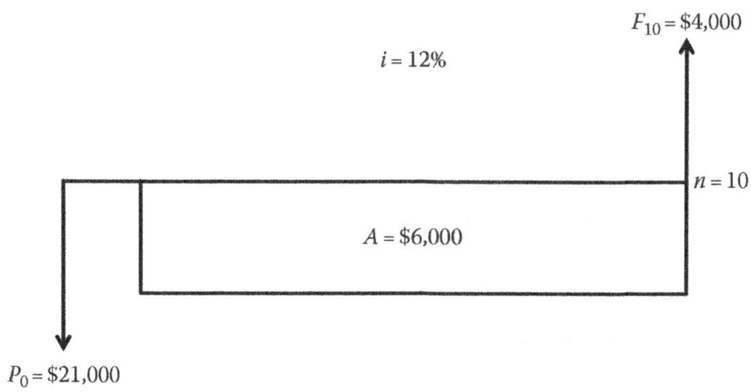

FIGURE 10.1 Cash flow diagram for purchasing vehicles for Example 10.2.

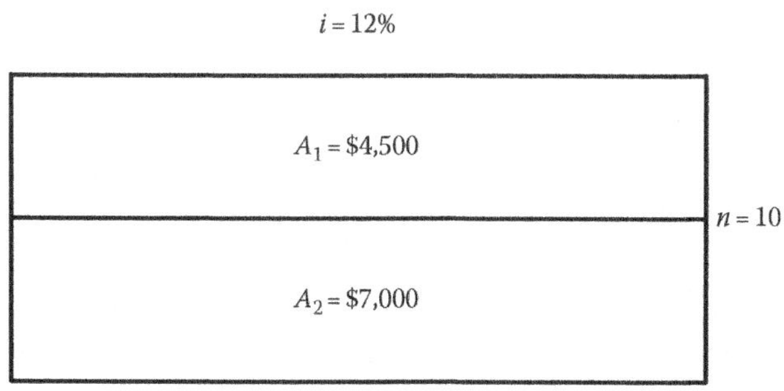

FIGURE 10.2 Cash flow diagram for leasing vehicles for Example 10.2.

Solution

Purchasing vehicles (defender)

$$\begin{aligned}
\text{EUAC}_D &= P_0(A/P,i,n) + F_{10}(A/F,i,n) + A \\
&= -\$21{,}000.00(A/P,12,10) + \$4{,}000.00(A/F,12,10) - \$6{,}000.00 \\
&= -\$21{,}000.00(0.17698) + \$4{,}000.00(0.05698) - \$6{,}000.00 \\
&= -\$3{,}716.58 + \$227.92 - \$6{,}000.00 \\
&= -\$9{,}488.66 \text{ per vehicle}
\end{aligned}$$

$$\text{Total} = \frac{-\$9{,}488.66}{\text{Vehicle}} \times 8 \text{ vehicles} = -\$75{,}909.28$$

Leasing vehicles (challenger)

$$EUAC_C = -\$4,500.00 - \$7,000.00 = -\$11,400.00 \text{ per vehicle}$$

$$\text{Total} = \frac{\$11,400.00}{\text{Vehicle}} \times 8 \text{ vehicles} = -\$92,000.00$$

Therefore, the firm should continue to purchase vehicles rather than switching to leasing vehicles since the equivalent uniform annual cost for purcasing the vehicles is less than the equivalent uniform annual cost for leasing the vehicles −$75,909.28 < −$92,000.00.

Example 10.3 is a problem using replacement analysis techniques to determine whether to replace an existing pipeline.

Example 10.3

An oil pipeline in East Texas has been in service for five years. A petroleum engineer working for the oil company is evaluating whether to replace the pipeline with a new one. The data for the existing pipeline and the potential new pipeline are listed in Table 10.3. Using an interest rate of 6%, and net present worth analysis and then equivalent uniform annual worth analysis, the petroleum engineer determines whether the existing pipeline should be replaced with the proposed new pipeline. Figures 10.3 and 10.4 are the cash flow diagrams for the existing pipeline and the proposed new pipeline.

TABLE 10.3
Data for Pipeline Alternatives

Costs	Existing Pipeline	Proposed Pipeline
Initial cost	—	$300,000.00
Salvage value at time zero	$1,000,000.00 increases each replacement every 10 years by $500,000.00 per year	—
Yearly maintenance	$100,000.00 per year increasing by $50,000.00 per year for 10 years	$100,000.00 per year increasing by $10,000.00 per year
Life in years	10	40

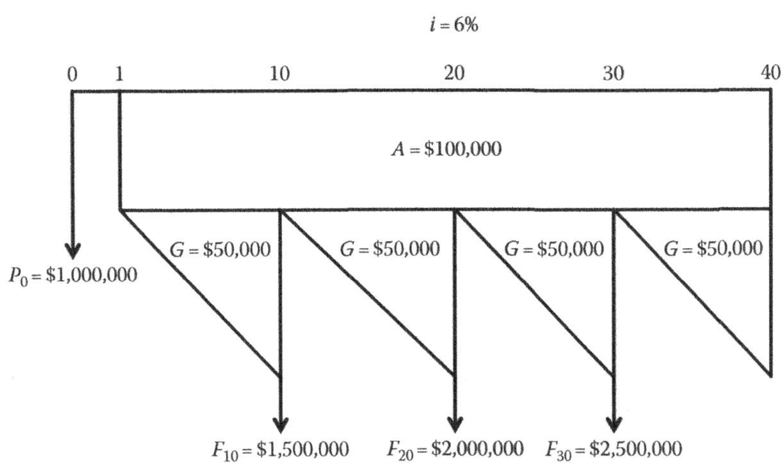

FIGURE 10.3 Cash flow diagram for the existing pipeline in Example 10.3.

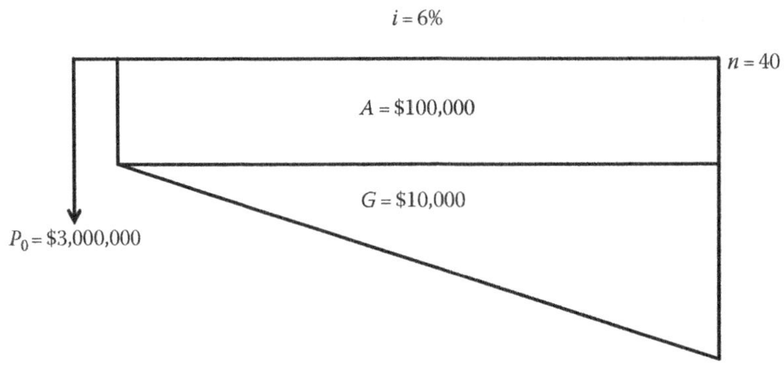

FIGURE 10.4 Cash flow diagram the proposed new pipeline in Example 10.3.

Solution

Existing pipeline

The salvage value for the existing pipeline is considered to be a cost of keeping the pipeline in service when analyzing the existing pipeline.

Calculate the net present worth of the existing pipeline and then convert the net present worth into an equivalent uniform annual cost:

$$
\begin{aligned}
\text{NPW}_D &= P_0 + F_{10}(P/F,i,n) + F_{20}(P/F,i,n) + F_{30}(P/F,i,n) + A(P/A,i,n) \\
&\quad + G(A/G,i,n)(P/A,i,n) \\
&= -\$1,000,000.00 - \$1,500,000.00(P/F,6,10) \\
&\quad - \$2,000,000.00(P/F,6,20) - \$2,500,000.00 \\
&\quad - \$100,000.00(P/A,i,40) - \$50,000.00(A/G,6,10)(P/A,i,40) \\
&= -\$1,000,000.00 - \$1,500,000.00(0.55839) \\
&\quad - \$2,000,000.00(0.31180) - \$2,500,000.00(0.17411) \\
&\quad - \$100,000.00(15.046) - \$50,000.00(4.0220)(15.046) \\
&= -\$1,000,000.00 - \$837,585.00 - \$623,600.00 - \$435,275.00 \\
&\quad - \$1,504,600.00 - \$3,025,750.60 \\
&= -\$7,426,810.60
\end{aligned}
$$

$$
\begin{aligned}
\text{EUAC}_D &= -\$7,426,810.60(A/P,6,40) = -\$7,426,810.60(0.06646) \\
&= -\$493,585.83
\end{aligned}
$$

Proposed pipeline

Calculate the net present worth of the proposed pipeline and then convert the net present worth into an equivalent uniform annual worth:

$$
\begin{aligned}
\text{NPW}_C &= P_0 + A(P/A,i,n) + G(P/G,i,n) \\
&= -\$3,000,000.00 - \$100,000.00(P/A,6,40) - \$10,000.00(P/G,6,40) \\
&= -\$3,000,000.00 - \$100,000.00(15.046) - \$10,000.00(185.95) \\
&= -\$3,000,000.00 - \$1,504,600.00 - \$1,859,500.00 \\
&= -\$6,364,100.00
\end{aligned}
$$

$$EUAC_D = -\$6,364,100.00(A/P,6,40) = -\$6,364,100.00(0.06646)$$
$$= -\$422,958.09$$

Therefore, the new pipeline should be installed since it has a lower equivalent uniform annual cost than the existing pipeline $-\$422,958.09 < -\$493,585.83$.

Example 10.4 provides another problem where equivalent uniform annual cost analysis techniques are used to compare two facilities to determine whether to replace the existing facility with a proposed new facility.

Example 10.4

An engineering office building has a current value of \$3,000,000.00. The building requires \$200,000.00 a year to operate and maintain the building. If the office building is retained, it will have an estimated resale value of \$15,000,000.00 in 40 years. Construction of a new building is being investigated to replace the existing building. The new building under consideration would cost \$4,000,000.00 and have maintenance and operating costs of \$100,000.00 per year. The estimated resale value of the proposed building would be \$10,000,000.00 in 40 years. The interest rate for financing the new building is 6%. Determine whether the existing building should be retained or whether it should be replaced by the proposed building using equivalent uniform annual cost analysis. Figures 10.5 and 10.6 are the cash flow diagrams for the existing and proposed buildings.

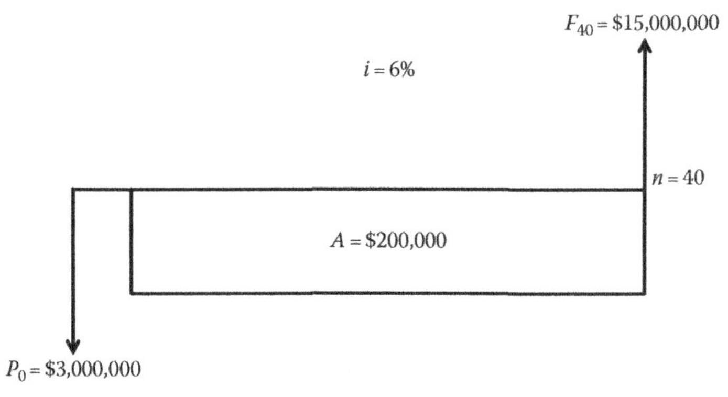

FIGURE 10.5 Cash flow diagram for the existing building in Example 10.4.

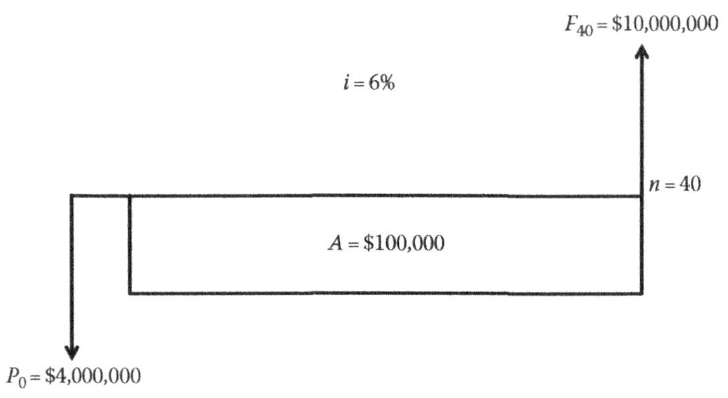

FIGURE 10.6 Cash flow diagram for the proposed building in Example 10.4.

Solution

Existing building

$$\begin{aligned}
\text{EUAC}_D &= \left[P_0 + A(P/A,i,n) + F_{40}(P/F,i,n)\right](A/P,i,n) \\
&= \left[-\$3{,}000{,}000.00 - \$200{,}000.00(P/A,6,40)\right. \\
&\quad \left. + \$15{,}000{,}000.00(P/F,6,40)\right](A/P,6,40) \\
&= \left[-\$3{,}000{,}000.00 - \$200{,}000.00(15.046)\right. \\
&\quad \left. + \$15{,}000{,}000.00(0.09722)\right](0.06646) \\
&= \left[-\$3{,}000{,}000.00 - \$3{,}009{,}200.00 + \$1{,}458{,}300.00\right](0.06646) \\
&= -\$4{,}550{,}900.00(0.06646) \\
&= -\$302{,}452.81
\end{aligned}$$

Proposed building

$$\begin{aligned}
\text{EUAC}_C &= \left[P_0 + A(P/A,i,n) + F_{40}(P/F,i,n)\right](A/P,i,n) \\
&= \left[-\$4{,}000{,}000.00 - \$100{,}000.00(P/A,6,40)\right. \\
&\quad \left. + \$10{,}000{,}000.00(P/F,6,40)\right](A/P,6,40) \\
&= \left[-\$4{,}000{,}000.00 - \$100{,}000.00(15.046)\right. \\
&\quad \left. + \$10{,}000{,}000.00(0.09722)\right](0.06646) \\
&= \left[-\$4{,}000{,}000.00 - \$1{,}504{,}600.00 + \$972{,}200.00\right](0.06646) \\
&= -\$4{,}532{,}400.00(0.06646) \\
&= -\$301{,}223.30
\end{aligned}$$

The new building has a slightly lower equivalent uniform annual cost than the existing building −\$301,223.30 < −\$302,452.81; therefore, the new building should be selected by the firm.

10.4 SUMMARY

This chapter provided an introduction to replacement analysis and the methods for determining whether to replace an existing facility, equipment, or asset with a proposed new facility, equipment, or asset. This chapter included definitions for terms used in conjunction with replacement analysis including replacements, augmentation, retirement, challengers, defenders, physical life, accounting life, service period, sunk cost, block replacement, reduced performance, alternate requirements, and obsolescence. Methods for determining when to replace a facility, equipment, or asset were included and the last part of this chapter provided example problems demonstrating the analysis process for mutually exclusive alternatives to determine when to replace an existing facility, equipment, or asset with a new asset.

KEY TERMS

Accounting life
Augmentation
Block replacements
Challenger
Defender
Economic life
Obsolescence
Ownership life
Physical life
Reduced performance
Replacement
Replacement analysis
Retirement
Service period
Sunk costs

PROBLEMS

10.1 Two maintenance plans are being considered by a city for a public arena. Table 10.4 lists the data for the two options. Using an interest rate of 8% and equivalent uniform annual worth analysis, determine which option should be selected by the city.

TABLE 10.4
Data for Maintenance Options for Public Arena

Costs or Disbursements	Maintenance Alternative 1	Maintenance Alternative 2
Initial cost	$10,000,000.00	$10,000,000.00
Operating and maintenance costs	$5,000,000.00	$1,300,000.00
Annual income	$6,000,000.00	$6,000,000.00 decreasing by $1,000,000.00 per year
Life in years	20	5

10.2 Using the data in Problem 10.2 and an interest rate of 20%, determine which alternative should be selected by the city.

10.3 Two processes are being analyzed for a water treatment facility. The first alternative costs $100,000.00, it has a life of 19 years, and the operating and maintenance costs are $10,000.00 per year increasing by $5,000.00 per year until year 10. The second process costs $300,000.00 and the operating and maintenance costs are $10,000.00 increasing by $1,000.00 per year for 40 years. Using net present worth analysis and an interest rate of 6%, determine which process should be selected by the firm.

10.4 Using the data in Problem 10.3 and the second process with operating and maintenance costs of $12,000.00 increasing by $1,500.00 per year for 40 years, determine which process should be selected by the firm.

10.5 An engineer working for a construction firm is determining whether to replace one of the bulldozers. The existing bulldozer could be sold for $40,000.00. If the firm decides to keep the bulldozer, it will decrease in value by $10,000.00 each year. Maintenance costs are $12,000.00 per year increasing by $1,400.00 each year. A new bulldozer costs $80,000.00 and the maintenance costs are $10,000.00 starting at year two and increasing by $10,000.00 per year until year five. The salvage value is $60,000.00 the first year and decreases by $10,000.00 each year. Using an interest rate of 15% determine whether the existing bulldozer should be kept for one more year.

10.6 An existing roadway has operating and maintenance costs of $200,000.00 per year. The concrete on the roadway could be sold to recyclers for $3,000,000.00. A new road has been proposed that would cost $4,000,000.00 and have operating and maintenance costs of $100,000.00 per year. The road will be financed using bonds paying 6% interest. The roadways are estimated to last 40 years, and the first alternative has a salvage value of $15,000,000.00 and the second alternative has a salvage value of $10,000,000.00. Using net present worth analysis, determine whether the county should replace the existing roadway.

10.7 If the roadways in Problem 10.6 will both have a salvage value of $20,000,000.00, determine which alternative the firm should select.

10.8 A construction company owns two excavators. The fair market value of the excavators is $420,000.00. Annual operating and maintenance costs are $120,000.00 and the salvage value in 10 years will be $80,000.00 for both excavators. The company is considering leasing excavators, and it needs to determine whether leasing excavators is more cost effective than owning them. The lease costs would be $140,000.00 per year. Using equivalent uniform annual worth analysis, a life of 10 years, and an interest rate of 12%, determine whether the construction company should lease the excavators.

10.9 Using the data in Problem 10.8 and an interest rate of 25%, determine whether the company should lease the vehicles.

10.10 An agriculture firm has a combine they have owned for three years. The combine currently has an equivalent uniform annual cost of $52,100.00, and an anticipated remaining life of five years due to technological advances available on newer combines. The firm is considering a replacement that would cost $250,000.00 and have a salvage value of $38,000.00. The operating and maintenance costs for the replacement will be $7,200.00 per year and the useful life is 12 years. Determine whether the existing machine should be replaced based on equivalent uniform annual worth analysis.

10.11 If the initial cost of the new combine in Problem 10.10 is $350,000.00 instead of $250,000.00, determine whether the firm should purchase the new combine.

10.12 Calculate the equivalent uniform annual cost of the new combine in Problem 10.10 using a life of five years rather than 12 years to determine if the agriculture firm should purchase the new combine.

10.13 A surveying firm is determining whether to purchase new surveying equipment. The current equipment is three years old and it costs the firm $95,000.00 a year to operate the equipment. The equipment could be sold for $35,000.00 in seven years. The new equipment costs $280,000.00, it has a life of 14 years, the operating and maintenance costs are $55,000.00 per year (the equipment requires less labor to operate), and the salvage value is $20,000.00. Using equivalent uniform annual cost analysis, determine the minimum amount the surveying firm should accept if they trade-in the existing equipment and buy the new equipment. Use an interest rate of 15%.

10.14 Using the data from Problem 10.13 and a life of 10 years instead of 14 years for the new surveying equipment, determine the trade-in value for the old equipment.

10.15 A heavy construction company has the option to augment one of its existing dump trucks with another one of the same age or the firm could purchase a newer, larger capacity dump truck. Table 10.5 provides the data for the existing dump truck and the two proposed dump truck options. Determine which option the company should select if the interest rate is 12% using equivalent uniform annual worth analysis.

TABLE 10.5
Data for Dump Truck Options

Costs or Disbursements	Existing Dump Truck	Additional Dump Truck with Same Capacity	Larger Capacity Dump Truck
Initial cost	$180,000.00	$580,000.00	$720,000.00
Operating and maintenance costs	$15,000.00	$15,000.00	$25,000.00
Salvage value	$5,100.00	$69,600.00	$72,000.00
Life in years	9	12	12

10.16 Solve for the net present worth of the dump truck options listed in Problem 10.15.

10.17 If the data in Problem 10.15 are used with an interest rate of 25%, determine which option the company should select using equivalent uniform annual cost analysis.

10.18 A mechanical system is being considered for replacement. There are two options being reviewed to determine if the existing system should be replaced now. The data for all three systems are listed in Table 10.6. Using equivalent uniform annual worth, an interest rate of 8%, and a life of 10 years, determine whether the mechanical system should be replaced by option 2 or option 3.

TABLE 10.6
Data for Mechanical Systems Options

Costs or Disbursements	Existing Mechanical System Alternative 1	Mechanical System Alternative 2	Mechanical System Alternative 3
Initial cost	—	$25,000.00	$38,000.00
Trade-in value	$9000.00 for alternative 2 $7000.00 for alternative 3		
Operating and maintenance costs	$3000.00	$4,000.00	$2,500.00
Labor costs	$5000.00	$5,000.00	
Salvage value	—	$1,000.00	$1,000.00
Life in years	2	12	20

10.19 Using the data in Problem 10.19, determine the equivalent uniform annual cost of option 1 if option 2 is selected and if the firm selects option 3.

10.20 Solve for the net present worth of the options for the mechanical system listed in Problem 10.18.

11 Breakeven Analysis Comparisons

This chapter introduces another method for comparing economic alternatives known as *breakeven analysis*. Breakeven analysis provides information to decision makers on the number of units of production per year where one alternative begins to have a higher equivalent uniform annual worth than another alternative or an existing alternative. This chapter includes a graph with a plot showing the breakeven point for units of production per year for two alternatives that helps illustrate graphically the breakeven point between two alternatives. To help explain breakeven comparisons, this chapter includes information on fixed and variable costs as they relate to the calculations performed for breakeven comparisons. The procedures for determining the breakeven point when comparing alternatives are explained and solved example problems and a case study are included, which demonstrate the process for calculating the breakeven point.

11.1 FIXED AND VARIABLE COSTS

When firms are producing some type of product, consuming resources, or there are operating and maintenance costs associated with the operations of the firm, then the type of comparisons managers of firms would perform between an existing and a proposed asset would be dependent on the variable costs associated with the project. *Variable costs* are any type of cost that varies with production. In addition to variable costs, projects also have *fixed costs* that do not vary with production. Fixed costs include capital costs such as construction costs, installation costs, and the cost of acquiring land or facilities. No matter whether firms are producing products there are still fixed costs.

In many instances, there is an inverse relationship between fixed capital costs and the variable costs associated with operating a facility. If a firm chooses to invest in more expensive equipment or assets (fixed costs), the equipment or assets are normally more efficient than less costly alternatives, and this in turn should result in reductions in yearly operating costs (variable costs).

11.2 LOCATING THE BREAKEVEN POINT

In order for firms to maximize their profits, engineers may be required to determine if the extra expense of purchasing higher cost equipment or assets is justified based on the units of production of the equipment or facility. The method for determining the point in time, in terms of units of production, where the purchase of new or more expensive equipment or assets is justified is called the *breakeven point*.

If production is below the breakeven point, then an existing facility or asset should be retained or the lower cost alternative should be purchased by a firm. If production is above the breakeven point, a firm is justified in purchasing new equipment or assets or purchasing higher cost equipment or assets.

The relationship of the breakeven point to variable units is best illustrated by the graph shown in Figure 11.1.

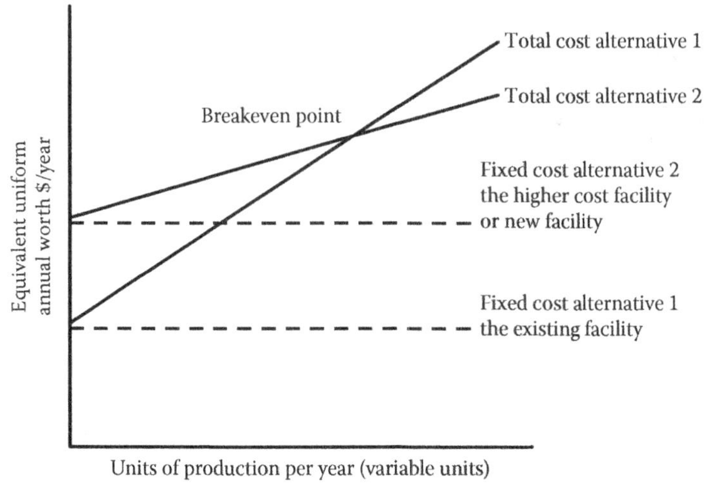

FIGURE 11.1 Graph of breakeven point for units of production per year.

The graph in Figure 11.1 shows the point where the variable costs for alternative 1 and alternative 2 intersect and this is the breakeven point. If the units of production are less than the number of units at the breakeven point, then the first alternative is selected, and if the units of production exceed the breakeven point, then alternative 2 is selected over alternative 1 since the equivalent uniform annual cost of alternative 2 would be less than alternative 1 in this range of production.

In addition to plotting the breakeven point on a graph, it may be determined numerically using net present worth or equivalent uniform annual worth analysis to express the total cost of each of the alternatives in terms of a function of the variable being analyzed in the problem. The method for locating the breakeven point involves several steps:

1. Calculate the equivalent uniform annual series of the capital cost (initial cost).
2. Determine the independent variable, which is normally in the form of annual units of production or consumption.
3. Develop the total annual cost equation, which is Equation 11.1:

$$\text{Total annual cost} = \text{Annual equivalent capital cost} + \text{Cost per variable unit} \times \text{Number of variable units per year} \quad (11.1)$$

4. Calculate the yearly production rate.

11.3 SOLVED EXAMPLE PROBLEMS

This section provides example problems demonstrating the process for calculating the breakeven point.

Example 11.1

Table 11.1 lists data for two industrial production machines a firm is considering purchasing for one of their plants. The company expects a return of 10% (MARR). Using the information in Table 11.1, determine the following: (1) how many tons per year need to be produced to justify the purchase of the more expensive machine? (2) If the company produces 20,000 tons per year, which machine should the company select to purchase? Figures 11.2 and 11.3 are the cash flow diagrams for production machines 1 and 2.

TABLE 11.1
Production Machine Data

Costs and Disbursements	Production Machine 1	Production Machine 2
Initial cost	\$80,000.00	\$230,000.00
Salvage value	—	\$40,000.00
Cost of operators	3 operators at \$8.00 per hour	1 operator at \$12.00 per hour
Production	6 tons per hour	8 tons per hour
Annual operating and maintenance cost	\$15,000.00	\$35,000.00
Life in years	5	10

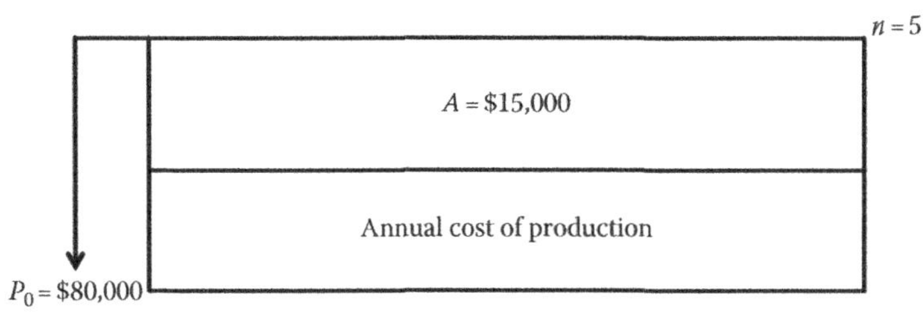

FIGURE 11.2 Cash flow diagram for production machine 1 in Example 11.1.

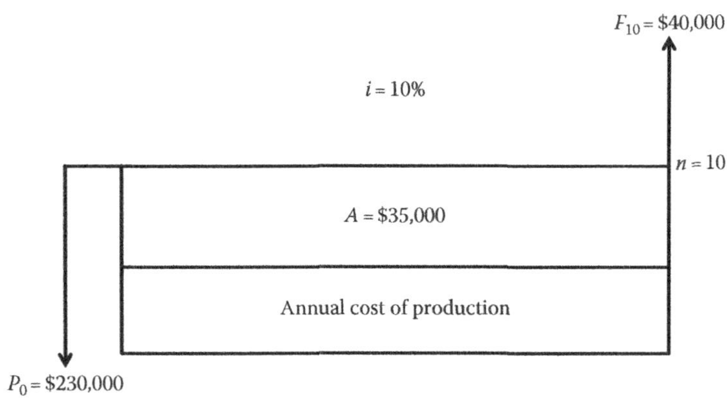

FIGURE 11.3 Cash flow diagram for production machine 2 in Example 11.1.

Solution

1. Solve for the number of tons per year to justify the more expensive machine.

Production machine 1

First, solve for the variable annual cost of production for machine 1:

$$\text{Variable annual cost} = 3\,\text{operators} \times \frac{1\ \text{hour}}{6\ \text{tons}} \times \frac{\$8.00}{\text{hour}} \times \frac{x\ \text{tons}}{\text{year}} = \frac{24}{6}x = 4x$$

where x is dollars per year of production

Second, solve for the annual cost of production for machine 1:

$$\text{Annual cost of production for machine 1} = P_0\left(A/P,i,n\right) + A + \text{variable cost}$$

$$= -\$80{,}000.00\left(A/P,10,5\right) - \$15{,}000.00 - 4x$$

$$= -\$80{,}000.00\left(0.26380\right) - \$15{,}000.00 - 4x$$

$$= -\$21{,}104.00 - \$15{,}000.00 - 4x$$

$$= -\$36{,}104.00 - 4x$$

Production machine 2

First, solve for the variable annual cost of production for machine 2:

$$\text{Variable annual cost} = \frac{\$12.00}{\text{hour}} \times \frac{1 \text{ hour}}{8 \text{ tons}} \times \frac{x \text{ tons}}{\text{year}} = \frac{12}{8}x$$

Second, solve for the annual cost of production for machine 2:

$$\text{Annual cost of production for machine 2} = P_0\left(A/P,i,n\right) + A + F_{10}\left(A/F,i,n\right) + \text{variable cost}$$

$$= -\$230{,}000.00\left(A/P,10,10\right) - \$35{,}000.00$$

$$+ \$40{,}000.00\left(A/F,10,10\right) - \frac{12}{8}x$$

$$= -\$230{,}000.00\left(0.16275\right) - \$35{,}000.00$$

$$+ \$40{,}000.00\left(0.06275\right) - \frac{12}{8}x$$

$$= -\$37{,}432.50 - \$35{,}000.00 + \$2{,}510.00 - \frac{12}{8}x$$

$$= -\$69{,}922.50 - 1.5x$$

To solve for (x), equate the annual cost of production for the two production machines:

$$-\$69{,}922.50 - 1.5x = -\$36{,}104.00 - 4x$$

$$-\$69{,}922.50 - (-\$36{,}104.00) = -4x - -1.5x$$

$$-\$33{,}818.50 = -2.5x$$

$$\frac{-\$33{,}818.50}{-2.5} = x$$

$$x = 13{,}527.40 \text{ tons per year}$$

Therefore, 13,527.40 is the number of tons that need to be produced to justify the purchase of machine 2.

2. Determine which machine should be purchased if the company produces 20,000 tons per year. Solve for which machine should be purchased if the annual production is 20,000 tons per year. The 20,000 tons per year is used in both equations as the value of (x) and the machine with the lowest equivalent uniform annual cost is selected for purchase:

$$\text{Machine 1 EUAC} = -\$36{,}104.00 - 3 \text{ operators} \times \frac{1 \text{ hour}}{6 \text{ tons}} \times \frac{\$8.00}{\text{hour}} \times \frac{20{,}000 \text{ tons}}{\text{year}}$$

$$= -\$36{,}104.00 - \$80{,}000.00$$

$$= -\$116{,}104.00$$

$$\text{Machine 2 EUAC} = -\$69{,}922.50 - \frac{\$12.00}{\text{hour}} \times \frac{1\text{ hour}}{8\text{ tons}} \times \frac{20{,}000\text{ tons}}{\text{year}}$$

$$= -\$69{,}922.50 - \$30{,}000.00$$

$$= -\$99{,}922.50$$

Therefore, Machine 2 should be selected since it has a lower equivalent uniform annual cost than machine 1: −\$99,922.50 < −\$116,104.00.

The principles that apply to determining the breakeven point for production machinery could also be applied to personal decisions such as whether to purchase a new vehicle or keep an existing vehicle.

Example 11.2 demonstrates how breakeven analysis techniques are applied when making a decision on whether to purchase a new vehicle.

Example 11.2

An aerospace engineer has recently graduated from college and he is considering buying a new vehicle. In order to determine whether the purchase of a new vehicle is justified in terms of economic viability, he has chosen to determine the breakeven point where the number of miles driven per year would justify the purchase of a new vehicle. The fixed costs for his current vehicle are \$7,500.00 per year, and they include yearly registration and automobile insurance. His variable costs associated with his existing vehicle are gas, tires, maintenance, and repairs, and they are \$0.82 per mile. For the new vehicle, the fixed costs will be \$12,000.00 per year since they will include a loan payment in addition to the existing fixed costs. The mileage costs for the new vehicle will be \$0.59 per mile. Find the breakeven point in terms of miles per year where the purchase of the new vehicle would be justified rather than keeping the existing vehicle.

Solution

First, calculate the equivalent uniform annual cost of the existing vehicle:

$$\text{EUAC of existing vehicle} = \text{Fixed costs} - \left(\text{Variable costs} \times \frac{x\text{ miles}}{\text{year}}\right)$$

$$= -\$7500.00 - \left(\frac{\$0.82}{\text{mile}} \times \frac{x\text{ miles}}{\text{year}}\right)$$

$$= -\$7500.00 - 0.82x$$

Second, calculate the equivalent uniform annual cost of the new vehicle:

$$\text{EUAC of new vehicle} = \text{Fixed costs} - \left(\text{Variable costs} \times \frac{x\text{ miles}}{\text{year}}\right)$$

$$= -\$12{,}000.00 - \left(\frac{\$0.59}{\text{mile}} \times \frac{x\text{ miles}}{\text{year}}\right)$$

$$= -\$12{,}000.00 - 0.59x$$

Third, to solve for (x), equate the existing vehicle to the new vehicle:

$$-\$7,500.00 - 0.82x = -\$12,000.00 - 0.59x$$

$$-\$7,500.00 + \$12,000.00 = 0.82x - 0.59x$$

$$\$45,000.00 = 0.23x$$

$$x = 19,565.52 \text{ miles per year}$$

If the aerospace engineer drives 19,565.52 or more miles per year the purchase of the new vehcile is justified.

Example 11.3 demonstrates using breakeven analysis techniques to help with a decision related to the purchase of heavy construction equipment.

Example 11.3

A field engineer working for a heavy construction company needs to determine whether the company should keep an existing backhoe or replace it with a new backhoe. The field engineer has collected data on the fixed and variable costs for each machine, and they are listed in Table 11.2. The interest rate is 10%. How many feet of trenches need to be excavated each year to justify the purchase of the new backhoe? Figures 11.4 and 11.5 are the cash flow diagrams for the existing and new backhoe.

TABLE 11.2
Data for Existing and New Backhoe

Costs and Disbursements	Existing Backhoe	New Backhoe
Initial cost	\$300,000.00 (foregone resale value)	\$200,000.00
Salvage value	\$50,000.00 at year 10	\$50,000.00 at year 10
Operating and maintenance costs	\$100.00 per hour increasing \$10.00 per hour each year	\$60.00 per hour increasing by \$6.00 per hour each year
Excavation rate	50 ft per hour	20 ft per hour
Life in years	10	10

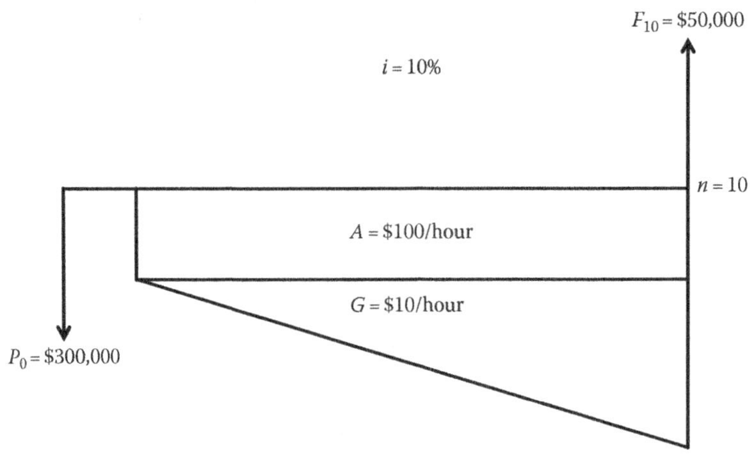

FIGURE 11.4 Cash flow diagram for existing backhoe in Example 11.3.

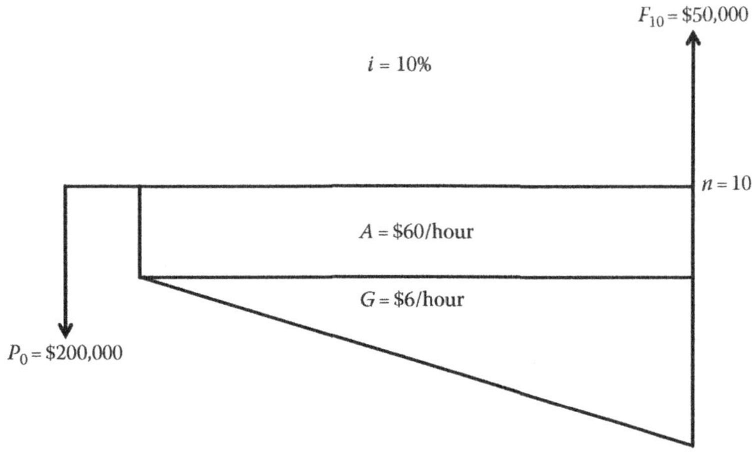

FIGURE 11.5 Cash flow diagram for new backhoe in Example 11.3.

Solution

First, calculate the equivalent uniform annual cost of the existing backhoe:

$$\text{EUAC}_D = P_0(A/P,i,n) + F_{10}(A/F,i,n) - \left[A + G(A/G,i,n)\right]\left(\frac{1\text{ hour}}{50\text{ ft}} \times \frac{x\text{ feet}}{\text{year}}\right)$$

$$= -\$300{,}000.0010(A/P,10,10) + \$50{,}000.00(A/F,10,10)$$

$$-\left[\frac{\$100.00}{\text{hour}} + \frac{\$10.00}{\text{hour}}(A/G,10,10)\right]\left(\frac{1\text{ hour}}{50\text{ ft}} \times \frac{x\text{ feet}}{\text{year}}\right)$$

$$= -\$300{,}000.00(0.16275) + \$50{,}000.00(0.06275)$$

$$-\left[\frac{\$100.00}{\text{hour}} + \frac{\$10.00}{\text{hour}}(3.7254)\right]\left(\frac{1\text{ hour}}{50\text{ ft}} \times \frac{x\text{ feet}}{\text{year}}\right)$$

$$= -\$48{,}825.00 + \$3{,}137.50 - \left(\frac{\$100.00}{\text{hour}} + \frac{\$37.25}{\text{hour}}\right)\left(\frac{1\text{ hour}}{50\text{ ft}} \times \frac{x\text{ feet}}{\text{year}}\right)$$

$$= -\$45{,}687.50 - 2.745x$$

Second, calculate the equivalent uniform annual cost of the new backhoe:

$$\text{EUAC}_C = P_0(A/P,i,n) + F_{10}(A/F,i,n) - \left[A + G(A/G,i,n)\right]\left(\frac{1\text{ hour}}{20\text{ ft}} \times \frac{x\text{ feet}}{\text{year}}\right)$$

$$= -\$200{,}000.00(A/P,10,10) + \$50{,}000.00(A/F,10,10)$$

$$-\left[\frac{\$60.00}{\text{hour}} + \frac{\$6.00}{\text{hour}}(A/G,10,10)\right]\left(\frac{1\text{ hour}}{20\text{ ft}} \times \frac{x\text{ feet}}{\text{year}}\right)$$

$$= -\$200{,}000.00(0.16275) + \$50{,}000.00(0.06275)$$

$$-\left[\frac{\$60.00}{\text{hour}} + \frac{\$6.00}{\text{hour}}(3.7254)\right]\left(\frac{1\text{ hour}}{20\text{ ft}} \times \frac{x\text{ feet}}{\text{year}}\right)$$

$$= -\$32{,}550.00 + \$3{,}137.50 - \left(\frac{\$60.00}{\text{hour}} + \frac{\$22.35}{\text{hour}}\right)\left(\frac{1\text{ hour}}{20\text{ ft}} \times \frac{x\text{ feet}}{\text{year}}\right)$$

$$= -\$29{,}412.50 - 4.1175x$$

To solve for (x), equate the equivalent uniform annual cost of the existing backhoe to the new backhoe:

$$-\$45,687.00 - 2.745x = -\$29,412.00 - 4.1175x$$

$$\$45,687.00 - \$29,412.00 = 4.1175 - 2.745x$$

$$\$16,275.00 = 1.3725x$$

$$x = 11,857.92 \text{ ft per year}$$

Therefore, if 11,857.92 ft per year or more of trenches are excavated the purchase of the new backhoe is justified.

This section provides a case study that illustrates using breakeven analysis in the oil industry.

Case Study 11.1 Breakeven Analysis

Breakeven analysis techniques are prevalent in major industries, such as the oil industry, where the drilling and refining of oil is dependent on the price of oil per barrel. When the price of a barrel of oil drops to \$50.00 per barrel from a high of over \$100.00 per barrel, many oil companies reduce their drilling operations or cease drilling in locations where it is difficult to drill for oil such as the Arctic region. All of the major oil companies, and many of the smaller companies, perform breakeven analysis to determine what is the minimum price per barrel oil needs to be selling for in order for the company to continue certain types of operations such as oil exploration activities, or hydraulic fracturing. Oil wells already drilled and producing do not require as high a price for a barrel of oil since their annual costs are lower than wells requiring exploration and drilling. Therefore, it is potential and new oil wells that are usually the first to be shut down when the price of oil per barrel drops too low. The following demonstrates how breakeven analysis techniques are implemented in the oil industry.

An oil company is trying to determine whether to expand operations on a series of oil wells currently being explored and drilled in the Arctic region. It initially cost the oil company \$230,000,000.00 to acquire the leases on the land where they hoped to drill for oil. The yearly exploration and drilling costs are \$35,000,000.00. The materials and equipment being used on this project would have a salvage value of \$40,000,000.00 at the end of 10 years. The oil company has developed a second option on how they would expand their exploration and drilling operations. The second option would have an initial cost of \$80,000,000.00, annual operating and maintenance costs of \$15,000,000.00, and no salvage value. If the oil field produces 500,000 barrels of oil per year, the interest rate is 10%, and the life of the wells is 10 years, determine the price per barrel to justify implementing the second option if the second option would produce an additional 500,000 barrels of oil per year and have a life of five years. Figures 11.6 and 11.7 are the cash flow diagrams for the two drilling options.

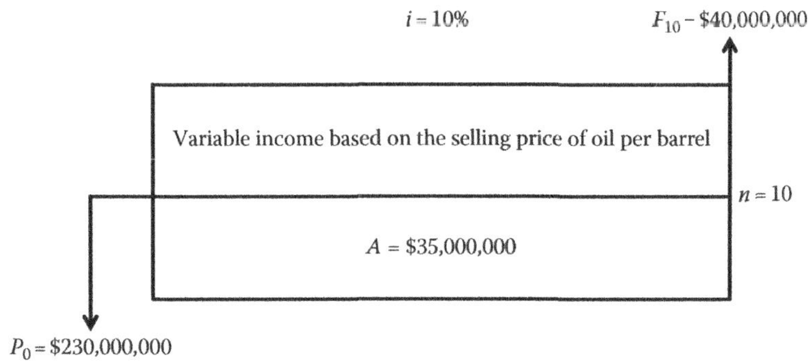

FIGURE 11.6 Cash flow diagram for original oil field production for Case Study 11.1.

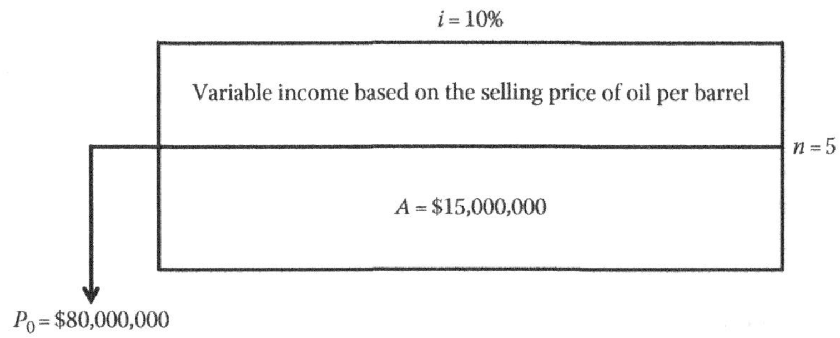

FIGURE 11.7 Cash flow diagram for proposed oil field production for Case Study 11.1.

SOLUTION

First, calculate the equivalent uniform annual cost of the cost of production for the original oil field:

$$\text{EUAC}_D = P_0(A/P,i,n) + A + F_{10}(A/F,i,n) + \left(\frac{500,000 \text{ barrels}}{\text{year}} \times x\right)$$

where x is dollars per barrel

$$= -\$230,000,000.00(A/P,10,10) - \$35,000,000.00$$

$$+ \$40,000,000.00(A/F,10,10) + 500,000x$$

$$= -\$230,000,000.00(0.16275) - \$35,000,000.00$$

$$+ \$40,000,000.00(0.06275) + 500,000x$$

$$= -\$37,432,500.00 - \$35,000,000.00 + \$2,510,000.00 + 500,000x$$

$$= -\$69,922,500.00 + 500,000x$$

Second, calculate the equivalent uniform annual cost of the proposed oil field production alternative:

$$\begin{aligned}
\text{EUAC}_D &= P_0\left(A/P,i,n\right)+A+\left(\frac{1{,}000{,}000\text{ barrels}}{\text{year}}\times x\right)\\
&= -\$80{,}000{,}000.00\left(A/P,10,5\right)-\$15{,}000{,}000.00+1{,}000{,}000x\\
&= -\$80{,}000{,}000.00\left(0.26380\right)-\$15{,}000{,}000.00+1{,}000{,}000x\\
&= -\$21{,}104{,}000.00-\$15{,}000{,}000.00+1{,}000{,}000x\\
&= -\$36{,}104{,}000.00+1{,}000{,}000x
\end{aligned}$$

To solve for (x), equate the equivalent uniform annual cost for the two options:

$$\begin{aligned}
&-\$69{,}922{,}500.00+500{,}000x=-\$36{,}104{,}000.00+1{,}000{,}000x\\
&\$69{,}922{,}500.00-\$36{,}104{,}000.00=1{,}000{,}000x-500{,}000x\\
&\$33{,}818{,}500.00=500{,}000x\\
&x=\$67.64\text{ per barrel}
\end{aligned}$$

Therefore, $67.64 per barrel is the breakeven point to justify the proposed oil field production option.

11.4 SUMMARY

This chapter introduced another method for analyzing alternatives using engineering economic analysis called breakeven analysis or breakeven comparison. Information was included in this chapter on fixed and variable costs as they relate to the calculations performed for breakeven comparisons. Methods for determining the breakeven point when comparing alternatives were reviewed and solved examples and a case study demonstrating the process for calculating the breakeven point between two alternatives were provided in the last section of this chapter.

KEY TERMS

Breakeven comparisons
Breakeven point
Fixed costs
Variable costs

PROBLEMS

11.1 An electrical engineer is purchasing new machines to manufacture electrical parts. The data for the two machines are listed in Table 11.3. In order to operate the machines, four laborers are required at a cost of $25.00 per hour per laborer. The machines produce 10,000 parts per day. The firm uses a minimum attractive rate of return of 15%. If the company charges $5.00 per part, how many parts need to be manufactured each year to justify purchasing the two machines?

TABLE 11.3
Data for Machines Manufacturing Electrical Parts

Costs or Disbursements	Electrical Parts Machine 1	Electrical Parts Machine 2
Initial cost	$180,000.00	$120,000.00
Operating and maintenance costs	$60,000.00	$50,000.00
Overhaul	$30,000.00 at year 3	—
Salvage value	$20,000.00	−$5,000.00
Life in years	6	4

11.2 An engineer is evaluating the purchase of a dump truck costing $300,000.00 that will have operating and maintenance costs of $1.00 per mile. The dump truck will have a life of 15 years and no salvage value. The existing dump truck could be sold for $100,000.00, and it has operating and maintenance costs of $1.50 per mile. The existing truck will have a life of five years and no salvage value. The interest rate is 10%. Calculate the breakeven point in terms of miles per year required to justify the purchase of the new dump truck.

11.3 If the interest rate is 20%, calculate the breakeven point for the data in Problem 11.2.

11.4 A construction company owns its own concrete plant. The plant cost $7,400,000.00 and the operating and maintenance costs are $41.04 per yd^3. If concrete is selling for $90.00 per yd^3, how much concrete does the company need to sell to break even if the minimum attractive rate of return is 20%?

11.5 If the interest rate is 9%, use the data from Problem 11.4 to calculate how much concrete the company needs to sell to break even.

11.6 An aerospace firm will be hiring a new engineer for $60,000.00 per year plus $60,000.00 in benefits. The firm has a fixed cost of $1,480,000.00 per year and variable costs of $22,400.00 per contract. The average income per contract is $38,600. How many more contracts have to be acquired by the firm for the company to break even if they hire the new engineer?

11.7 For the data in Problem 11.6 if the average cost per contact is reduced to $20,800.00 and the firm will have 100 contracts, should the firm hire the new engineer?

11.8 A construction firm is currently digging postholes for fence posts using manual labor that costs $10.00 per hour, and they are able to dig four holes per person per hour. The company is considering the purchase of an automatic posthole digger at a cost of $10,000.00 that is able to dig five holes per hour. The automatic hole digger has a life of 10 years with no salvage value. The automatic digger requires one operator at a cost $16.00 per hour and operating costs are $4.00 per hour. The interest rate is 10%. How many holes need to be dug to break even if the firm purchases the automatic digger?

11.9 If the labor costs increase by $1.00 per hour per year in Problem 11.8 for laborers and equipment operators, what is the breakeven point?

11.10 A chemical company is considering upgrading a manual chemical analysis process to an automated process. The automated process cost $69,000.00 and it has a salvage value of $12,000.00 at the end of 10 years. The automated process requires one operator that costs $36.00 per hour. The automated process is able to analyze 24 liters of chemicals per hour. Operating and maintenance costs will be $10,500.00 per year. The manual analysis process cost $24,000.00 and it has a life of five years with no salvage value. The manual system is able to analyze 18 liters per hour. The manual process requires three people per hour to perform the analysis at a cost of $24.00 per person per hour. Operating and maintenance costs are $1,500.00 per year. The interest rate is 10%. How many liters per year need to be analyzed to justify purchasing the automatic analysis process?

11.11 If the interest rate is 20%, use the data from Problem 11.10 to calculate how many liters per year need to be analyzed to justify the purchase of the automatic analysis process.

11.12 A hazardous waste testing lab is evaluating two options for a testing machine. The first testing machine cost $50,000.00, it has operating and maintenance costs of $26,000.00 per year, a life of six years, and it cost $10.00 per hour to pay the operator. The second testing machine cost $35,000.00, operating and maintenance costs are $10,000.00 per year, the life is six years, and it requires three operators at a cost of $13.30 per operator per hour. If the interest rate is 10%, how many samples need to be tested per year to justify the purchase of the higher cost alternative?

11.13 A transportation engineering firm is analyzing whether to replace their existing fleet of vehicles with new vehicles. The fixed cost per year for the existing fleet is $75,000.00 and the variable cost is $8.30 per mile. The new vehicles have a total fixed cost of $120,000.00 per year and the variable cost is $5.90 per mile. What is the breakeven point in terms of miles per year to justify the purchase of the new vehicles?

11.14 A processing plant is evaluating one of their machines to determine whether to replace it with a new machine. The data for the two machines are listed in Table 11.4. If the interest rate is 10%, how many feet of product needs to be produced per year to justify replacing the existing machine with the new machine?

TABLE 11.4
Data for Process Plant Machines

Costs or Disbursements	Existing Process Plant Machine	New Process Plant Machine
Initial cost	$600,000.00 (foregone resale value)	$400,000.00
Operating and maintenance costs	$200.00 per hour increasing by $20.00 per hour each year	$120.00 per hour increasing by $12.00 per hour each year
Processing rate	$100 ft per hour	40 ft per hour
Salvage value	$100,000.00	$100,000.00
Life in years	10	10

11.15 If the interest rate is 20%, use the data from Problem 11.14 to determine how many feet of product needs to be produced per year to justify replacing the existing machine with the new machine?

11.16 A construction firm is considering the purchase of a new scraper to replace an existing scraper. The data for the existing scraper and the new scraper are listed in Table 11.5. If the interest rate is 10%, determine the number of cubic yards per year that have to be excavated to justify the purchase of the new scraper.

TABLE 11.5
Data for Existing and New Scraper

Costs and Disbursements	Existing Scraper	New Scraper
Initial cost	$3,000,000.00 (foregone resale value)	$2,000,000.00
Salvage value	$500,000.00 at year 10	$50,0000.00 at year 10
Operating and maintenance costs	$1,000.00 per hour increasing by $100.00 per hour each year	$600.00 per hour increasing by $60.00 per hour each year
Excavation rate	500 ft per hour	200 ft per hour
Life in years	10	10

11.17 A manufacture of metal parts is reviewing data to determine whether they should replace an existing computer network control (CNC) machine. The data for the existing machine and a new machine are listed in Table 11.6. If the interest rate is 10%, determine how many metal parts need to be produced per year to justify the purchase of the new CNC machine.

TABLE 11.6
Computer Network Control (CNC) Machine Data

Costs and Disbursements	Existing CNC Machine	New CNC Machine
Initial cost	\$800,000.00 (foregone resale value)	\$2,300,000.00
Salvage value	—	\$400,000.00
Cost of operators	3 operators at \$80.00 per hour	1 operator at \$120.00 per hour
Production	60 metal plates per hour	80 metal plates per hour
Annual operating and maintenance cost	\$150,000.00	\$350,000.00
Life in years	5	10

11.18 If the cost for the operator for machine 2 in Problem 11.17 is \$160.00 per hour, determine how many metal parts need to be produced per year to justify the purchase of the new CNC machine.

11.19 A manager of a process engineering firm is considering installing solar panels on one of their office buildings. Table 11.7 lists the data for the solar panels and for the cost of electricity if it is purchased from a utility company. Using an interest rate of 10%, how many hours per year does the electrical system need to be used to reach the breakeven point that justifies the installation of the solar panels.

TABLE 11.7
Data for Utility Company Electricity and Solar Panels

Costs and Disbursements	Utility Company Electricity	Solar Panel Electricity
Initial cost	\$160,000.00	\$400,000.00
Salvage value	—	\$5,000.00
Electricity cost	\$12.00 per hour increasing by \$0.24 per hour	\$4.00 per hour
Life in years	15	15

11.20 Using the data in Problem 11.19 and an interest rate of 25%, determine the breakeven point to justify the installation of the solar panels.

12 Benefit/Cost Ratio Economic Evaluations

Another engineering economic technique for evaluating potential projects is *benefit/cost ratio economic analysis* also referred to as *discounted profitability index, profitability index, profit investment ratio,* or *value investment ratio.* Benefit/cost ratio economic analysis is used in conjunction with present worth and equivalent uniform annual worth analysis techniques since the costs and benefits for each alternative need to be in equivalent terms before they are compared to determine which alternative has the highest benefit/cost ratio.

This chapter defines the terms used in benefit/cost ratio economic evaluations and explains the procedures for conducting benefit/cost ratio economic evaluations by including specific steps for performing a benefit/cost ratio economic evaluation. Example problems are provided to help illustrate the process for calculating benefit/cost ratios.

12.1 DEFINITIONS AND TERMS USED IN BENEFIT/COST RATIO ECONOMIC EVALUATIONS

This section defines the terms used in benefit/cost ratio economic evaluations and explains how they are integrated into the formulas for calculating *benefit/cost ratios.* There are two types of costs to owners associated with projects: (1) capital costs and (2) maintenance costs. Sections 12.1.1 and 12.1.2 explain these two types of costs in relation to benefit/cost ratio economic evaluations, and Sections 12.1.3 and 12.1.4 cover the benefits associated with benefit/cost ratio economic evaluations.

12.1.1 Benefit/Cost Ratio Costs

In the formulas for calculating benefit/cost ratios, there are three different terms representing different types of costs and they are the following:

1. C_f—Equivalent capital cost (construction, acquisition, or other costs) of a proposed future facility provided as an equivalent uniform annual cost
2. C_p—Equivalent capital worth of an existing facility provided as an equivalent uniform annual worth (This is also the current salvage value of an existing facility.)
3. $C_n = C_f - C_p$—Net capital cost required if a new facility replaces an existing facility

12.1.2 Benefit/Cost Ratio Maintenance Costs

There are three types of maintenance costs in benefit/cost ratios and they are the following:

1. M_f—Equivalent annual operating and maintenance costs for a proposed future facility
2. M_p—Equivalent annual operating and maintenance costs of an existing facility
3. $M_n = M_f - M_p$—Net operating and maintenance costs of a proposed (future) facility minus the existing (present) facility operating and maintenance costs

12.1.3 Benefit/Cost Ratio Benefits

The following are the types of benefits included in benefit/cost ratios:

1. B_n or U_n—Net annual benefits or savings in cost occurring due to improvements in safety procedures and decreased expenses
2. $U_n = U_f - U_p$—Where U_p is the user benefits of a present facility and U_f is the user benefits of the future facility. The net user benefit U_n is the annual equivalent benefit to an owner.
3. $U_n = U_p - U_f$—Where the benefits are derived from a reduction in the yearly cost U_p to the owner through a new facility U_f

12.1.4 Benefits and Disbursements

When evaluating projects using benefit/cost ratios, one of the first steps is to determine what are the benefits and the disbenefits as defined by the following:

1. *Benefits* are advantages accruing to an owner. For public projects, the owner is considered to be the public. Benefits on public projects include cost savings, less wear and tear on vehicles, lower fuel consumption, safer roadways, time reductions, and so forth.
2. *Disbenefits* are disadvantages to the owner and they are subtracted from the benefits, not added to the costs.

12.2 BENEFIT/COST RATIO ECONOMIC ANALYSIS

Benefit/cost ratio economic analysis techniques are mainly for evaluating public sector projects, but they are also sometimes used by private owners and investors when they are considering investments. Benefit/cost ratios provide a comparison of the benefits and costs associated with a project and they are used to determine whether the benefits will be greater than the costs, which would result in a benefit/cost ratio greater than one. Since public projects are not built to generate profits, the benefits are usually in the form of reduced costs to users of facilities. One example is providing a road that reduces the distance drivers have to travel to a particular destination. Other examples of public projects providing benefits to citizens are bridges, public buildings, tunnels, upgrading technologies such as fiber optic cables, and water treatment facilities.

If only one alternative is being evaluated using benefit/cost ratio analysis techniques, the alternative is compared to the do nothing alternative. When there are two alternatives, they are compared to each other. Before benefit/cost ratio comparisons are performed, all of the costs and benefits need to be converted to either a present worth or an equivalent uniform annual worth so all of the values being compared are being analyzed based on equivalent terms. The basic benefit/cost ratio formula is Equation 12.1.

$$B/C = \frac{\text{Benefits} - \text{Disbenefits}}{\text{Costs}} \tag{12.1}$$

Equation 12.1 is used when the B/C ratio of one facility is being calculated to determine if the benefits are greater than the costs, as shown in Equations 12.2 and 12.3.

$$B/C \geq 1.0 \text{ project is justified} \tag{12.2}$$

$$B/C \leq 1.0 \text{ project is not justified} \tag{12.3}$$

In addition to the basic B/C ratio, there are two other B/C ratios used when comparing alternatives and they are the *conventional B/C* and *modified B/C*.

12.2.1 Conventional Benefit/Cost Ratio

The conventional B/C ratio is Equation 12.4.

$$\text{Conventional } B/C = \frac{\text{Net savings to users}}{\text{Owner's net capital cost + Owner's net operating and maintenance costs}} \tag{12.4}$$

Equation 12.5 is the formula for the conventional B/C ratio written using the terms introduced in Section 12.1.

$$\text{Conventional } B/C = \frac{U_n}{C_n + M_n} \quad \text{or} \quad = \frac{B_n}{C_n + M_n} = \frac{(U_f - U_p)}{(C_f - C_p) + (M_f - M_p)} \tag{12.5}$$

12.2.2 Modified Benefit/Cost Ratio

The modified B/C uses the same data as the conventional B/C, but the net operating and maintenance costs M_n are considered to be disbenefits and not costs, as they are in the conventional B/C ratio. The modified B/C ratio is Equation 12.6.

$$\text{Modified } B/C = \frac{U_n - M_n}{C_n} \quad \text{or} \quad = \frac{B_n - M_n}{C_n} = \frac{(U_f - U_p) - (M_f - M_p)}{C_f - C_p} \tag{12.6}$$

12.2.3 Incremental Benefit/Cost Ratio

If there are two or more alternatives being evaluated, then the incremental $\Delta B/\Delta C$ ratio is calculated using Equation 12.7 and the difference between the two alternatives to determine whether the $\Delta B/\Delta C \geq 1.0$:

$$\Delta B/\Delta C = \frac{B_f - B_p}{C_f - C_p} \tag{12.7}$$

If the $\Delta B/\Delta C \geq 1.0$, then the higher initial cost alternative is selected as the most beneficial alternative. If the $\Delta B/\Delta C \leq 1.0$, then the lower initial cost alternative is selected rather than the higher initial cost alternative because the incremental increase in the cost of the higher initial cost alternative is not justified in economic terms.

12.3.4 Steps for Performing Benefit/Cost Ratio Economic Analysis

The following steps outline the process for performing a benefit/cost ratio economic analysis:

1. Determine which of the elements being analyzed are benefits:
   a. Benefits are advantageous elements expressed in dollars accruing to the owner (public).
   b. Disadvantages are disbenefits and they are not included with the costs but are subtracted from the benefits.
2. Determine which of the elements being analyzed are costs:
   a. Costs include items such as construction costs and operating and maintenance expenses.
   b. If a government project is being evaluated, the expenses are incurred by the appropriate government agency.

3. Before calculating the benefit/cost ratio, convert all dollar values to equivalent amounts:
   a. Using the present worth formula or
   b. Using the equivalent uniform annual worth formula
4. If only one proposal is being considered, compare it to the do nothing alternative. If there are two or more alternatives being considered, compare them in order of the lowest to highest initial cost:
   a. If the $\Delta B/\Delta C \geq 1.0$, the extra benefits of the higher initial cost alternative justify the selection of the higher initial cost alternative.
   b. If the $\Delta B/\Delta C \leq 1.0$, the extra benefits of the higher initial cost alternative do not justify the selection of the higher initial cost alternative.
5. The capitalized cost of an existing facility is its salvage value for comparison purposes:
   a. The salvage value is the amount a firm would receive if the facility were purchased or demolished and its contents and materials are sold.
   b. Do not subtract the salvage value from the proposed facility since it is a cost of the existing facility because it is the amount not realized if the facility is retained and continues in service.
6. Benefit/cost ratios will be in one of the two forms:
   a. $B/C = \dfrac{\text{Present worth of benefits}}{\text{Present worth of costs}}$
   b. $B/C = \dfrac{\text{Equivalent uniform annual benefit}\left(\text{EUAB}\right)}{\text{Equivalent uniform annual cost}\left(\text{EUAC}\right)}$

12.3 SOLVED EXAMPLE PROBLEMS

This section provides several example problems that demonstrate calculating benefit/cost ratios. Example 12.1 compares two alternatives using the basic B/C ratio.

Example 12.1

A systems engineer is considering the purchase of a new device that will save his firm money on an annual basis by speeding up the production process. There are two devices being considered and both would save the firm money. Both devices cost $100,000.00, but the first one will save the firm $30,000.00 per year for five years and the second one will save $40,000.00 in the first year and the savings will decline by $5,000.00 each year until year five. If the firm uses an interest rate of 7%, determine which device should be purchased by calculating the benefit/cost ratio of each alternative using present worth analysis techniques. Figures 12.1 and 12.2 are the cash flow diagrams for the two devices.

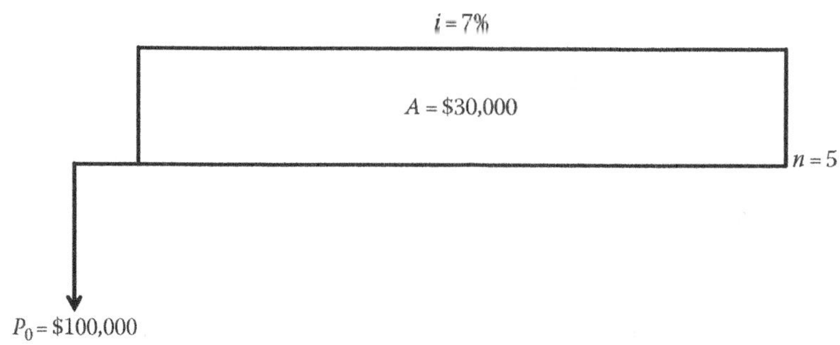

FIGURE 12.1 Cash flow diagram for the systems engineering device 1 in Example 12.1.

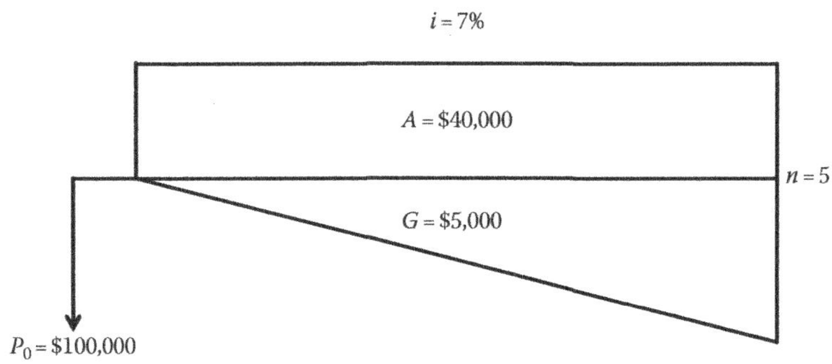

FIGURE 12.2 Cash flow diagram for the systems engineering device 2 in Example 12.1.

Solution

Present worth of costs for device 1 = \$100,000.00

$$\text{Present worth of benefits for device 1} = \$30,000.00(P/A,7,5)$$
$$= \$30,000.00(4.1002)$$
$$= \$123,006.00$$
$$B/C = \frac{\$123,006.00}{\$100,000.00} = 1.23$$

Present worth of costs for device 2 = \$100,000.00

$$\text{Present worth of benefits for device 2} = \$40,000.00(P/A,7,5)$$
$$-\$5,000.00(P/G,7,5)$$
$$= \$40,000.00(4.1002)$$
$$-\$5,000.00(7.6466)$$
$$= \$164,008.00 - \$38,233.00$$
$$= \$125,775.00$$
$$B/C = \frac{\$125,775.00}{\$100,000.00} = 1.26$$

Therefore, since device 2 has a higher B/C ratio than device 1 select device 2, $1.26 > 1.23$

Note: Since both of the devices under consideration have the same initial cost, it is not possible to compare them using the $\Delta B/\Delta C$ ratio.

Example 12.2 demonstrates calculating the $\Delta B/\Delta C$ ratio for two alternatives.

Example 12.2

Two metal punching machines are being evaluated for purchase by an industrial engineer. The first punching machine has an initial cost of \$200,000.00 and a salvage value at the end of six years of \$50,000.00. This machine would provide the firm with an annual benefit of \$95,000.00. The second machine would cost \$700,000.00 and have a salvage value of \$150,000.00 at the end of 12 years. The second machine provides an annual benefit of \$120,000.00. The interest rate is 10%.

Calculate the benefit/cost ratio for each individual punching machine, and then using incremental benefit/cost ratio analysis techniques and equivalent uniform annual cost analysis, determine which punching machine should be selected by the firm. Figures 12.3 and 12.4 are the cash flow diagrams for the punching machines.

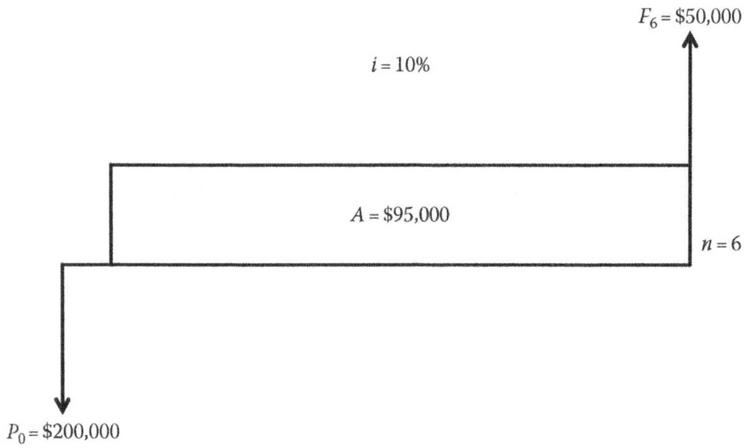

FIGURE 12.3 Cash flow diagram for punching machine 1 in Example 12.2.

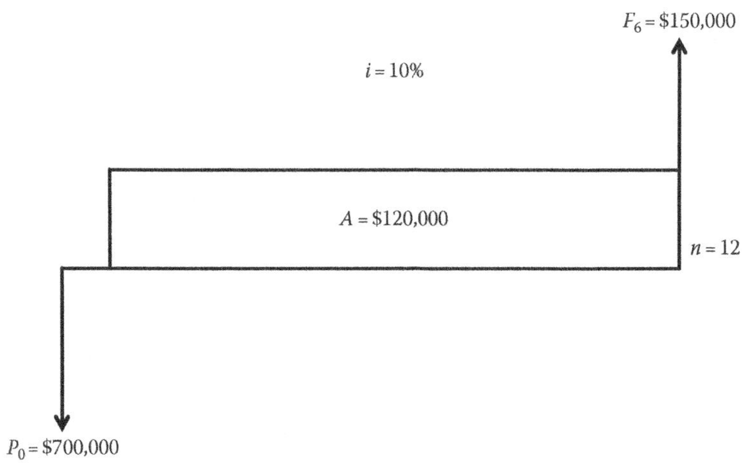

FIGURE 12.4 Cash flow diagram for punching machine 2 in Example 12.2.

Solution

The salvage value is considered a reduction in costs not a benefit.

Punching machine 1

$$\begin{aligned}
\text{EUAC}_1 &= P_0(A/P,10,6) - F_6(A/F,10,6) \\
&= -\$200{,}000.00(A/P,10,6) - \$50{,}000.00(A/F,10,6) \\
&= -\$200{,}000.00(0.22961) - \$50{,}000.00(0.12961) \\
&= -\$45{,}922.00 - \$6{,}480.50 \\
&= -\$52{,}402.50
\end{aligned}$$

$$\text{EUAB}_1 = \$95{,}000.00$$

$$B/C = \frac{\$95{,}000.00}{\$52{,}402.50} = 1.81$$

Punching machine 2

$$\begin{aligned}
\text{EUAC}_2 &= P_0(A/P,10,6) - F_{12}(A/F,10,6) \\
&= -\$700{,}000.00(A/P,10,12) - \$150{,}000.00(A/F,10,12) \\
&= -\$700{,}000.00(0.14676) - \$150{,}000.00(0.04676) \\
&= -\$102{,}732.00 - \$7{,}014.00 \\
&= -\$109{,}746.00.00
\end{aligned}$$

$$\text{EUAB}_2 = \$120{,}000.00$$

$$B/C = \frac{\$120{,}000.00}{\$109{,}746.00} = 1.09$$

Calculate the incremental benefit/cost ratio for the difference between the two metal punching machines

$$\Delta B/\Delta C = \frac{\text{benefits of machine 2} - \text{benefits of machine 1}}{\text{cost of machine 2} - \text{cost of machine 1}} = \frac{\$120{,}000.00 - \$95{,}000.00}{\$109{,}746.00 - \$52{,}402.50}$$

$$= \frac{\$25{,}000.00}{\$57{,}343.50} = 0.44$$

Therefore, since the $\Delta B/\Delta C$ of $0.44 \leq 1.0$ choose punching machine 1

Example 12.3 uses benefit/cost ratios and incremental benefit/cost ratios to compare three alternatives.

Example 12.3

A county is considering three locations for a new dam. The three alternatives for the dams for the three locations cost \$25,000,000.00, \$30,000,000.00, and \$32,000,000.00. Currently flood damage amounts to \$20,000,000.00 per year. If the new dams are built, the flood damage will be reduced to \$17,000,000.00; \$16,000,000.00; and \$15,500,000.00 per year, respectively. Determine which dam should be built based on benefit/cost ratio analysis by determining the individual benefit/cost ratios using an interest rate of 5% and a life of 10 years and then calculate the incremental benefit/cost ratios.

Solution

First, calculate the equivalent uniform annual cost of the initial cost for each of the three dam alternatives:

$$EUAC_{1\,of\,initial\,cost} = \$25,000,000.00(A/P,5,10) = \$25,000,000.00(0.12950)$$
$$= \$3,327,500.00$$

$$EUAC_{2\,of\,initial\,cost} = \$30,000,000.00(A/P,5,10) = \$30,000,000.00(0.12950)$$
$$= \$3,885,000.00$$

$$EUAC_{3\,of\,initial\,cost} = \$32,000,000.00(A/P,5,10) = \$32,000,000.00(0.12950)$$
$$= \$4,144,000.00$$

Second, calculate the benefit for each alternative:

$$Benefit_1 = \$20,000,000.00 - \$17,000,000.00 = \$3,000,000.00$$
$$Benefit_2 = \$20,000,000.00 - \$16,000,000.00 = \$4,000,000.00$$
$$Benefit_3 = \$20,000,000.00 - \$15,500,000.00 = \$4,500,000.00$$

Third, calculate the benefit/cost ratios for each alternative:

$$B/C_1 = \frac{\$3,000,000.00}{\$3,237,500.00} = 0.93$$

$$B/C_2 = \frac{\$4,000,000.00}{\$3,885,000.00} = 1.03$$

$$B/C_3 = \frac{\$4,500,000.00}{\$4,144,000.00} = 1.09$$

Finally, calculate the incremental benefit/cost ratios:

$$\Delta B/\Delta C_{2-1} = \frac{\$4,000,000.00 - \$3,000,000.00}{\$3,885,000.00 - \$3,237,500.00} = \frac{\$1,000,000.00}{\$647,500.00} = 1.54$$

Since the $\Delta B/\Delta C_{2-1} \geq 1.0$ retain alternative 2 and compare it to alternative 3

$$\Delta B/\Delta C_{3-2} = \frac{\$4,500,000.00 - \$4,000,000.00}{\$4,144,000.00 - \$3,885,000.00} = \frac{\$500,000.00}{\$259,000.00} = 1.93$$

Therefore, since the $\Delta B/\Delta C_{3-2} \geq 1.0$ select alternative 3.

Example 12.4 uses the conventional benefit/cost ratio to determine which of two alternatives should be built.

Example 12.4

For Example 12.3, if the maintenance costs for alternative 2 are \$200,000.00 per year and \$250,000.00 per year for alternative 3, calculate the conventional benefit/cost ratio for the difference between alternatives 3 and 2.

Solution

$$\text{Conventional } \Delta B/\Delta C_{3-2} = \frac{(U_f - U_p)}{(C_f - C_p) + (M_f - M_p)}$$

$$= \frac{(\$4{,}500{,}000.00 - \$4{,}000{,}000.00)}{(\$4{,}144{,}000.00 - \$3{,}885{,}000.00) + (\$250{,}000.00 - \$200{,}000.00)}$$

$$= \frac{\$500{,}000.00}{\$259{,}000.00 + \$50{,}000.00}$$

$$= \frac{\$500{,}000.00}{\$309{,}000.00} = 1.62$$

Therefore, since the $\Delta B/\Delta C_{3-2} \geq 1.0$ select alternative 3.

Example 12.5 demonstrates the process for using the modified benefit/cost ratio to analyze two alternatives.

Example 12.5

Calculate the modified benefit/cost ratio of alternative 3 compared to alternative 2 using the data from Examples 12.3 and 12.4.

Solution

$$\text{Modified } \Delta B/\Delta C = \frac{U_n - M_n}{C_n} \quad \text{or} \quad = \frac{B_n - M_n}{C_n} = \frac{(U_f - U_p) - (M_f - M_p)}{C_f - C_p}$$

$$= \frac{(\$4{,}500{,}000.00 - \$4{,}000{,}000.00) - (\$250{,}000.00 - \$200{,}000.00)}{\$4{,}144{,}000.00 - \$3{,}885{,}000.00}$$

$$= \frac{\$500{,}000.00 - \$50{,}000.00}{\$259{,}000.00}$$

$$= \frac{\$450{,}000.00}{\$259{,}000.00} = 1.74$$

Therefore, since the $\Delta B/\Delta C_{3-2} \geq 1.0$ select alternative 3.

Example 12.6

A city is planning to build a public swimming pool costing \$3,000,000.00 that will have a life of 25 years. The annual maintenance cost for the swimming pool will be \$1,000.00 per year increasing by \$200.00 per year. There will be no salvage value. An average benefit is assigned of \$1.50 per use and there will be an average of 75 users per hour, 8 hours per day, 316 days per year. Calculate the modified benefit/cost ratio using a rate of return of 6%.

Solution

First, calculate the total yearly benefit:

$$\text{Yearly benefit} = \frac{\$1.50}{\text{Use}} \times \frac{75 \text{ users}}{\text{Hour}} \times \frac{8 \text{ hours}}{\text{Day}} \times \frac{316 \text{ days}}{\text{Year}} = \$284,400.00$$

Second, calculate the equivalent uniform annual cost of the initial cost of the swimming pool:

$$\text{EUAC} = \$3,000,000.00(A/P,6,25) = \$3,000,000.00(0.07823) = \$224,690.00$$

Third, calculate the equivalent uniform annual cost of the gradient:

$$\text{EUAC}_G = \$200.00(A/G,6,25) = \$200.00(9.0722) = \$1,814.44$$

Fourth, add the equivalent uniform annual cost of the gradient to the yearly uniform series maintenance cost:

$$\text{EUAC}_M = \$1,000.00 + \$1,814.44 = \$2,814.44$$

Finally, calculate the modified benefit/cost ratio:

$$\text{Modified } B/C = \frac{U_n - M_n}{C_n} = \frac{\$284,400.00 - \$2,814.44}{\$224,690.00} = \frac{\$281,585.56}{\$224,690.00} = 1.25$$

Case Study 12.1 demonstrates calculating conventional and modified B/C ratios.

Case Study 12.1 Conventional and Modified *B/C* Ratios

A city is trying to determine whether it should install a new traffic signal at a major intersection. Table 12.1 provides the data collected on the costs and benefits of the existing traffic signal and the proposed traffic signal. Using an interest rate of 6% and a life of five years for both traffic signals, determine whether the proposed traffic signal should be installed by the city using both the conventional and modified incremental benefit/cost ratios.

TABLE 12.1
Data for Traffic Signal Alternatives

Costs and Benefits	Existing Traffic Signal	Proposed Traffic Signal
Initial cost	—	$68,670.00
Maintenance costs	$3,000.00 per month	$60,000 per year
User costs for accidents	$95,200.00 per year	$3,000.00 per year
Acceleration and deceleration costs (average cost to stop)	$0.025	$0.025
Number of vehicles	80,000 per day	40,000 per day

SOLUTION

$$\text{Cost to stop for existing signal} = \frac{\$0.025}{\text{Vehicle}} \times \frac{80,000 \text{ vehicles}}{\text{Day}} \times \frac{365 \text{ days}}{\text{Year}}$$

$$= \$730,000 \text{ per year}$$

$$\text{Cost to stop for proposed signal} = \frac{\$0.025}{\text{Vehicle}} \times \frac{40{,}000 \text{ vehicles}}{\text{Day}} \times \frac{365 \text{ days}}{\text{Year}}$$

$$= \$365{,}000 \text{ per year}$$

$$\text{EUAC}_E \text{ of existing signal initial cost} = 0$$

$$\text{EUAC}_P \text{of proposed signal initial cost} = \$68{,}670.00(A/P,6,5)$$

$$= \$68{,}670.00(0.23740)$$

$$= \$16{,}302.26$$

$$\text{Total benefits for existing signal} = \$730{,}000.00 + \$95{,}280.00 = \$825{,}280.00$$

$$\text{Total benefits for proposed signal} = \$365{,}000.00 + \$3{,}000.00 = \$368{,}000.00$$

$$\text{Total mainteance cost for existing signal} = \frac{\$3{,}000.00}{\text{Month}} \times \frac{12 \text{ months}}{\text{Year}} = \$36{,}000.00$$

$$\text{Total mainteance cost for proposed signal} = \$60{,}000.00$$

$$\text{Conventional } \Delta B/\Delta C = \frac{U_n}{C_n + M_n} = \frac{U_f - U_p}{(C_f - C_p) + (M_f - M_p)}$$

$$= \frac{\$825{,}280.00 - \$368{,}000.00}{(\$16{,}302.26 - 0) + (\$60{,}000.00 - \$36{,}000.00)}$$

$$= \frac{\$457{,}280.00}{\$16{,}302.26 + \$24{,}000.00} = \frac{\$457{,}280.00}{\$40{,}302.26} = 11.35$$

$$\text{Modified } \Delta B/\Delta C = \frac{U_n - M_n}{C_n} = \frac{(U_f - U_p) - (M_f - M_p)}{C_f - C_p}$$

$$= \frac{(\$825{,}280.00 - \$368{,}000.00) - (\$60{,}000.00 - \$36{,}000.00)}{(\$16{,}302.46.00 - 0)}$$

$$= \frac{\$427{,}280.00 - \$24{,}000.00}{\$16{,}302.26} = \frac{\$403{,}280.00}{\$16{,}302.26} = 24.74$$

Therefore, since the $\Delta B/\Delta C \geq 1.0$ for both methods the new traffic signal should be installed by the city.

12.4 SUMMARY

This chapter defined the terms related to benefit/cost ratio economic evaluations and explained the procedures for calculating benefit/cost ratios. It also described specific steps for performing a benefit/cost ratio economic evaluation. Example problems and a case study were included to illustrate the procedures for calculating benefit/cost ratios.

KEY TERMS

Benefit/cost ratios
Benefits
Conventional *B/C*
Benefit/cost ratio analysis
Disbenefits
Discounted profitability index
Modified *B/C*
Profit investment ratio
Profitability index
Value investment ratio

PROBLEMS

12.1 A state highway department is comparing two routes for a new roadway. The first route costs \$16,000,000.0 and will have annual benefits of \$2,000,000.00 to the local community. The second route costs \$24,000,000.00 and provides \$2,500,000.00 in benefits. Using an interest rate of 8% and a life of 20 years, calculate the benefit/cost ratio for each alternative.

12.2 A public works department is evaluating two alternatives for a wastewater treatment plant. The first alternative costs \$100,000,000.00, the operating and maintenance costs are \$350,000.00 per year, and the user cost is \$4,500,000.00 per year. The second alternative costs \$150,000,000.00, the operating and maintenance costs are \$550,000.00 per year, and the user cost is \$2,000,000.00 per year. If the interest rate is 5% and the life is 30 years, which alternative should be selected based on incremental benefit/ratio analysis?

12.3 A company is evaluating a new program for installing underground piping. The new program will have an additional cost of \$3,000,000.00 and it will save \$1,000,000.00 a year for 10 years. In order to fund the program, \$400,000.00 a year would be used from other programs. Using an interest rate of 6%, determine whether the program is justified using benefit/cost ratio analysis.

12.4 Three projects are being compared using benefit/cost ratio analysis. Table 12.2 contains the data for the projects under consideration. Determine the benefit/cost ratios for each project and the incremental benefit/cost ratios.

TABLE 12.2
Data for Project Options

Costs or Disbursements	Project Option 1	Project Option 2	Project Option 3
Initial cost	\$150,000.00	\$200,000.00	\$250,000.00
Annual benefit	\$375,000.00	\$460,000.00	\$500,000.00

12.5 Two highway routes are being proposed to replace an existing route. Table 12.3 provides the data for the existing route and the proposed route. Use an interest rate of 6% and a life of 15 years to calculate the conventional incremental benefit/cost ratio for route 1 compared to the existing route and for route 2 compared to the existing route.

TABLE 12.3
Data for Existing and Proposed Highway Routes

Costs or Disbursements	Existing Route	Proposed Route 1	Proposed Route 2
Initial cost	—	\$10,000.00	\$10,000.00
Operating and maintenance costs	\$25,000,000.00	\$27,000,000.00	\$24,000,000.00
User cost per year	\$20,000,000.00	\$16,500,000.00	\$19,500.00
Annual savings over existing route	—	\$481,000.00	\$481,000.00

12.6 Compare the two proposed routes in Problem 12.5 using the modified incremental benefit/cost ratio.

12.7 A biomedical engineer is analyzing the purchase of a new machine using benefit/cost ratios. The new machine costs \$100,000.00 and it will have a benefit of \$400,000.00 in 60 years. The interest rate is 1%. Calculate the benefit/cost ratio using equivalent uniform annual worth analysis.

12.8 A second alternative is being compared to the new machine in Problem 12.7. The second alternative costs \$100,000.00 and has a benefit of \$140,000.00 now. Calculate the benefit/cost ratio for the second machine.

12.9 Use the data in Problems 12.7 and 12.8 with an interest rate of 7% to determine whether to select alternative 1 or 2 using benefit/cost ratios.

12.10 A county is comparing an existing water slide to a replacement water slide. The data for the two alternatives under consideration are listed in Table 12.4. The interest rate is 6% and the life is 25 years. Compare the two water slide options using the conventional incremental benefit/cost ratio.

TABLE 12.4
Data for Water Slide Alternatives

Costs or Disbursements	Water Slide Alternative 1	Water Slide Alternative 2
Uses per year	20,000,000.00	80,000,000.00
Income per use	\$0.50	\$0.25
Maintenance cost	\$20,000,000.00	\$20,000,000.00
Capital cost	—	\$50,000,000.00

12.11 Compare the two water slides in Problem 12.10 using the modified incremental benefit/cost ratio.

12.12 A government is building a hydroelectric power plant costing \$240,000,000.00. It will be able to sell the power for \$9,000,000.00 per year increasing by \$20,000.00 per year. The local area will save \$6,500,000.00 per year due to reduced flood control measures. Another benefit to the local area is irrigation that will cost \$6,000,000.00 per year increasing by \$12,000.00 per year. A park is also part of the project and it will create user benefits of \$3,000,000.00 per year increasing by \$5,000.00 per year. Operating and maintenance costs will be \$2,000,000.00 per year increasing by \$10,000.00 per year. A disbenefit of the project is loss of production for the surrounding land of $1,000,000.00 a year increasing by $50,000.00 per year. The interest rate is 6% and the project has a life of 50 years. Calculate the benefit/cost ratio for this project.

12.13 Calculate the modified benefit/cost ratio for the hydroelectric power plant in Problem 12.12.

12.14 Calculate the conventional benefit/cost ratio for the hydroelectric power plant in Problem 12.12.

12.15 A county in New Mexico is analyzing whether to continue to use gravel on a road or to pave the road. The data for one mile of gravel road and the proposed paved road are listed in Table 12.5. The interest rate is 7%. If the roadway is 10 miles long and it has a life of 40 years, calculate the benefit/cost ratio for the existing gravel road and for the proposed paved road.

TABLE 12.5
Data for One Mile of Gravel and Paved Roads

Costs or Disbursements	Gravel Roadway	Proposed Paved Roadway
Capital investment	$2,000.00 every year	$38,955.46 once every 40 years
Maintenance cost	$1,887.44 per mile increasing by $50.00 per year	$13.45 per mile increasing by $5.00 per year
Road user cost	$6,500.00 increasing by $100.00 per year	$2,000.00 increasing by $25.00 per year

12.16 Calculate the modified incremental benefit/cost ratio for the data in Problem 12.15.

12.17 Calculate the conventional incremental benefit/cost ratio for the data in Problem 12.15

12.18 Two pollution testing processes are being evaluated for purchase by a county. The first process has an initial cost of $45,000.00, a salvage value of $9,000.00, annual maintenance costs of $10,000.00, and a life of 20 years. The second process under consideration costs $90,000.00, it will not have a salvage value, the yearly maintenance costs will be $7,000.00, and it will have a life of 30 years. If the interest rate is 12%, determine which process should be selected based on the individual benefit/cost ratios.

12.19 Calculate the conventional incremental benefit/cost ratio for the data in Problem 12.18.

12.20 Calculate the modified incremental benefit/cost ratio for the data in Problem 12.18.

13 Depreciation

This chapter introduces depreciation and the formulas for calculating the depreciation that is deducted from taxable income when calculating U.S. federal income taxes. This chapter provides definitions for depreciation and other related terms, discusses the components considered when calculating deprecation, and covers the methods for calculating the four most prevalent types of depreciation—production, straight line, declining balance, and sum-of-the-years digits.

13.1 DEFINITIONS FOR DEPRECIATION TERMS

Depreciation only exists as a means of reducing the income taxes businesses are obligated to pay to the U.S. federal and state governments since income taxes are paid on the net income of a firm minus business expenses and depreciation. Individuals owning rental property are able to depreciate it and deduct the yearly depreciation from the income earned on the rental property or other income since rental property is considered to be a business asset by the Internal Revenue Service (IRS).

Sections 13.1.1 through 13.1.6 provide definitions for terms related to depreciation.

13.1.1 Depreciation

Depreciation is defined by the IRS as a decrease in value of the assets of a business. The IRS allows businesses to calculate yearly depreciation for their assets and deduct the depreciation from their gross income. Depreciation is only calculated on business assets, not personal assets.

13.1.2 Deterioration

Assets deteriorate over time when they are being used for their intended purpose. When an asset wears out, or it no longer performs its intended function as well as when it was first purchased, then this is *deterioration* in economic terms.

13.1.3 Obsolescence

Assets still functional, but the function they perform could be performed in a more efficient manner by other assets, are considered to be obsolete. The *obsolescence* of existing assets may be caused by technological improvements in new products that supersede existing products.

13.1.4 Book Value

The *book value* of an asset represents the current value of the asset as determined by the IRS. Book value is the original cost of the asset minus the depreciation to date. Book values are calculated at the end of each year and they are the book value at the beginning of the year minus the depreciation for the year. The book value of an asset at a particular year is the original cost of the asset minus depreciation to date. Book values only apply when after-tax engineering economic analysis techniques are being used in evaluations.

13.1.5 Market Value

Market value is the amount of money realized if an asset is sold on the open market. The market value of an asset could be different from its book value since some assets sell for more or less on the open market than the book value determined by the techniques prescribed by the IRS for calculating book value.

13.1.6 Land Value

It is important to note that land is not a depreciable asset since the IRS considers land to not decrease in value. Therefore, before calculating the value of the depreciation for facilities, the value of the land is subtracted from the value of the facility.

13.2 COMPONENTS CONSIDERED WHEN CALCULATING DEPRECIATION

The three major components considered when calculating the yearly depreciation of assets are covered in Sections 13.2.1 through 13.2.3.

13.2.1 Allowable Depreciation

The IRS sets the type of assets subject to depreciation and this information is available in IRS publication 534. Form 4552—Depreciation and Amortization—is the IRS form that explains the process for depreciating assets.

13.2.2 Useful Life of Assets

The IRS determines the *useful life* of assets, which is the number of years over which an asset may be depreciated by a business. Examples of useful lives are listed in Table 13.1.

TABLE 13.1
Examples of the Useful Life of Assets as set by the IRS (2015)

Asset	Useful Life (Years)
Office furniture	10
Automobiles	3
Trucks	4
Apartments	40
Houses	18

13.2.3 Depreciation of Real Property

Houses and apartments are considered to be *real property* by the IRS and they may only be depreciated if they are business property. Tenants, or owners who occupy their homes, are not able to depreciate them since depreciation only applies to business property. No personal property may be depreciated for tax purposes.

13.3 METHODS FOR CALCULATING DEPRECIATION

There are four main methods for calculating depreciation on a yearly basis. Once depreciation is determined for a particular year using one of these four methods, it is deducted from the gross income of a firm, along with other business expenses, to arrive at the taxable income. Members of firms use one of the four methods for calculating depreciation and the selection of which method to use is based on the type of firm, the business performed, the amount of yearly income, the type of asset, and whether a firm wants to deduct more depreciation during the first few years of owning an asset while the asset is generating more income than in future years.

The following are the four methods for calculating depreciation discussed in Sections 13.3.1 through 13.3.4:

1. Production
2. Straight line
3. Declining balance
4. Sum-of-the-years digits (SOYD)

The IRS allows for one-time changes in the method used for calculating depreciation for certain types of assets such as real property. For additional information on this option, see IRS publication 534.

13.3.1 Production Depreciation

Production depreciation is based on the number of units of production (output) and the useful life of an asset in terms of production such as units, tons, feet, meters, cubic yards, cubic meters, hours, or mileage. Production depreciation is commonly used by construction or manufacturing firms, where the depreciable assets are heavy construction equipment or large processing equipment. Examples of the measurements used for determining the useful life of an asset when calculating production depreciation are the following:

- 10,000 hours for mobile equipment
- 200,000 hours for transport vehicles
- 1,000,000 yards (914,400 meters) of production for manufacturing or process plants

The formula for production depreciation per unit of production is Equation 13.1.

$$\text{Production depreciation} = \frac{\text{Number of units}}{\text{Life in units of production}} \times (\text{Cost} - \text{salvage value}) \tag{13.1}$$

In this equation, the salvage value is subtracted from the cost since an asset should not be depreciated below the salvage value. If an asset is depreciated below the salvage value and then sold for more than its depreciated book value, there will be recaptured depreciation on the amount realized on the sale above the depreciated value and this amount is subject to taxes.

Example 13.1 uses Equation 13.1 to calculate the production depreciation of an asset.

Example 13.1

A construction firm purchases a new bulldozer for \$160,000.00 that will be operated for six hours per day, five days per week, and 52 weeks per year. The bulldozer will have a salvage value of \$10,000.00 in 10 years. Determine the production depreciation.

Solution

Use Equation 13.1 to calculate the deprecation for the bulldozer:

$$\text{Yearly production} = \frac{6 \text{ hours}}{\text{Day}} \times \frac{5 \text{ days}}{\text{Week}} \times \frac{52 \text{ weeks}}{\text{Year}}$$

$$= 1{,}560 \text{ hours per year}$$

$$\text{Production depreciation} = \frac{\text{Number of hours}}{\text{Life in hours of production}} \times (\text{Cost} - \text{salvage value})$$

$$= \frac{1{,}560 \text{ hours}}{10{,}000 \text{ hours}} \times (\$160{,}000.00 - \$10{,}000.00)$$

$$= 0.156 \times \$150{,}000.00$$

$$= \$23{,}400.00$$

If the yearly use is not known for construction equipment, an average of 2,000 hours per year is used for the purpose of calculating yearly depreciation.

Another example calculating production depreciation is Example 13.2.

Example 13.2

A manufacturing plant purchased a new processing machine for $100,000,000.00. The processing machine will have a salvage value of $30,000,000.00 at the end of its useful life. The processing machine produces 50,000 yards (45,720 meters) per year and its useful life is 1,000,000 yards (914,400 meters). Using the yards of production for manufacturing plants, calculate the depreciation for the processing machine.

Solution

$$\text{Production depreciation (yards)} = \frac{\text{Number of yards}}{\text{Life in yards of production}} \times (\text{Cost} - \text{salvage value})$$

$$= \frac{50{,}000 \text{ yards}}{1{,}000{,}000 \text{ yards}} \times (\$100{,}000{,}000.00 - \$30{,}000{,}000.00)$$

$$= 0.05 \times \$70{,}000{,}000.00$$

$$= \$3{,}500{,}000.00$$

$$\text{Production depreciation (meters)} = \frac{\text{Number of meters}}{\text{Life in meters of production}} \times (\text{Cost} - \text{salvage value})$$

$$= \frac{45{,}720 \text{ meters}}{914{,}400 \text{ meters}} \times (\$100{,}000{,}000.00 - \$30{,}000{,}000.00)$$

$$= 0.05 \times \$70{,}000{,}000.00$$

$$= \$3{,}500{,}000.00$$

Example 13.3 also calculates production depreciation.

Example 13.3

The owner of a manufacturing plant bought a new conveying system. It cost $10,000,000 and it processes 1,900,000 units per year. It has a salvage value of $1,500,000.00 and its useful life is 12,000,000 units. Determine the production depreciation.

Solution

$$\text{Production depreciation} = \frac{\text{Number of units}}{\text{Life in units of production}} \times (\text{Cost} - \text{salvage value})$$

$$= \frac{1{,}900{,}000 \text{ units}}{12{,}000{,}000 \text{ units}} \times (\$10{,}000{,}000.00 - \$1{,}500{,}000.00)$$

$$= 0.1583 \times \$8{,}500{,}000.00 = \$1{,}345{,}550.00$$

13.3.2 Straight Line Depreciation

The most frequently used method for calculating yearly depreciation is *straight line depreciation*. When using straight line depreciation, the book value of an asset decreases linearly since the yearly depreciation is the same for every year the asset is being depreciated for tax purposes. The formula for straight line depreciation is Equation 13.2.

$$\text{Straight line depreciation} = \frac{P - F}{n} \tag{13.2}$$

where

P is the present value (cost)
F is the future value (salvage value)
n is the useful life

Note: The IRS may set the useful life in this equation; therefore, the useful life is subject to change and it should be verified by checking with the IRS when the useful life is being used for tax purposes.

The present value, which is the initial cost of an asset, includes the cost of the asset, taxes, delivery charges, installation costs, and any other costs incurred in purchasing and installing the equipment or asset. The salvage value is the amount realized by the sell of the asset or equipment at the end of its useful life, or whenever the asset is sold, minus any costs associated with removing or dismantling the asset.

When using straight line depreciation to calculate book values, the annual depreciation is multiplied by (m) the number of years of service and this is subtracted from the initial cost. The formula for the book value of an asset when using straight line depreciation is Equation 13.3.

$$BV_m = P - (m \times D) \tag{13.3}$$

where

BV_m is the book value after (m) years
P is the present value (cost)
m is the number of years of depreciation
D is the annual depreciation

The following examples demonstrate calculating straight line depreciation and book value.

Example 13.4

A manager of an electrical engineering firm purchases a new device costing \$160,000.00. The device will have a salvage value of \$10,000.00 after five years. What are the yearly straight line depreciation and the book value at the end of three years?

Solution

$$\text{Straight line depreciation} = \frac{P-F}{n} = \frac{\$160{,}000.00 - \$10{,}000.00}{5} = \$30{,}000.00 \text{ per year}$$

$$BV_m = P - (m \times D_m) = \$160{,}000.00 - (3 \times \$30{,}000.00) = \$70{,}000.00$$

Example 13.5

The purchasing manager for a nuclear power company purchases a new turbine for one of its generators for $1,100,000,000.00. The turbine has a useful life of 10 years and a salvage value of $400,000,000.00. What is the yearly straight line depreciation?

Solution

$$\text{Straight line depreciation} = \frac{P-F}{n} = \frac{\$1{,}100{,}000{,}000.00 - \$400{,}000{,}000.00}{10}$$

$$= \$70{,}000{,}000.00$$

Example 13.6

A civil engineer working for a design firm purchases a new 3D printer for $9,000.00. The IRS allows depreciating a printer over five years. At the end of five years, the printer will have a salvage value of $700.00. Calculate the straight line depreciation and the book value for each of the five years over which the printer will be depreciated for tax purposes.

Solution

$$\text{Straight line depreciation} = \frac{P-F}{n} = \frac{\$9{,}000.00 - \$700.00}{5} = \$1{,}660.00 \text{ per year}$$

Table 13.2 includes the straight line depreciation and the book values calculated using Equations 13.2 and 13.3 for all five years of the useful life of the printer.

TABLE 13.2
Straight Line Depreciation and Book Values for Example 13.6

Year	Book Value before Depreciation	Straight Line Depreciation	Book Value after Depreciation
1	$9,000.00	$1,660.00	$7,340.00
2	$7,340.00	$1,660.00	$5,680.00
3	$5,680.00	$1,660.00	$4,020.00
4	$4,020.00	$1,660.00	$2,360.00
5	$2,360.00	$1,660.00	$700.00

Note: The book value for the last year after depreciation should be equal to the salvage value.

Example 13.7

A tool and die company owns several models of lathes that have a total cost of \$15,000,000.00, a salvage value of \$2,750,000.00, and a life of five years. Develop a table of depreciation and book values for the lathes.

Solution

$$\text{Straight line depreciation} = \frac{P-F}{n} = \frac{\$15{,}000{,}000.00 - \$2{,}750{,}000.00}{5} = \$2{,}450{,}000.00$$

Table 13.3 lists the depreciation and book values for the lathes.

TABLE 13.3
Straight Line Depreciation and Book Values for Example 13.7

Year	Book Value before Depreciation	Straight Line Depreciation	Book Value after Depreciation
1	\$15,000,000.00	\$2,450,000.00	\$12,550,000.00
2	\$12,550,000.00	\$2,450,000.00	\$10,100,000.00
3	\$10,100,000.00	\$2,450,000.00	\$7,650,000.00
4	\$7,650,000.00	\$2,450,000.00	\$5,200,000.00
5	\$5,200,000.00	\$2,450,000.00	\$2,750,000.00

Note: The book value for the last year after depreciation should not be below the salvage value.

13.3.3 Declining Balance (Accelerated Cost Recovery System) Depreciation

In addition to the straight line depreciation, the IRS also allows *accelerated cost recovery depreciation* methods for depreciating assets and deducting them from taxable income since most assets actually lose more value during the first few years of service. In the *declining balance depreciation* method, the annual depreciation is 2.0, 1.5, or 1.25 times the current book value of the asset divided by the total economic life. If 2.0 times the current book value is used, then it is *double declining balance* depreciation and it doubles the amount of depreciation compared to using straight line depreciation. If 1.5 times the book value is used, then the depreciation would be 1.5 times the depreciation calculated using straight line depreciation.

As with straight line depreciation, when using declining balance depreciation, the book value is not allowed to be less than the salvage value. If additional depreciation is deducted from taxable income beyond this limit, then the IRS requires a firm to pay taxes on the amount of the salvage value minus the book value when the asset is sold and this is called recaptured depreciation. Since by depreciating the asset beyond its salvage value the company realizes a tax savings, then if the asset is sold at a value that has been recaptured, the business is required to pay taxes on the recaptured amount.

The IRS allows firms to use declining balance depreciation and then switch to straight line depreciation before the depreciation is less than the amount of depreciation available using straight line depreciation. A firm may only change from using declining balance depreciation to straight line depreciation once during the life of an asset. After the company changes depreciation methods to straight line, the amount of depreciation will be a set amount each year of its remaining life.

Equation 13.4 is the formula for calculating double declining balance depreciation:

$$\text{Double declining balance depreciation} = \frac{2}{n}BV \tag{13.4}$$

Equation 13.5 is the formula for calculating the book value when using declining balance depreciation and Equation 13.6 summarizes Equation 13.5:

$$BV_m = P - \text{Depreciation to date} \tag{13.5}$$

$$BV_m = BV_{m-1} - R\left(P - \sum D\right) \tag{13.6}$$

where

BV_m is the book value ($BV_m > F$)

R is the depreciation rate $\left(\frac{2}{n}, \frac{1.5}{n}, \text{ or } \frac{1.25}{n}\right)$

P is the present value

$\sum D$ is the sum of the depreciation to date

The types of declining balance depreciation the IRS allows are listed in Table 13.4, along with the category of assets for each type of depreciation.

TABLE 13.4
Internal Revenue Service Allowed Declining Balance Depreciation Rates

Type of Declining Balance Depreciation	Category of Asset
$\frac{2}{n}BV$	All new property except real estate
$\frac{1.5}{n}BV$	All used property and real estate
$\frac{1.25}{n}BV$	Used rental residential property

The following examples calculate double declining balance depreciation and book values.

Example 13.8

Calculate the depreciation and book values for the data provided in Example 13.6 using double declining balance depreciation.

Solution

$$\text{Double declining balance depreciation} = \frac{2}{n}(BV)$$

Table 13.5 shows the calculations for the depreciation and book values.

TABLE 13.5
Double Declining Balance Depreciation and Book Values for Example 13.8

Year	Book Value before Depreciation	Depreciation for Year	Book Value after Depreciation
1	\$9,000.00	$\frac{2}{5}(\$9,000.00-0)=\$3,600.00$	$\$9,000.00-\$3,600.00 = \$5,400.00$
2	\$5,400.00	$\frac{2}{5}(\$9,000.00-\$3,600.00) = \$2,160.00$	$\$5,400.00-\$2,160.00 = \$3,240.00$
3	\$3,240.00	$\frac{2}{5}(\$9,000.00-\$3,600.00) -\$2,160.00=\$1,296.00$	$\$3,240.00-\$1,296.00 = \$1,944.00$
4	\$1,944.00	$\frac{2}{5}(\$9,000.00-\$3,600.00 -\$2,160.00-\$1,296.00)=\$777.60$	$\$1,944.00-\$777.60 = \$1,166.40$
5	\$1,166.40	$\frac{2}{5}(\$9,000.00-\$3,600.00-\$2,160.00 -\$1,296.00-\$777.60)=\$466.56$	$\$1,166.40-\$466.40 = \$700.00$

The allowable depreciation needs to be calculated and compared to the total depreciation to ensure that an asset is only being depreciated down to the salvage value:

$$\text{Allowable depreciation} = P - F = \$9,000.00 - \$700.00 = \$8,300.00$$

$$\text{Total depreciation} = \$3,600.00 + \$2,160.00 + \$1,296.00 + \$777.60 + \$466.56 = \$8,300.00$$

Example 13.9

Calculate the depreciation and book values using 1.5 declining balance depreciation for an asset costing \$50,000.00, with a salvage value of \$5,000.00, and a life of four years.

Solution

$$1.5\text{ Declining balance depreciation} = \frac{1.5}{n}(BV)$$

Table 13.6 shows the calculations for the depreciation and book values.

TABLE 13.6
1.5 Declining Balance Depreciation and Book Value for Example 13.9

Year	Book Value before Depreciation	Depreciation for Year	Book Value after Depreciation
1	\$50,000.00	$\frac{1.5}{4}$(\$50,000.00 − 0) = \$18,750.00	\$50,000.00 − \$18,750.00 = \$31,250.00
2	\$31,250.00	$\frac{1.5}{4}$(\$50,000.00 − \$18,750.00) = \$11,718.75	\$31,250.00 − \$11,718.75 = \$19,531.25
3	\$19,531.25	$\frac{1.5}{4}$(\$50,000.00 − \$18,750.00 − \$11,718.75) = \$7,324.22	\$19,531.25 − \$7,324.22 = \$12,207.03
4	\$12,207.03	$\frac{1.5}{4}$(\$50,000.00 − \$18,750.00 − \$11,718.75 − \$7,324.22) = \$4,577.64	\$12,207.03 − \$4,557.64 = \$7,629.39

Verify that the total depreciation does not exceed the allowable depreciation:

$$\text{Allowable depreciation} = P - F = \$50{,}000.00 - \$5{,}000.00 = \$45{,}000.00$$

$$\begin{aligned}\text{Total depreciation} &= \$18{,}750.00 + \$11{,}718.75 + \$7{,}324.23 + \$4{,}577.64 \\ &= \$42{,}370.62\end{aligned}$$

Example 13.10

Calculate the depreciation and book values for the data from Example 13.7 using double declining balance depreciation.

Solution

Table 13.7 lists the deprecation and book values for Example 13.10.

$$\text{Double declining balance depreciation} = \frac{2}{n}\,(BV)$$

TABLE 13.7
Double Declining Balance Depreciation and Book Values for Example 13.10

Year	Book Value before Depreciation	Depreciation for Year	Book Value after Depreciation
1	\$15,000,000.00	$\frac{2}{5}$(\$15,000,000.00 − 0) = \$6,000,000.00	\$15,000,000.00 − \$6,000,000.00 = \$9,000,000.00
2	\$9,000,000.00	$\frac{2}{5}$(\$15,000,000.00 − \$6,000,000.00) = \$3,600,000.00	\$9,000,000.00 − \$3,600,000.00 = \$5,400,000.00
3	\$5,400,000.00	$\frac{2}{5}$(\$15,000,000.00 − \$6,000,000.00 − \$3,600,000.00) = \$2,160,000.00	\$5,400,000.00 − \$2,160,000.00 = \$3,240,000.00
4	\$3,240,000.00	$\frac{2}{5}$ (\$15,000,000.00 − \$6,000,000.00 − \$3,600,000.00 − \$2,160,000.00) = \$1,296,000.00	\$3,240,000.00 − \$1,296,000.00 = \$1,944,000.00 This value is below the salvage value, only allowed \$490,000.00 in depreciation
5	\$2,750,000.00	\$0	\$2,750,000.00

13.3.4 Sum-of-the-Years Digits Depreciation

Sum-of-the-years digits (SOYD) depreciation is another accelerated cost recovery method for calculating depreciation that allows for more depreciation in the earlier years during the life of an asset. Sum-of-the-years digits depreciation uses the sum of the years one through (n) and the total number of years of the life of the asset. The depreciation is calculated by multiplying the initial cost of the asset minus the salvage value ($P-F$) by the remaining number of years divided by the sum-of-the-years digits. The formulas for SOYD depreciation are Equations 13.7 and 13.8.

$$\text{SOYD depreciation} = \frac{\text{Years remaining}}{\text{Sum-of-the-years digits (SOYD)}}(P-F) \tag{13.7}$$

$$= \frac{m}{\frac{n(n+1)}{2}}(P-F) \tag{13.8}$$

where

m is the remaining years
n is the total number of years
$\text{SOYD} = \frac{n(n+1)}{2}$
P is the present value (initial cost)
F is the future value (salvage value)

The book value when using sum-of-the-years digits depreciation for any given year is calculated using Equation 13.9.

$$BV_m = P - \left[\frac{m\left(n - \left(\frac{m}{2}\right) + 0.5\right)}{\text{SOYD}} \right](P - F) \tag{13.9}$$

Example 13.11 uses Equation 13.6 to calculate the sum-of-the-years digits depreciation for two different time frames.

Example 13.11

Calculate the sum-of-the-years digits for n = 5 years and n = 10 years.

Solution

For n = 5 years,

$$\text{SOYD} = \frac{n(n+1)}{2} = \frac{5(5+1)}{2} = \frac{5(6)}{2} = \frac{30}{2} = 15$$

or

$$\text{SOYD} = 5 + 4 + 3 + 2 + 1 = 15$$

For n = 10 years,

$$\text{SOYD} = \frac{n(n+1)}{2} = \frac{10(10+1)}{2} = \frac{10(11)}{2} = \frac{110}{2} = 55$$

or

$$\text{SOYD} = 10 + 9 + 8 + 7 + 6 + 5 + 4 + 3 + 2 + 1 = 55$$

Example 13.12 uses Equation 13.6 to calculate sum-of-the-years digits depreciation and book values.

Example 13.12

Using the data from Example 13.5, calculate the sum-of-the-years digits depreciation and book values for five years.

Solution

Table 13.8 shows the calculations for depreciation and the book values for Example 13.12:

$$\text{SOYD depreciation} = \frac{\text{Years remaining}}{\text{Sum-of-the-years digits (SOYD)}}(P - F)$$

TABLE 13.8
Sum-of-the-Years Digits Depreciation and Book Values for Example 13.12

Year	Book Value before Depreciation	SOYD Depreciation for Year	Book Value after Depreciation
1	\$9,000.00	$\frac{5}{15}$(\$9,000.00 – \$700.00) = \$2,766.67	\$9,000.00 – \$2,766.67 = \$6,233.33
2	\$6,233.33	$\frac{4}{15}$(\$9,000.00 – \$700.00) = \$2,213.33	\$6,233.33 – \$2,213.33 = \$4020.00
3	\$4,020.00	$\frac{3}{15}$(\$9,000.00 – \$700.00) = \$1,660.00	\$4,020.00 – \$1,660.00 = \$2,360.00
4	\$2,360.00	$\frac{2}{15}$(\$9,000.00 – \$700.00) = \$1,106.67	\$2,360.00 – \$1,106.67 = \$1,253.33
5	\$1,253.33	$\frac{1}{15}$(\$9,000.00 – \$700.00) = \$553.33	\$1,253.33 – \$553.33 = \$700.00

Note: The book value for the last year after depreciation is \$700.00 and it should be equal to the salvage value, which is \$700.00.

Example 13.13 provides another problem solving for depreciation using sum-of-the-years digits digits depreciation.

Example 13.13

Calculate the depreciation and book values using sum-of-the-years digits deprecation for the first three years of the life of an asset if it has an initial cost of \$250,000.00, a salvage value of \$40,000.00, and a life of eight years using Equation 13.8.

Solution

Calculate the sum-of-the-years digits and then the sum-of-the-years digits depreciation:

$$\text{SOYD} = \frac{n(n+1)}{2} = \frac{8(8+1)}{2} = \frac{8(9)}{2} = \frac{72}{2} = 36$$

or

$$\text{SOYD} = 8+7+6+5+4+3+2+1 = 36$$

Calculate the sum-of-the-years digits depreciation for years one through three using Equation 13.8

$$\text{SOYD depreciation} = \frac{m}{\frac{n(n+1)}{2}}(P-F)$$

$$\text{SOYD depreciation year 1} = \frac{8}{36}(\$250{,}000.00 - \$40{,}000.00)$$

$$= 0.22222 \times \$210{,}000.00 = \$46{,}666.20$$

$$\text{SOYD depreciation year 2} = \frac{7}{36}(\$250{,}000.00 - \$40{,}000.00)$$
$$= 0.19444 \times \$210{,}000.00 = \$40{,}832.40$$

$$\text{SOYD depreciation year 3} = \frac{6}{36}(\$250{,}000.00 - \$40{,}000.00)$$
$$= 0.16667 \times \$210{,}000.00 = \$35{,}000.70$$

Use Equation 13.9 to calculate the book value for years one through three:

$$BV_1 = BV - \frac{8}{36}(P - F)$$
$$= \$250{,}000.00 - \frac{8}{36}(\$250{,}000.00 - \$40{,}000.00)$$
$$= \$250{,}000.00 - (0.2222 \times \$210{,}000.00)$$
$$= \$250{,}000.00 - \$46{,}666.20$$
$$= \$203{,}333.80$$

$$BV_2 = BV_1 - \frac{7}{36}(P - F)$$
$$= \$203{,}333.80 - \frac{7}{36}(\$210{,}000.00)$$
$$= \$203{,}333.80 - (0.19444 \times \$210{,}000.00)$$
$$= \$203{,}333.80 - \$40{,}832.40$$
$$= \$162{,}501.40$$

$$BV_3 = BV_2 - \frac{6}{36}(P - F)$$
$$= \$162{,}501.40 - \frac{6}{36}(\$210{,}000.00)$$
$$= \$162{,}501.40 - (0.16667 \times \$210{,}000.00)$$
$$= \$162{,}501.40 - \$35{,}000.70$$
$$= \$127{,}500.70$$

Example 13.14

Calculate the sum-of-the-years digits depreciation and book values using the data in Problem 13.7.

Solution

Table 13.9 contains the depreciation and book values for Example 13.14:

$$\text{SOYD depreciation} = \frac{\text{Years remaining}}{\text{Sum-of-the-years digits (SOYD)}}(P - F)$$

TABLE 13.9
Sum-of-the-Years Digits Depreciation and Book Values for Example 13.14

Year	Book Value before Depreciation	SOYD Depreciation for Year	Book Value after Depreciation
1	\$15,000,000.00	$\frac{5}{15}$(\$15,000,000.00 – \$2,750,000.00) =4,083,333.33	\$15,000,000.00 – \$4,083,333.33 = \$10,916,666.67
2	\$10,916,666.67	$\frac{4}{15}$(\$15,000,000.00 – \$2,750,000.00) = \$3,266,667.67	\$10,916,666.67 – \$3,266,667.67 = \$7,650,000.00
3	\$7,650,000.00	$\frac{3}{15}$(\$15,000,000.00 – \$2,750,000.00) = \$2,450,000.00	\$7,650,000.00 – \$2,450,000.00 = \$5,200,000.00
4	\$5,200,000.00	$\frac{2}{15}$(\$15,000,000.00 – \$2,750,000.00) = \$1,633,333.33	\$5,200,000.00 – \$1,633,333.33 = \$3,566,666.67
5	\$3,566,666.67	$\frac{1}{15}$(\$15,000,000.00 – \$2,750,000.00) = \$816,666.67	\$3,566,666.67 – \$816,666.67 = \$2,750,000.00

Appendix C provides spreadsheet formulas for solving for all four types of depreciation.

Chapter 14 explains incorporating depreciation into taxes and after-tax rate of return analysis.

13.4 SUMMARY

This chapter introduced the concept of depreciation and four methods for calculating depreciation for tax purposes—production, straight line, declining balance, and sum-of-the-years digits. This chapter also provided definitions for depreciation and related terms and discussed the components considered when calculating deprecation. The last part of this chapter covered the methods for calculating the four different types of depreciation and provided examples demonstrating how to calculate each type of depreciation.

KEY TERMS

Accelerated cost recovery depreciation
Book value
Declining balance depreciation
Depreciation
Deterioration
Double declining balance depreciation
Market value

Obsolescence
Production depreciation
Real property
Straight line depreciation
Sum-of-the-years digits depreciation
Useful life

PROBLEMS

13.1 The owner of a construction company purchased an excavator for $300,000.00 and it will have a salvage value of $80,000.00. The excavator moves 90,000 tons of gravel in the first year and it has a useful life of 1,000,000 tons. What is the depreciation using the production method?

13.2 A manager of a processing plant purchased a new processing machine for $110,000,000.00. The equipment will have a salvage value of $10,000,000.00 and the useful life of the processing machine is 150,000,000 units. If the processing machine produces 15,000,000 units in the first year, what is the deprecation using the production method?

13.3 The owner of a construction company purchases a scraper for $6,000,000,000.00. The scraper will not have a salvage value. Its useful life is 20,000,000 tons. If it moves 46,000 tons in the first year, what is the depreciation using the production method?

13.4 An agricultural engineer purchases a truck that will be used for company business. The truck cost $52,000.00 and will have a salvage value of $8,000.00. The life of the truck is 10,000 hours. If the truck is used 2,700 hours per year, what is the production depreciation?

13.5 A petroleum engineer purchases a new liquefied natural gas exchanger for $80,000,000.00. It will be used for 20 years and be sold for a salvage value of $12,000,000.00. Calculate the straight line depreciation for the exchanger.

13.6 Calculate the double declining balance depreciation and book values for the first three years for the exchanger in Problem 13.5 using the format in Table 13.5.

13.7 Calculate the sum-of-the-years digits depreciation and book values for the data in Problem 13.5 for 3 years using the format in Table 13.8.

13.8 Calculate the double declining balance depreciation and book values for an asset costing $80,000.00 with a salvage value of $10,000.00 that will be depreciated over 10 years.

13.9 Calculate the first four years of double declining balance depreciation for an asset costing $80,000.00 that has a salvage value at year 7 of $24,000.00.

13.10 A new machine has an initial cost of $23,000,000.00, a salvage value of $2,000,000.00, and a life of 12 years. Determine which year the depreciation charge for straight line depreciation will exceed the depreciation for double declining balance depreciation.

13.11 A biomedical engineer purchases an electron microscope for $4,000,000.00. The life of the electron microscope is eight years and the salvage value is 10% of the purchase price. Using straight line depreciation, determine the salvage value, annual depreciation, and the book value at year four.

13.12 A $1,000,000.00 mechanical testing machine is being used by a firm and depreciated over five years. The testing machine will be sold at the end of five years for $240,000.00. Calculate the annual depreciation using straight line depreciation and the book value at the end of year four.

13.13 Calculate the depreciation and book value for the first year for the data in Problem 13.12 using double declining balance depreciation.

13.14 Calculate the depreciation for the data in Problem 13.12 using sum-of-the-years digits depreciation.

13.15 A civil engineering firm has a soil testing lab that purchases a new piece of testing equipment costing $250,000.00 that will not have a salvage value. The equipment will last five years. Calculate the book value for each of the five years using straight line depreciation.

13.16 Calculate the book value for each of the five years for the data in Problem 13.15 using double declining balance depreciation.

13.17 Calculate the sum-of-the-years digits depreciation for each of the five years for the data in Problem 13.16.

13.18 An asset costing $250,000.00 with a salvage value of $80,000.00 is being depreciated over three years. Calculate the depreciation for the third year using straight line, double declining balance, and sum-of-the-years digits depreciation.

13.19 A materials engineer is depreciating an asset costing $175,000.00 that will have a salvage value of $35,000.00 at the end of four years. Calculate the 1.5 declining balance depreciation.

13.20 An electrical engineer purchases a new numerical controller. It costs $455,000.00 and the salvage value will be $65,000.00 at the end of four years. Calculate the sum-of-the-years digits book value for all four years that the firm depreciates the controller.

14 Taxes and After-Tax Economic Analysis

The subject of engineering economic analysis would not be complete without the inclusion of a discussion on the effects of U.S. federal and state taxes on the time value of money, rates of return, and economic decisions. Decisions made by engineers and managers are affected by tax implications; therefore, taxes are factored into engineering economic analysis problems to determine the after-tax rate of return on the investments entered into by corporate entities and individuals.

This chapter focuses on U.S. federal income taxes, after-tax cash flows and their effect on corporate economic decisions, and the processes used for calculating after-tax rates of return. This chapter also includes information on individual taxes to illustrate using engineering economic analysis techniques when calculating individual taxes. The last part of the chapter discusses mortgages and briefly introduces the different types of mortgages available from lending institutions.

14.1 CORPORATE TAXES

This section introduces terms related to corporate tax calculations and formulas for calculating U.S. federal *corporate income taxes.*

14.1.1 Gross and Taxable Income

The U.S. federal income tax system is administered by the *Internal Revenue Service* (IRS) and this is the administrative agency responsible for developing tax guidelines, forms, and publications and administering tax laws. The IRS also sets federal corporate income tax rates and the deductions firms may subtract from their income to reduce their yearly taxes. The tax rates for corporations and individuals frequently change; therefore, IRS publications should be consulted to ensure the appropriateness of the income tax rates being used for a particular year.

The amount of federal corporate income taxes paid by businesses is based on the taxable income of a business and the corporate tax rate for the appropriate year. The starting point for determining taxable corporate income is the *adjusted gross income*, which includes all of the revenue earned by a firm during the year. The IRS allows firms to subtract all ordinary and necessary expenditures to conduct business from their adjusted gross income. These expenditures do not include capital expenditures because capital expenditures are recovered through depreciation, the second item subtracted by businesses from adjusted gross income. Equation 14.1 is the formula for calculating the *taxable income* of a business:

$$\text{Taxable business income} = \text{Adjusted gross income} - \text{All expenditures (except capital expenditures)} - \text{Depreciation} \qquad (14.1)$$

where

Adjusted gross income is all income from revenue generating sources
Expenditures are all costs incurred while transacting business
Depreciation is calculated using one of the four IRS approved methods

Firms could also have an operating loss rather than income and an operating loss is a net loss rather than a net profit.

14.1.2 Capital Gains and Losses

In addition to income from revenue-generating sources, firms may also generate *capital gains*, which result when an asset or real property is sold for an amount higher than its current book value. The amount of money realized on the sale of the asset or real property is a capital gain. The formula for calculating a capital gain is Equation 14.2.

$$\text{Capital gain} = \text{Selling price} - \text{Adjusted basis of purchase price (original price} + \text{cost of all renovations)} \quad (14.2)$$

There are two types of capital gains rates and they are based on the length of time a firm or individual holds onto an asset. There is one set of rates for assets held less than one year and a second set of rates for assets held more than one year. The capital gains tax rates for 2015 were the following:

> The tax rate on most net capital gains is no higher than 15% for most taxpayers. Some or all of the net capital gain may be taxed at 0% if someone is in the 10% or 15% ordinary income tax brackets. However, a 20% tax rate on net capital gain applies to the extent that a taxpayer's taxable income exceeds the thresholds set for the 39.6% ordinary tax rate ($413,200.00 for single, $464,850.00 for married filing jointly or qualifying widow[er], $439,000.00 for head of household, and $232,425.00 for married filing separately).
>
> There are a few other exceptions where capital gains may be taxed at rates greater than 15%:
>
> 1. The taxable part of a gain from selling section 1202 qualified small business stock is taxed at a maximum 28% rate.
> 2. Net capital gains from selling collectibles (such as coins or art) are taxed at a maximum 28% rate.
> 3. The portion of any unrecaptured section 1250 gain from selling section 1250 real property is taxed at a maximum 25% rate.
>
> **Note**: Net short-term capital gains (assets held less than one year) are subject to taxation as ordinary income at graduated tax rates (Internal Revenue Service, January 4, 2016).

A company could also have a *capital loss* if an asset or real property is sold for an amount lower than the book value at the time of the sale. Equation 14.3 is the formula for calculating a capital loss:

$$\text{Capital loss} = \text{Book value (cost} - \text{depreciation to date)} - \text{Selling price} \quad (14.3)$$

Capital losses could be carried forward for three years or carried backward for five years using amended tax returns for previous years. Capital losses are carried forward or backward when there is insufficient income to deduct the loss against in the year the loss is incurred by a firm or an individual taxpayer.

The IRS also sets guidelines on whether a business is a business or a hobby. "The IRS presumes that an activity is carried on for profit if it makes a profit during at least three of the last five tax years, including the current year—at least two of the last seven years for activities that consist primarily of breeding, showing, training or racing horses" (Internal Revenue Service, August 17, 2012). If it is a hobby, then an individual is not able to deduct business expenses or depreciate assets.

14.1.3 Recaptured Depreciation

As was mentioned in Sections 13.3.1 and 13.3.3, taxes are also charged on *recaptured depreciation*. Recaptured depreciation occurs when assets or real property are sold for more than the current book value. The amount realized over and above the current book value is the amount taxed by the IRS at the ordinary income tax rate up to the original cost of the asset. If an asset is sold for more than its original cost, the amount realized above the original cost is taxed at the appropriate capital gains rate. Equation 14.4 is the formula for calculating recaptured depreciation:

$$\text{Recaptured depreciation} = \text{Selling price of asset} - \text{Book value (original cost} - \text{depreciation to date)} \quad (14.4)$$

Example 14.1 demonstrates calculating recaptured depreciation for tax purposes.

Example 14.1

The owner of a construction firm purchases a piece of heavy construction equipment for \$500,000.00. The equipment is depreciated over three years. At the end of three years the book value is \$200,000.00 and the equipment is sold for \$450,000.00. (1) What is the amount of the recaptured depreciation taxed by the IRS and (2) what would be the amount taxed as recaptured taxes at the appropriate capital gains rate if the equipment is sold for \$650,000.00?

Solution

1. The recaptured depreciation is calculated using Equation 14.4:

$$\begin{aligned}\text{Recaptured depreciation} &= \text{Selling price of asset} - \text{Book value (original cost} - \text{depreciation to date)} \\ &= \$450{,}000.00 - \$200{,}000.00 = \$250{,}000.00\end{aligned}$$

2. The recaptured depreciation and capital gains taxes are calculated using Equations 14.4 and 14.2:

$$\begin{aligned}\text{Recaptured depreciation} &= \$500{,}000.00 - \$200{,}000.00 \\ &= \$300{,}000.00 \text{ taxed at the ordinary tax rate}\end{aligned}$$

$$\begin{aligned}\text{Capital gains} &= \$600{,}000.00 - \$500{,}000.00 \\ &= \$100{,}000.00 \text{ taxed at the appropriate capital gains rate}\end{aligned}$$

14.1.4 Taxes

The amount a firm pays in *federal corporate income taxes* depends on the amount of income the firm generates and the amount they are able to deduct as business expenses and depreciation. Equation 14.5 is the formula for calculating federal corporate income taxes.

$$\text{Federal corporate income taxes} = (\text{Gross income} - \text{Business expenses} - \text{Depreciation}) \times \text{Tax rate} \quad (14.5)$$

Managers try to reduce corporate taxes by maximizing their deductions for business expenses and depreciation. A firm with a federal income tax rate of 38% before deducting any business expenses and depreciation could reduce their federal income taxes to a lower income tax rate after deducting business expenses and depreciation. Therefore, some firms are not paying the federal income tax rates published by the IRS for their income level.

The *effective federal corporate income tax rate* is the amount of federal taxes divided by the total revenue, which is Equation 14.6.

$$\text{Effective federal corporate income tax rate} = \frac{\text{Federal income taxes}}{\text{Adjusted gross income}} \quad (14.6)$$

The federal corporate income tax rates for 2015 are shown in Table 14.1.

TABLE 14.1
Federal Income Tax Rate Schedule for Corporations for 2015

If Taxable Income Is Over	But Not Over	Tax Is	Of the Amount Over
$0	$50,000.00	15%	$0
$50,000.00	$75,000.00	$7,500.00 + 25%	$50,000.00
$75,000.00	$100,000.00	$13,750.00 + 34%	$75,000.00
$100,000.00	$335,000.00	$22,250.00 + 39%	$100,000.00
$335,000.00	$10,000,000.00	$113,900.00 + 34%	$335,000.00
$10,000,000.00	$15,000,000.00	$3,400,000.00 + 35%	$10,000,000.00
$15,000,000.00	$18,333,333.00	$5,150,000.00 + 38%	$15,000,000.00
$18,333,333.00	—	35%	0

Source: Internal Revenue Service, Form 1120 U.S. Corporate Income Tax Return—Instructions, U.S. Government Printing Office, Washington, DC, January 21, 2016, https://www.irs.gov/pub/irs-pdf/i1120.pdf, accessed on January 31, 2016.

In addition to federal income taxes, firms and individuals may also have to pay state and city taxes. When all of the tax rates for federal, state, and city taxes are combined, this is the *effective tax rate*.

Example 14.2 demonstrates calculating federal corporate income taxes.

Example 14.2

How much would a firm pay in federal corporate income taxes for 2015 if their adjusted gross income was $475,000.00 and they had no deductions for business expenses or depreciation? What is the effective federal corporate income tax rate for this firm?

Solution

Use Table 14.1 to determine the federal corporate income taxes for Example 14.2.

For $335,000.00 the tax rate is $113,900.00. For the amount over $335,000.00, which is $475,000.00 minus $335,000.00 or $140,000.00, the tax rate is 34%. Therefore, the total tax is

$$\begin{aligned}\text{Federal corporate income taxes} &= \$113{,}900.00 + 0.34(\$140{,}000.00)\\ &= \$113{,}900.00 + \$47{,}600.00\\ &= \$161{,}500.00\end{aligned}$$

The effective federal corporate income tax rate is calculated using Equation 14.6:

$$\text{Effective federal corporate income tax rate} = \frac{\text{Federal income taxes}}{\text{Adjusted gross income}}$$

$$= \frac{\$161{,}500.00}{\$475{,}000.00} = 0.34 \text{ or } 34\%$$

The effective corporate income tax rate is the same as the corporate income tax rate for this problem because the firm does not have any deductions for business expenses or depreciation.

If a firm has business expenses and depreciation that they are able to deduct from their adjusted gross income then it reduces their taxable income.

Example 14.3 illustrates reducing federal corporate income taxes by deducting business expenses and depreciation.

Example 14.3

A firm has an adjusted gross income of $475,000.00, business expenses of $121,000.00, and depreciation of $65,000.00. What would be the federal corporate income taxes for 2015 and the effective federal corporate income tax rate?

Solution

First, calculate the taxable business income using Equation 14.1:

$$\begin{aligned}\text{Taxable business income} &= \text{Adjusted gross income} - \text{All expenditures (except capital expenditures)} - \text{Depreciation} \\ &= \$475{,}000.00 - \$121{,}000.00 - \$65{,}000.00 \\ &= \$289{,}000.00\end{aligned}$$

Second, calculate the federal corporate income taxes using Table 14.1:

$22,250.00 on the first $100,000.00

39% on the amount over $100,000.00

Amount over $100,000.00 = $289,000.00 − $100,000.00 = $189,000.00

$$\begin{aligned}\text{Federal corporate income taxes} &= \$22{,}250.00 + 0.39(\$189{,}000.00) \\ &= \$22{,}250.00 + \$73{,}710.00 \\ &= \$95{,}960.00\end{aligned}$$

Third, calculate the effective federal corporate income tax rate using Equation 14.6:

$$\text{Effective federal corporate income tax rate} = \frac{\text{Federal income taxes}}{\text{Adjusted gross income}}$$

$$= \frac{\$95{,}960.00}{\$475{,}000.00} = 0.202 = 20.2\%$$

14.2 AFTER-TAX CASH FLOW

Once firms are able to calculate their taxes, then the taxes are incorporated into engineering economic analysis evaluations to determine the *after-tax cash flow* (ATCF). After-tax cash flows are used when calculating the after-tax rate of return, net present worth, future worth, and equivalent uniform annual worth. The calculations for solving for these values use the same equations introduced in the previous chapters, but taxes are subtracted from the revenue before any of the formulas are applied to calculate these values.

Up until this chapter, all of the cash flow problems have been *before-tax cash flows* (BTCFs). At this point, all of the cash flows become ATCFs since in the United States ATCFs are the accurate method for determining engineering economic values. The principal elements of ATCFs are the following:

1. Before-tax cash flows
2. Depreciation
3. Recaptured depreciation
4. Capital gains
5. Taxable income (adjusted gross income – business expenses – depreciation)
6. After-tax cash flow (before tax cash flow – taxes)

Table 14.2 provides a spreadsheet format for tabulating the elements required for developing ATCFs.

TABLE 14.2
Before-Tax and After-Tax Cash Flow Format

Year	Before-Tax Cash Flow (BTCF)	Depreciation	Taxable Income (BTCF–Depreciation)	Income Taxes (Taxable Income × Tax Rate)	After-Tax Cash Flow (BTCF–Income Taxes)

Example 14.4 demonstrates solving for the after-tax rate of return.

Example 14.4

The owner of a chemical processing firm pays $300,000.00 for a new piece of machinery. The firm is able to earn $280,000.00 from the machinery in years one and two. The machinery will have a salvage value of $75,000.00 at year five. Calculate the after-tax rate of return for years one and two using straight line depreciation and a corporate tax rate of 38%. Figure 14.1 is the before-tax cash flow diagram for the chemical processing machinery.

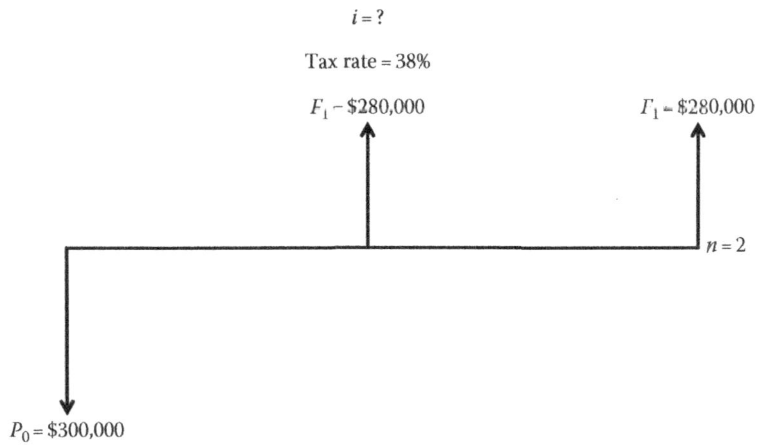

FIGURE 14.1 Before-tax cash flow diagram for the processing machinery in Example 14.4.

Solution

First, calculate the straight line depreciation for years one and two using Equation 13.2:

$$\text{Straight line depreciation} = \frac{P-F}{n} = \frac{\$300{,}000.00 - \$75{,}000.00}{5} = \frac{\$225{,}000.00}{5}$$

$$= \$45{,}000.00$$

Second, develop Table 14.3 using the format in Table 14.2 with before-tax cash flow, depreciation, taxable income, income taxes, and the after-tax cash flow for the first two years.

TABLE 14.3
Before-Tax and After-Tax Cash Flows for Example 14.4

Year	BTCF	Straight Line Depreciation	Taxable Income (BTCF − Depreciation)	Income Taxes (Taxable Income × Tax Rate)	ATCF (BTCF−Income Taxes)
0	−$300,000.00	—	—	—	−$300,000.00
1	+$280,000.00	$45,000.00	$280,000.00−$45,000.00 = $235,000.00	$235,000.00 × 0.38 = $89,300.00	$280,000.00−$89,300.00 = $190,700.00
2	+$280,000.00	$45,000.00	$280,000.00−$45,000.00 = $235,000.00	$235,000.00 × 0.38 = $89,300.00	$280,000.00−$89,300.00 = $190,700.00

Third, draw the after-tax cash flow diagram, as shown in Figure 14.2.

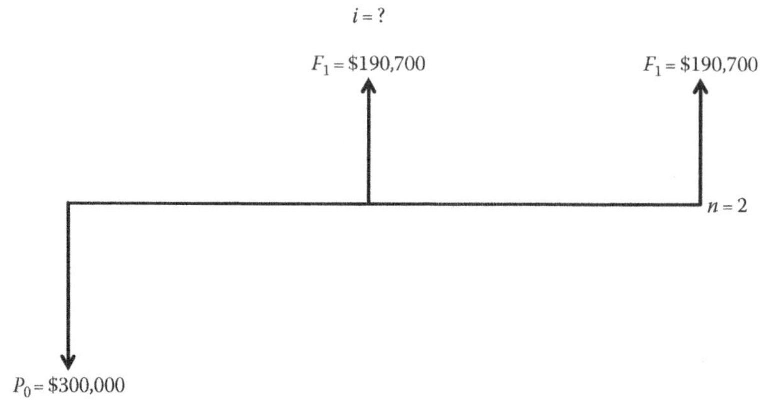

FIGURE 14.2 After-tax cash flow diagram for the processing machinery in Example 14.4.

Fourth, develop the net present worth equation for the after-tax cash flow:

$$NPW = P_0 + A(P/A, i, n)$$

$$= -\$300{,}000.00 + \$190{,}700.00(P/A, i, 2)$$

Fifth, use trial and error in the net present worth equation to calculate the unknown rate of return.

Try 15%:

$$= -\$300{,}000.00 + \$190{,}700.00(1.6257) = -\$300{,}000.00 + \$310{,}020.99$$

$$= \$10{,}020.99$$

Try 20%:

$$= -\$300{,}000.00 + \$190{,}700.00(1.5277) = -\$300{,}000.00 + \$291{,}332.39$$

$$= -\$8{,}667.61$$

Sixth, use interpolation and Table 14.4 to calculate the rate of return:

TABLE 14.4
Table for Developing Interpolation Problem for Unknown Rate of Return with n = 2 Years for Example 14.4

	ROR	Net Present Worth	
	15%	10,020.99	*a*
d	Unknown ROR	0	*b*
	20%	−8,667.61	

$$\text{ROR} = c + \left(\frac{a}{b}\right)d$$

$$\text{ROR} = 15 + \left(\frac{10{,}020.99 - 0}{10{,}020.99 - (-8{,}667.61)}\right) \times (20 - 15)$$

$$= 15 + \left(\frac{10{,}020.99}{18{,}688.60}\right) \times 5$$

$$= 15 + (0.536209 \times 5)$$

$$= 15 + 2.6810$$

$$= 17.68\%$$

Example 14.5 calculates the after-tax equivalent uniform annual worth.

Example 14.5

An electrical engineer purchases a new transformer for $250,000.00. The transformer will have a salvage value of $50,000.00 in four years. With the new transformer, the firm will increase its yearly profits by $175,000.00 for four years. The firm uses sum-of-the-years digits depreciation, the interest rate is 7%, and the corporate income tax rate is 34%. Determine the after-tax cash flow and the equivalent uniform annual worth? Figure 14.3 is the before-tax cash flow diagram for the new transformer.

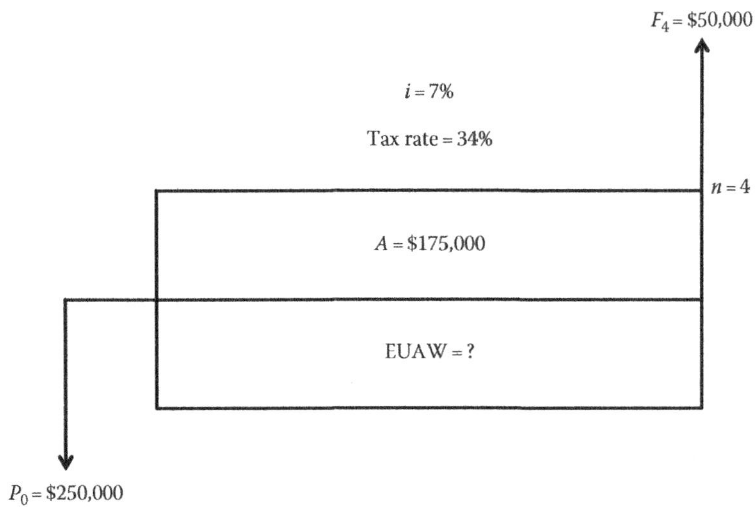

FIGURE 14.3 Before-tax cash flow diagram for the new transformers in Example 14.5.

Solution

First, solve for the sum-of-the-years digits depreciation using Equation 13.6 for the four years:

$$\text{SOYD depreciation} = \frac{m}{\frac{n(n+1)}{2}}(P-F) = \frac{m}{\text{SOYD}}(P-F)$$

$$\text{SOYD} = 4+3+2+1 = 10$$

$$\text{SOYD depreciation year } 1 = \frac{4}{10}(\$250{,}000.00 - \$50{,}000.00)$$
$$= 0.40 \times \$200{,}000.00 = \$80{,}000.00$$

$$\text{SOYD depreciation year } 2 = \frac{3}{10}(\$250{,}000.00 - \$50{,}000.00)$$
$$= 0.30 \times \$200{,}000.00 = \$60{,}000.00$$

$$\text{SOYD depreciation year } 3 = \frac{2}{10}(\$250{,}000.00 - \$50{,}000.00)$$
$$= 0.2 \times \$200{,}000.00 = \$40{,}000.00$$

$$\text{SOYD depreciation year } 4 = \frac{1}{10}(\$250{,}000.00 - \$50{,}000.00)$$
$$= 0.1 \times \$200{,}000.00 = \$20{,}000.00$$

Second, develop Table 14.5 using the format in Table 14.2 with before-tax cash flow, depreciation, taxable income, taxes, and the after-tax cash flow.

TABLE 14.5
Before-Tax and After-Tax Cash Flows for Example 14.5

Year	BTCF	SOYD Depreciation	Taxable Income (BTCF−Depreciation)	Income Taxes (Taxable Income × Tax Rate)	ATCF (BTCF−Income Taxes)
0	−\$250,000.00	—	—	—	−\$250,000.00
1	+\$175,000.00	\$80,000.00	\$175,000.00 − \$80,000.00 = \$95,000.00	\$95,000.00 × 0.34 = \$32,300.00	\$175,000.00 − \$32,300.00 = \$142,700.00
2	+\$175,000.00	\$60,000.00	\$175,000.00 − \$60,000.00 = \$115,000.00	\$115,000.00 × 0.34 = \$39,100.00	\$175,000.00 − \$39,100.00 = \$135,900.00
3	+175,000.00	\$40,000.00	\$175,000.00 − \$40,000.00 = \$135,000.00	\$135,000.00 × 0.34 = \$45,900.00	\$175,000.00 − \$45,900.00 = \$129,100.00
4	+175,000.00	\$20,000.00	\$175,000.00 − \$20,000.00 = \$155,000.00	\$155,000.00 × 0.34 = \$52,700.00	\$175,000.00 − \$52,700.00 = \$122,300.00
4	+50,000.00				+ \$50,000.00

Third, draw the after-tax cash flow diagram, which is shown in Figure 14.4.

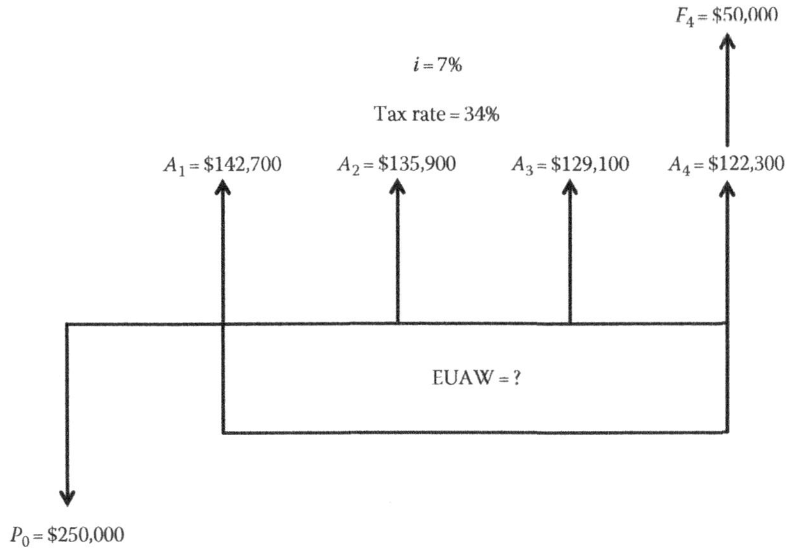

FIGURE 14.4 After-tax cash flow diagram for the new transformers in Example 14.5.

Fourth, develop the equivalent uniform annual worth equation using the after-tax cash flow and calculate the equivalent uniform annual worth:

$$\begin{aligned}
\text{EUAW} &= P_0(A/P,i,n) + \sum A + F_4(A/F,i,n) \\
&= -\$250{,}000.00(A/P,7,4) + \$142{,}700.00 + \$135{,}900.00 + \$129{,}100.00 \\
&\quad + \$122{,}300.00 + \$50{,}000.00(A/F,7,4) \\
&= -\$250{,}000.00\ (0.29523) + \$142{,}700.00 + \$135{,}900.00 + \$129{,}100.00 \\
&\quad + \$122{,}300.00 + \$50{,}000.00(0.22523) \\
&= -\$73{,}807.50 + \$142{,}700.00 + \$135{,}900.00 + \$129{,}100.00 \\
&\quad + \$122{,}300.00 + \$11{,}261.50 \\
&= \$467{,}454.00
\end{aligned}$$

14.3 INDIVIDUAL TAXES

The IRS administers *individual income taxes* and determines the income tax rates each year. This section covers individual taxable income, exemptions, standard deductions, itemized deductions, and individual income tax rates.

14.3.1 Individual Taxable Income

The formula for individual taxable income is Equation 14.7.

$$\text{Individual taxable income} = \text{Adjusted gross income} - \text{Deductions for exemptions} \\ - \text{Standard deduction or itemized deduction} \quad (14.7)$$

The adjusted gross income for tax purposes includes any type of income earned during the year. The following are the main elements of adjusted gross income:

- Wages
- Taxable interest
- Ordinary dividends from stocks
- Qualified dividends from stocks
- Taxable refunds, credits, and offsets of state and local taxes
- Alimony received
- Business income or loss
- Capital gains or losses
- Individual retirement account (IRA) distribution
- Pensions and annuities
- Rental real estate, royalties, partnerships, S corporations (a S corporation is a special type of corporation created through the IRS. An eligible domestic corporation may avoid double taxation—once to the corporation and again to the shareholders—by electing to be a S corporation)
- Farm income or loss
- Social security benefits (taxable amount)
- Other income

The adjusted gross income allows for certain types of deductions to income such as the following:

- Education expenses
- Certain business expenses of reservists, performing artists, and fee-based government officials
- Health savings account deduction
- Deductible part of self-employment tax
- Self-employed SEP (simplified employee pension plan), SIMPLE (a method for small businesses to contribute to their employees and their retirement plants), and qualified plans
- Self-employed health insurance deduction
- Penalty for early withdrawal of savings
- Alimony paid
- Individual retirement account (IRA) deduction
- Student loan interest deduction
- Tuition and fees
- Domestic production activities deduction

14.3.2 Exemptions

Each person being claimed on income taxes is allowed a deduction for *exemptions* and the amount of the exemption is set each year by the IRS. For 2015, the personal exemption was \$4000.00.

14.3.3 Standard Deduction

If an individual, or a couple, does not itemize deductions, then he or she (they) is (are) allowed to subtract a ***standard deduction*** from their adjusted gross income and Table 14.6 lists the standard deductions for 2015.

TABLE 14.6
2015 Standard Deductions for Federal Income Taxes

Filing Status	Standard Deduction
Single	\$6,300.00
Married filing jointly	\$12,600.00
Married filing separately	\$6,300.00
Head of household	\$9,250.00
Qualifying widow(er)	\$12,600.00

Source: Internal Revenue Service, 2015 Federal tax rates, personal exemptions, and standard deductions, U.S. Government Printing Office, Washington, DC, 2015, https://www.irs.com/articles/2015-federal-tax-rates-personal-exemptions-and-standard-deductions, accessed on November 9, 2015.

Example 14.6 demonstrates calculating individual taxable income using standard deductions.

Example 14.6

A single engineer earns \$65,000.00 a year and she received a bonus at the end of the year of \$5,000.00. She does not itemize deductions; therefore, she uses the standard deduction. What is her federal taxable income for 2015?

Solution

Use Table 14.6 to determine the appropriate standard deduction and Equation 14.7 to calculate the taxable income;

$$\begin{aligned}\text{Individual taxable income} &= \text{Adjusted gross income} - \text{Deductions for exemptions} \\ &\quad - \text{Standard deduction or itemized deduction} \\ &= \$65{,}000.00 + \$5{,}000.00 - \$4{,}000.00 - \$6{,}300.00 \\ &= \$59{,}700.00\end{aligned}$$

14.3.4 Itemized Deductions

If an individual has over \$6,000.00, a married couple \$12,600.00, head of household \$9,250.00, or a qualifying widow(er) \$12,600.00 in deductions, then he or she (they) should itemize their deductions rather than claiming the standard deduction. The following are the types of items included in the category of *itemized deductions*:

- Home mortgage interest (principal resident plus home improvement loans and one second residence)
- Points on home mortgage loans

- Mortgage insurance premiums
- Medical and dental expenses over 10% of the adjusted gross income (if born after January 2, 1949 it is 7.5%)
- Vehicle registration fees
- Property taxes
- Professional dues
- Investment interest
- Charitable contributions
- Casualty or theft
- Tax preparation fees
- Safety deposit box fee
- Unreimbursed employee expenses (job travel, job-related education, vehicles, insurance, office equipment, office supplies, home office)

There are also tax credits that may change from year to year (see IRS.gov).

It is important to note that state and local taxes are a deduction on federal taxes, but federal taxes are not deducted from state or local taxes. In some cities, there may be city taxes, and even if someone does not live in the city but they work in the city, they may be required to pay city taxes.

By far the largest deduction on personal income taxes is home mortgage interest. The engineering economic analysis techniques discussed in the previous chapters assist in calculating mortgage payments and the amount of the mortgage payment that is the principal and interest. Engineering economic analysis formulas are also incorporated into amortization schedules, which are a month-by-month representation of mortgage payments, interest, and principal. Amortization schedules are a tabulation of the periodic payment (A) required to pay off the principal (P) over (n) periods with a fixed or adjustable interest rate (i). Case study 14.1 demonstrates calculating payments and amortization schedules and provides the first 10 months of the amortization schedules for a 30-year loan and a 15-year loan.

Case Study 14.1 Home Mortgage Amortization Schedules

A homeowner needs to determine the amount of his yearly mortgage payments that he is able to deduct from his taxable income. In order to determine the amount of the deduction, the homeowner has to first determine the monthly payments and the amount of the monthly payments that is the interest since he is only able to deduct the interest on the mortgages, not the entire mortgage payment. The homeowner develops two *amortization schedules*, one for a 30-year mortgage for his primary residence and one for a 15-year mortgage for his second home, to help him calculate his mortgage interest deduction based on the following information:

Cost of each home = \$392,000.00

Down payments of 5% for each home $= \$392{,}000.00 \times 0.05 = \$19{,}600.00$

Balanced owed after down payment for each home $= \$392{,}000.00 - \$19{,}600.00 = \$372{,}400.00$

Amount financed through a mortgage for each home = \$372,400.00

Interest rate = 4.5%

SOLUTION

First, calculate the monthly interest rate and total number of months:

$$\text{Monthly interest rate} = \frac{4.5\%/\text{year}}{12\ \text{months/year} \times 100} = 0.00375\%\ \text{per month}$$

$$\text{For the 15 year mortgage: } n = 12\ \text{months} \times 15\ \text{years} = 180\ \text{months}$$

$$\text{For the 30 year mortgage: } n = 12\ \text{months} \times 30\ \text{years} = 360\ \text{months}$$

Second, calculate the monthly payment using the equivalent uniform annual cost equation for 30 and 15 years:

$$\text{EUAC} = P_0\left[\frac{i(1+i)^n}{(1+i)^n - 1}\right]$$

$$\text{EUAC}_{15} = P_0\left[\frac{0.00375(1+0.00375)^{180}}{(1+0.00375)^{180} - 1}\right] = \$372{,}400.00\left(\frac{0.007356}{0.961555}\right)$$

$$= \$372{,}400.00(0.007650)$$

$$= \$2{,}848.90$$

$$\text{EUAC}_{30} = P_0\left[\frac{0.00375(1+0.00375)^{360}}{(1+0.00375)^{360} - 1}\right] = \$372{,}400.00\left(\frac{0.014429}{2.847698}\right)$$

$$= \$372{,}400.00(0.005067)$$

$$= \$1{,}886.95$$

Third, develop the amortization schedule for the 30-year mortgage and list the results in Table 14.7. The beginning balance for each month is the unpaid balance from the previous month.

TABLE 14.7
Amortization Schedule for 30-Year Mortgage

Month	Beginning Balance	Payment	Interest Per Month (Beginnning Balance × 0.00375)	Principal (Payment – Interest)	Unpaid Balance (Beginning Balance – Principal)
1	\$372,400.00	\$1,886.95	(\$372,400.00 × 0.00375) = \$1,396.50	\$1,886.95 – \$1,396.50 = \$490.45	\$372,400.00 – \$490.45 = \$371,909.55
2	\$371,909.55	\$1,886.95	(\$371,909.55 × 0.00375) = \$1,394.66	\$1,886.95 – \$1,394.66 = \$492.29	\$371,909.55 – \$492.29 = \$371,417.26
3	\$371,417.26	\$1,886.95	\$371,417.26 × 0.00375 = \$1,392.81	\$1,886.95 – \$1,392.81 = \$494.14	\$371,417.26 – \$494.14 = \$370,923.12
4	\$370,923.12	\$1,886.95	\$370,417.26 × 0.00375 = \$1,389.06	\$1,886.94 – \$1,389.06 = \$497.88	\$370,923.12 – \$497.88 = \$370,425.24
5	\$370,425.25	\$1,886.95	\$370,425.25 × 0.00375 = \$1,389.09	\$1,886.96 – \$1,389.09 = \$497.87	\$370,425.25 – \$497.87 = \$369,927.38
6	\$369,927.38	\$1,886.95	\$369,927.25 × 0.00375 = \$1,387.23	\$1,886.96 – \$1,387.23 = \$499.73	\$369,927.38 – \$499.73 = \$369,427.65
7	\$369,427.65	\$1,886.95	\$369,427.65 × 0.00375 = \$1,385.35	\$1,886.96 – \$1,385.35 = \$501.61	\$369,427.65 – \$501.61 = \$368,926.04
8	\$368,926.04	\$1,886.95	\$358,926.04 × 0.00375 = \$1,383.47	\$1,886.96 – \$1,383.47 = \$503.49	\$368,926.04 – \$503.49 = \$368,422.55
9	\$368,442.55	\$1,886.95	\$368,442.55 × 0.00375 = \$1,381.66	\$1,886.96 – \$1,381.66 = \$505.30	\$368,442.55 – \$505.30 = \$367,927.25
10	\$367,927.25	\$1,886.95	\$367,927.25 × 0.00375 = \$1,379.73	\$1,886.95 – \$1,379.73 = \$507.22	\$367,927.25 – \$507.22 = \$367,420.03

Fourth, develop the amortization schedule for the 15-year mortgage and list the results for the first 10 months in Table 14.8.

TABLE 14.8
Amortization Schedule for 15-Year Mortgage

Month	Beginning Balance	Payment	Interest Per Month (Beginning Balance × 0.00375)	Principal (Payment − Interest)	Unpaid Balance (Beginning Balance − Principal)
1	\$372,400.00	\$2,848.90	(\$372,400.00 × 0.00375) = \$1,396.50	\$2,848.90 − \$1,396.50 = \$1,452.40	\$372,400.00 − \$1,452.40 = \$370,947.60
2	\$370,947.60	\$2,848.90	(\$370,947.60 × 0.00375) = \$1,391.05	\$2,848.90 − \$1,391.05 = \$1,457.85	\$370,947.60 − \$1,457.85 = \$369,489.75
3	\$369,489.75	\$2,848.90	\$369,489.75 × 0.00375 = \$1,385.59	\$2,848.90 − \$1,385.59 = \$1,463.31	\$369,489.75 − \$1,463.31 = \$368,026.64
4	\$368,026.64	\$2,848.90	\$368,026.64 × 0.00375 = \$1,380.10	\$2,848.90 − \$1,380.10 = \$1,468.80	\$368,026.64 − \$1,468.80 = \$366,557.84
5	\$366,557.84	\$2,848.90	\$366,557.84 × 0.00375 = \$1,374.59	\$2,848.90 − \$1,374.59 = \$1,474.31	\$366,557.83 − \$1,474.31 = \$365,083.52
6	\$365,083.53	\$2,848.90	\$365,083.52 × 0.00375 = \$1,369.06	\$2,848.90 − \$1,369.06 = \$1,479.84	\$365,083.53 − \$1,479.84 = \$363,603.69
7	\$363,603.69	\$2,848.90	\$363,603.69 × 0.00375 = \$1,363.51	\$2,848.90 − \$1,363.51 = \$1,485.39	\$363,603.69 − \$1,485.39 = \$362,118.30
8	\$362,118.30	\$2,848.90	\$362,118.30 × 0.00375 = \$1,357.94	\$2,848.90 − \$1,357.94 = \$1,490.96	\$362,118.30 − \$1,490.96 = \$360,627.34
9	\$360,627.34	\$2,848.90	\$360,627.34 × 0.00375 = \$1,352.35	\$2,848.90 − \$1,352.35 = \$1,496.55	\$360,627.34 − \$1,496.55 = \$359,130.79
10	\$359,130.79	\$2,848.90	\$359,130.80 × 0.00375 = \$1,346.74	\$2,848.90 − \$1,346.74 = \$1,502.16	\$359,130.80 − \$1,502.16 = \$357,628.64

Case Study 14.1 helps illustrate the different amounts of interest paid on 15- and 30-year loans. For a purchase price of \$392,000.00 at 4.5% interest, a homeowner pays \$138,969.19 in interest if he repays the mortgage in 15 years. For the same loan amount, a homeowner will pay \$301,149.19 in interest over 30 years—a difference of \$162,180.07 in interest. The difference between the monthly payments of \$1,886.05 for the 30-year mortgage and \$2,848.90 for the 15-year mortgage is \$962.85. The homeowner is able to deduct the additional interest for the 15-year mortgage on his taxes. Even if a homeowner has a 30-year mortgage, he or she may pay an extra amount each month or make extra mortgage payments each year, and this helps to repay the principal faster, which in turn reduces the total amount of interest paid over the life of the mortgage. Extra payments each month or year help to more rapidly repay the principal since all of the extra payments are applied to repaying the principal.

14.3.5 Personal Income Tax Rates

In the same manner as federal corporate income tax rates, *individual tax rates* are figured using seven tax brackets that may change from year to year. The amount of taxes owed depends on the level of income and the filing status of the person who will be paying the taxes. Tables 14.9 through 14.12 list the individual tax rates for 2015 for the different filing statuses.

TABLE 14.9
Federal Income Tax Rates for Single Filers

Taxable Income	Tax Rate
\$0–\$9,225.00	10%
\$9,226.00–\$37,450.00	\$922.50 plus 15% of the amount over \$9,225.00
\$37,451.00–\$90,750.00	\$5,156.25 plus 25% of the amount over \$37,450.00
\$90,751.00–\$189,300.00	\$18,481.25 plus 28% of the amount over \$90,750.00
\$189,301.00–\$411,500.00	\$46,075.25 plus 33% of the amount over \$189,300.00
\$411,501.00–\$413,200.00	\$119,401.25 plus 35% of the amount over \$411,500.00
\$413,201.00 or more	\$119,996.25 plus 39.6% of the amount over \$413,200.00

Source: Internal Revenue Service, 2015 Federal tax rates, personal exemptions, and standard deductions, U.S. Government Printing Office, Washington, DC, 2015, https://www.irs.com/articles/2015-federal-tax-rates-personal-exemptions-and-standard-deductions, accessed on November 9, 2015.

TABLE 14.10
Federal Income Tax Rates for Married Filing Jointly or Qualifying Widow(er)

Taxable Income	Tax Rate
\$0–\$18,450.00	10%
\$18,451.00–\$74,900.00	\$1,845.00 plus 15% of the amount over \$18,450.00
\$74,901.00–\$151,200.00	\$10,312.50 plus 25% of the amount over \$74,900.00
\$151,201.00–\$230,450.00	\$29,387.50 plus 28% of the amount over \$151,200.00
\$230,451.00–\$411,500.00	\$51,577.50 plus 33% of the amount over \$230,450.00
\$411,501.00–\$464,850.00	\$111,324.00 plus 35% of the amount over \$411,500.00
\$464,851.00 or more	\$129,996.50 plus 39.6% of the amount over \$464,850.00

Source: Internal Revenue Service, 2015 Federal tax rates, personal exemptions, and standard deductions, U.S. Government Printing Office, Washington, DC, 2015, https://www.irs.com/articles/2015-federal-tax-rates-personal-exemptions-and-standard-deductions, accessed on November 9, 2015.

TABLE 14.11
Federal Income Tax Rates Married Filing Separately

Taxable Income	Tax Rate
\$0–\$9,225.00	10%
\$9,226.00–\$37,450.00	\$922.50 plus 15% of the amount over \$9,225.00
\$37,451.00–\$75,600.00	\$5,156.25 plus 25% of the amount over \$37,450.00
\$75,601.00–\$115,225.00	\$14,693.75 plus 28% of the amount over \$75,600.00
\$115,226.00–\$205,750.00	\$25,788.75 plus 33% of the amount over \$115,225.00
\$205,751.00–\$232,425.00	\$55,662.00 plus 35% of the amount over \$205,750.00
\$232,426.00 or more	\$64,998.25 plus 39.6% of the amount over \$232,425.00

Source: Internal Revenue Service, 2015 Federal tax rates, personal exemptions, and standard deductions, U.S. Government Printing Office, Washington, DC, 2015, https://www.irs.com/articles/2015-federal-tax-rates-personal-exemptions-and-standard-deductions, accessed on November 9, 2015.

TABLE 14.12
Federal Income Tax Rate for Head of Household

Taxable Income	Tax Rate
\$0–\$13,150.00	10%
\$13,151.00–\$50,200.00	\$1,315.00 plus 15% of the amount over \$13,150.00
\$50,201.00–\$129,600.00	\$6,872.50 plus 25% of the amount over \$50,200.00
\$129,601.00–\$209,850.00	\$26,772.50 plus 28% of the amount over \$129,600.00
\$209,851.00–\$411,500.00	\$49,192.50 plus 33% of the amount over \$209,850.00
\$411,501.00–\$439,000.00	\$115,737.00 plus 35% of the amount over \$411,500.00
\$439,001.00 or more	\$125,362.00 plus 39.6% of the amount over \$439,000.00

Source: Internal Revenue Service, 2015 Federal tax rates, personal exemptions, and standard deductions, U.S. Government Printing Office, Washington, DC, 2015, https://www.irs.com/articles/2015-federal-tax-rates-personal-exemptions-and-standard-deductions, accessed on November 9, 2015.

In addition to the tax brackets listed in Tables 14.9 through 14.12, the IRS also provides federal income tax tables where taxpayers are able to locate their taxable income and find the amount owed in taxes rather than having to use the tax rate tables and calculate the amount of taxes owed on their taxable income.

The effective federal individual income tax rate is calculated using Equation 14.8.

$$\text{Effective federal individual income tax rate} = \frac{\text{Individual income taxes}}{\text{Adjusted gross income}} \tag{14.8}$$

Examples 14.7 and 14.8 demonstrate using the federal adjusted gross income tax rate tables to calculate individual income taxes and the effective federal individual income tax rates.

Example 14.7

The single engineer from Example 14.6 is using Table 14.9 to calculate her federal income taxes. Her income for the year was \$65,000.00 and she received a bonus of \$5000.00. What is the total amount she owes in federal incomes taxes for 2015 and what is the effective federal individual income tax rate?

Solution

Use Table 14.6 to determine the appropriate standard deduction and Equation 14.7 to calculate the taxable income:

$$\begin{aligned}\text{Individual taxable income} &= \text{Adjusted gross income} - \text{Deductions for exemptions}\\ &\quad - \text{Standard deduction or itemized deduction}\\ &= \$65{,}000.00 + \$5{,}000.00 - \$4{,}000.00 - \$6{,}300.00\\ &= \$59{,}700.00\end{aligned}$$

Calculate the federal individual income taxes using Table 14.9:

\$5,156.25 plus the 25% of the amount over \$37,450.00

$$\text{Income taxes} = \$5{,}156.25 + 0.25(\$59{,}700.00 - \$37{,}450.00)$$
$$= \$5{,}156.25 + 0.25(\$22{,}250.00)$$
$$= \$5{,}156.25 + \$5{,}562.50$$
$$= \$10{,}718.75$$

Solve for the effective federal individual income tax rate using Equation 14.7:

$$\text{Effective federal individual income tax rate} = \frac{\text{Personal income taxes}}{\text{Adjusted gross income}}$$
$$= \frac{\$10{,}718.75}{\$70{,}000.00} = \$0.153 \text{ or } 15.3\%$$

In addition, the engineer will owe state and local taxes (if local taxes are required in her place of residence.

Example 14.8

If a married couple earns $200,000.00 a year, they file a joint return, and use the standard deduction, what amount would they owe in federal income taxes for 2015 and what is the effective federal individual income tax rate?

Solution

Use Table 14.6 to determine the appropriate standard deduction and Table 14.10 for married couples filing jointly to determine the federal income taxes:

$29,387.50 plus 28% of the amount over $151,200.00

$$\text{Taxable income} = \text{Adjusted gross income} - \text{Deductions for exemptions}$$
$$- \text{Standard deduction or itemized deduction}$$
$$= \$200{,}000.00 - 2(\$4{,}000.00) - \$12{,}600.00$$
$$= \$179{,}400.00$$

$$\text{Income taxes} = \$29{,}387.50 + 0.28(\$179{,}400.00 - \$151{,}200.00)$$
$$= \$29{,}387.50 + 0.28(\$28{,}200.00)$$
$$= \$29{,}387.50 + \$7{,}896.00$$
$$= \$37{,}283.50$$

Solve for the effective federal individual income tax rate using Equation 14.6:

$$\text{Effective federal individual income tax rate} = \frac{\text{Personal income taxes}}{\text{Adjusted gross income}}$$
$$= \frac{\$37{,}283.50}{\$200{,}00.00} = 0.1864 \text{ or } 18.64\%$$

Case Study 14.2 provides information demonstrating the variations in average tax rates based on whether standard deductions or itemized deductions are claimed by a taxpayer. This case study also illustrates deducting mortgage interest to lower individual income taxes.

Case Study 14.2 After-Tax Net Income

Table 14.13 provides sample scenarios for different personal tax situations. The first two scenarios represent calculations for determining the amount of taxes paid by a taxpayer who claims one exemption and uses the standard deduction. The third scenario includes tax information for a single taxpayer who owns a home and itemizes deductions including home mortgage interest. The fourth scenario is another single taxpayer who owns a primary residence along with a rental house and deductions are itemized including the depreciation on the rental house.

The data for the third scenario are the following:

- Home purchase price—\$201,199.00
- Land value—\$20,000.00
- Down payment—\$40,000.00
- Loan mortgage amount—\$161,199.00
- Yearly interest rate—9%
- Monthly interest—$\dfrac{9\%/\text{month}}{12 \text{ months/year}} = 0.0075\%$
- $n = 12$ months × 30 years = 360 months
- Annual payments—$P\left[\dfrac{i(1+i)^n}{(1+i)^n - 1}\right]$

$$= \$161{,}199.00\left[\frac{0.0075(1+0.0075)^{360}}{(1+0.0075)^{360}-1}\right] = \$1{,}297.00 \text{ per month}$$

- Yearly interest on the loan—\$14,000.00
- Yearly property taxes—\$2,000.00

The data for the fourth scenario are the following:

- Purchase price of homes—\$402,398.00 total for two homes
- Land value—\$39,816.00 total for two homes
- Down payment—\$69,000.00 total for two homes
- Loan mortgage amount—\$333,398.00 total for two homes
- Yearly interest rate—9%
- Monthly interest—$\dfrac{9\%/\text{month}}{12 \text{ months/year}} = 0.0075\%$
- $n = 12$ months × 30 years = 360 months
- Annual payments—$P\left[\dfrac{i(1+i)^n}{(1+i)^n - 1}\right]$

$$= \$333{,}398.00\left[\frac{0.0075(1+0.0075)^{360}}{(1+0.0075)^{360}-1}\right] = \$2{,}682.59 \text{ per month}$$

- Yearly interest on loans—\$28,000.00 total
- Property taxes—\$4,000.00 total
- Depreciation on rental home—$\dfrac{\$362{,}582.00 - \$0}{22} = \$16{,}481.00$ per year
- Additional deductions including rental property business expenses \$35,880.00

TABLE 14.13
After-Tax Net Income Calculations for Four Scenarios

	Single with Standard Deduction	Single with Higher Income and Standard Deduction	Single Itemized Deductions with One Home	Single Itemized Deductions with a Home and a Rental Home
Adjusted gross income	$50,000.00	$120,000.00	$120,000.00	$120,000.00
Exemptions	$4,000.00	$4,000.00	$4,000.00	$4,000.00
Standard deduction	$6,300.00	$6,300.00	—	
Itemized deduction (loan interest and property taxes)	—	—	$16,0000.00	$28,000.00 + $4,000.00 + $35,880 = $67,880.00
Depreciation				$16,484.00
Taxable income	$50,000.00 − $4,000.00 − $6,300.00 = $39,700.00	$120,000.00 − $4,000.00 − $6,400.00 = $109,700.00	$120,000.00 − $4,000.00 − $16,000.00 = $100,00.00	$120,000.00 − $4,000.00 − $67,880.00.00 − $16,481.00 = $31,639.00
Taxes (from tax tables)	$5,516.25 + 0.25($39,700.00 − $37,350.00) = $5,718.75	$18,481.25 + 0.28($109,700.00 − $90,750.00) = $23,787.25	$18,481.25 + 0.28($100,00.00 − $90,750.00) = $21,071.25	$922.50 + 0.15($31,639.00 − $9,225.00) = $4,284.60
Effective federal personal income tax rate	$\frac{\$5,718.75}{\$50,000.00} = 11.44\%$	$\frac{\$23,787.25}{\$120,000.00} = 19.83\%$	$\frac{\$21,071.25}{\$120,000.00} = 17.56\%$	$\frac{\$4,284.60}{\$120,000.00} = 3.6\%$
Income after federal taxes	$50,000.00 − $5,718.75 = $44,281.28	$120,000.00 − $23,787.25 = $96,212.75	$120,000.00 − $21,071.25 = $98,928.75	$120,000.00 − $4,284.60 = $115,715.40
Tax savings	—	—	$23,787.25 − $21,071.25 = $2,716.00.00	$23,787.25 − $4,284.60 = $19,502.65
Federal taxes	$5,718.75	$23,787.25	$21,071.25	$4,284.40
State taxes	$1,500.00	$2,500.00	$2,500.00	$2,500.00
Social Security taxes	$50,000.00 × 0.0765 = $3,825.00	$120,000.00 × 0.0765 = $9,180.00	$120,000.00 × 0.0765 = $9,180.00	$120,000.00 × 0.0765 = $9,180.00
Total taxes	$11,043.75	$35,467.25	$32,751.25	$15,964.40
Yearly net income after taxes	$50,000.00 − $11,043.75 = $38,956.25	$120,000.00 − $35,467.25 = $84,532.75	$120,000.00 − $32,751.25 = $87,248.75	$120,000.00 − $15,964.40 = $104,035.60
Monthly net income after taxes	$\frac{\$38,956.25}{12\text{ months}} = \$3,246.35$	$\frac{\$84,532.75}{12\text{ months}} = \$7,044.39$	$\frac{\$87,248.75}{12\text{ months}} = \$7,270.73$	$\frac{\$104,035.60}{12\text{ months}} = \$8,669.63$
Increase in monthly income			$7,270.73 − $7,044.39 = $226.34	$8,696.63 − $7,044.39 = $1,625.40

SOLUTION

Table 14.13 provides the solutions to Case study 14.2.

As the calculations in Case Study 14.2 demonstrate, with one primary residence and another rental property being depreciated, there is an increase in the monthly income of $1,625.40 due to the reduction in taxes from deducting the mortgage interest and depreciation on the rental property. A taxpayer is able to realize the increase in monthly income by claiming more deductions on the W-4 form he or she completes for their employer. When more deductions are listed on a W-4 form, an employer reduces the amount withheld from each paycheck for taxes; therefore, the monthly income increases since less taxes are being withheld from each paycheck.

14.4 MORTGAGES

This section covers several different types of mortgages including fixed, variable, renegotiable, adjustable, shared appreciation, graduated payment, negative amortization, graduated payment adjustable, reverse annuity, and buy down. This section also discusses balloon payments and prepayment penalties as they relate to mortgages.

Mortgages are a special type of loan providing evidence of debt. Mortgage documents include information about the amount of the loan, interest rate, term of the loan, and date of repayment, along with additional terms of the mortgage. Mortgages provide lenders with interest in the property until the loan is repaid. A mortgage is a lien on the property and remains with the property even if there is a transfer of ownership (30 states) or a mortgage transfers title to the lender until the note is repaid (20 states).

The institution holding the first mortgage has the first claim in bankruptcy proceedings after taxes and government claims. There could also be second mortgages on real property and these may be referred to as equity loans.

In order to secure a mortgage, a potential homeowner has to pass a credit check and provide verification of income, as well as two years of federal income tax returns. Some lenders may require additional documentation of income or assets. There are several costs associated with securing a mortgage called closing costs. Typical closing costs might include the following:

- *Points*—A percentage of the loan paid to the lending institution to lower the interest rate
- *Loan origination fee*—Usually 1%–3% of the loan value to cover the costs of the lending institution preparing and financing the loan
- Fees for the deed, abstract, title searches, title insurance, appraisal, credit report, tax service, underwriting, wire transfers, settlement and closing statements, endorsements, conveyances, and prepaid hazard insurance and mortgage insurance

The only items a homeowner may deduct related to mortgages when they are filing their federal income taxes are monthly interest on the loan, mortgage insurance, property taxes, and points the year he or she obtained the loan. Mortgage insurance is normally required on all mortgages when the down payment is less than 20% and it protects the lender against loan default. Once a home appreciates to a value where there is 20% equity in a home, a homeowner may apply to have the mortgage insurance removed from the loan.

14.4.1 Fixed Rate and Other Types of Mortgages

Fixed rate mortgages are loans where the interest rate is fixed for the entire term of the loan, which is normally either 15 or 30 years and these are *traditional mortgages*.

In addition to fixed rate mortgages, there are other options for financing real estate, which are *alternative mortgage instruments* and examples of how these types of mortgages are categorized are the following:

1. Mortgages restructuring the repayment schedule so the initial payments are less. Examples of this type of mortgage include
   a. Variable
   b. Graduated payment
   c. Buy down
2. Mortgages where the interest rate varies based on an index tied to inflation or other rates. This type of mortgage allows lenders to have a constant real rate of return when there is inflation. Examples of this type of mortgage are
   a. Adjustable rate
   b. Renegotiable
   c. Shared appreciation
3. There are also hybrid mortgages such as
   a. Reverse annuity
   b. Graduated payment adjustable

The types of mortgages listed are briefly explained in Sections 14.4.1 through 14.4.9.

14.4.2 Variable Rate Mortgages

Variable rate mortgages are any type of mortgage where the interest rate varies rather than being a fixed rate. The important factors related to variable rate mortgages that should be known before acquiring this type of a loan include

- The index the rate is tied to—the rate changes when this index changes.
- How often does the rate change?
- Are there limits on the incremental changes?
- How much the total rate may change during the life of the loan?

With variable rate mortgages, a homeowner would initially have a low monthly payment that increases rapidly if the index the interest rate is tied to rises quickly.

14.4.3 Adjustable Rate Mortgages

Adjustable rate mortgages were created in 1981 for federal savings and loan institutions. This type of mortgage has few regulatory restrictions. The terms of these loans that do not have any restrictions include

- Frequency of rate changes
- Maximum rate change
- Aggregate maximum rate change
- The percentage of the portfolio of an institution comprised of these loans
- No requirements for lenders to offer a fixed rate mortgage alternative
- No restrictions on using negative amortization (see Section 14.4.6)

The only requirements for these types of loans are that the index the interest rate is tied to be verifiable and the lender does not control the index. Typical indices for this type of mortgage are three to six month treasury bill rates.

14.4.4 Shared Appreciation Mortgages

For *shared appreciation mortgages*, lenders offer the mortgage at rates below market interest rates, but the lender will share in the net appreciation of the property. The lender receives their share of the appreciation when the earliest of the following occurs:

1. At the maturity of the mortgage.
2. If the property is transferred or sold to someone else.
3. The loan is repaid in full.

14.4.5 Graduated Payment Mortgages

In *graduated payment mortgages*, the interest rate is fixed during the entire life of the loan, but the payments start out low and increase by a constant gradient over the life of the loan. Having lower initial payments makes it easier for homebuyers to qualify for the loan, but they need to be aware of the amount of the increase in payments each year and the number of years the payments will increase.

14.4.6 Negative Amortization Mortgages

Negative amortization mortgages are also another way of initially reducing the amount of payments, but the payments are usually insufficient to pay all of the interest due. The interest not repaid is added to the principal. The principal increases rather than decreases over the life of the loan and a homeowner will be responsible for repaying the entire loan including the additional principal at some point in time.

14.4.7 Graduated Payment Adjustable Rate Mortgages

Graduated payment adjustable rate mortgages are similar to graduated payment mortgages except the interest rate may vary with market interest rates similar to the types of fluctuations that occur in adjustable rate mortgages.

14.4.8 Reverse Annuity Mortgages

Reverse annuity mortgages were designed for senior citizens to provide a means of augmenting their income during retirement. These mortgages use the equity in a home to provide the homeowner with a steady income stream and then the lender ends up owning the home at the end of the life of the person with the reverse annuity mortgage. This is a home mortgage where an elderly homeowner(s) is allowed a long-term loan in the form of monthly income against his or her home equity as collateral, repayable when the home is eventually sold. There are age restrictions for this type of mortgage: the youngest borrower has to be over 62 years of age to qualify for a Housing and Urban Development or Fannie Mae loan. Some states set the minimum age for privately financed reverse annuity mortgages.

14.4.9 Buy Down Mortgages

Buy down mortgages occur when someone makes a large payment to the lending institution and the lender applies the buy down funds towards reducing the interest rate for a specified period of time, which in turn lowers the payments during the specified time period. The buy down could be made by a builder, the seller, or the purchaser.

14.4.10 Balloon Payments

In any type of mortgage, the lending institution could include a clause whereby the total amount of the remaining balance on a loan is due and payable as a lump sum on a specified future date or when a specific event occurs. The person securing the loan should know before signing the loan documents whether the loan includes a *balloon payment* clause.

14.4.11 Prepayment Penalties

Another clause that may be added to any type of mortgage is a *prepayment penalty* clause. If this clause is included in a loan, the borrower is required to pay a monetary penalty if he or she repays the loan before the end of the loan period. If the borrower sells the house and uses the proceeds to repay the loan, he or she will have to pay the prepayment penalty in addition to the amount owed on the loan. A prepayment penalty would also apply if a borrower refinances a house.

14.5 SUMMARY

This chapter discussed the after-tax effects on corporate economic decisions and the procedures for calculating after-tax rates of return. This chapter also provided information on individual taxes to demonstrate the incorporation of engineering economic analysis into individual tax calculations. This chapter explained mortgages and their effect on the taxes of individuals and corporations. Two case studies were included—one on developing home mortgage amortization schedules and another addressing after-tax net incomes. Different types of home mortgages were explained in the last section of this chapter.

KEY TERMS

Adjustable rate mortgages
Adjusted gross income
After-tax cash flow
Alternative mortgage instruments
Amortization schedule
Balloon payments
Before-tax cash flow
Buy down mortgages
Capital gains
Capital losses
Effective federal corporate income tax rate
Effective federal individual income tax rate
Effective tax rate
Exemptions
Federal corporate income taxes
Fixed rate mortgages
Graduated payment adjustable rate mortgages
Graduated payment mortgages
Individual taxable income
Internal revenue service
Itemized deductions
Loan origination fee
Negative amortization mortgages
Personal income taxes

Personal tax rates
Points
Prepayment penalty
Recaptured depreciation
Reverse annuity mortgage
Shared appreciation mortgages
Standard deduction
Taxable income
Variable rate mortgages

PROBLEMS

14.1 An agricultural engineering firm has been depreciating a combine for five years using straight line depreciation. The combine originally cost $700,000.00 and the estimated salvage value is $125,000.00. (1) If the firm sells the combine for $350,000.00, what will be the recaptured depreciation? (2) Calculate the federal income taxes on the sale of the combine.

14.2 Calculate the federal income taxes on the combine in Problem 14.1 if it is sold for $900,000.00.

14.3 A materials engineering firm has an adjusted gross income of $3,700,000.00, business expenses of $2,300,000.00, and depreciation of $500,000.00. Calculate the federal income taxes owed by the firm.

14.4 A consulting engineering firm has earned $733,000.00 during the past year, and there were business expenses of $292,000.00 and no depreciation. Calculate the federal income taxes owed by the firm.

14.5 A process engineering firm has an adjusted gross income of $2,700,000.00 and business expenses of $1,100,000.00. The firm has assets worth $4,500,000.00 that they depreciate using double declining balance depreciation over eight years. The assets are in their second year of depreciation. Calculate the federal income taxes for this firm.

14.6 A mechanical engineering firm has an adjusted gross income of $17,200,000.00. The business expenses are $14,725,000.00, they depreciate assets that cost $9,000,000.00, and the assets will have a salvage value in 10 years of $1,000,000.00. They use sum-of-the-years digit depreciation. The asset is in its second year of depreciation. Calculate the federal income taxes for the firm.

14.7 A manufacturing company has a combined income of $90,000,000.00 and $62,000,000.00 in business expenses. The firm depreciates their machinery using production depreciation. The machinery was purchased for $162,000,000.00 and has a salvage value of $32,000,000.00. The machinery produced 150,000,000 units this year and the life in number of units is 1,500,000,000. Calculate the amount this firm has to pay in federal income taxes.

14.8 What is the effective federal income tax rate for the information in Problem 14.8?

14.9 What is the effective federal income tax rate for the information in Problem 14.3?

14.10 What is the effective federal income tax rate for the information in Problem 14.6?

14.11 A systems engineering firm has assets being depreciated using straight line depreciation for three years. The assets cost $750,000.00 and the salvage value is $292,000.00. The company has an income of $900,000.00 for each of the three years and their federal income tax rate is 34%. Calculate the after-tax cash flow for this firm.

14.12 Calculate the net present worth for the after-tax cash flow in Problem 14.11 using an interest rate of 4%.

14.13 Calculate the equivalent uniform annual worth of the after-tax cash flow in Problem 14.11 using an interest rate of 4%.

14.14 Calculate the rate of return for the after-tax cash flow in Problem 14.11 using net present worth analysis.

14.15 A petroleum engineering firm depreciates its assets over 5 years using double declining balance depreciation. The assets cost $45,000,000.00 and will have a salvage value of $2,000,000.00. The firm has an income of $30,000,000.00 per year. Calculate the after-tax cash flow for this firm over the next five years.

14.16 Calculate the net present worth of the after-tax cash flow in Problem 14.14 using an interest rate of 6%.

14.17 An individual has an adjusted gross income of $192,000.00. They are single and have itemized deductions of $41,750.00. Calculate the federal income taxes for this individual.

14.18 A married couple earns $292,000.00, they file jointly, and their itemized deductions are $67,500.00. Calculate the federal income taxes for the couple.

14.19 A married couple earns $63,000.00, and they file jointly, have five children, and do not itemize their deductions. Calculate the federal income taxes for the couple.

14.20 A married couple earns $187,000.00, they file jointly, they have two children, and the interest on their primary resident home mortgage was $24,000.00. They own a rental property and the interest on the mortgage is $16,000.00 and their business expenses related to the rental property are $8,700.00. Calculate the federal income taxes for the couple.

14.21 Based on the types of mortgages listed in Section 14.4, which type of a mortgage would you select and explain why you would select that type of mortgage.

REFERENCES

Internal Revenue Service. August 17, 2012. Business or hobby? Answer has implications for deductions. Washington, DC: U.S. Government Printing Office. Accessed on March 6, 2016. https://www.irs.gov/uac/Business-or-Hobby%3F-Answer-Has-Implications-for-Deductions.

Internal Revenue Service. 2015. 2015 Federal tax rates, personal exemptions, and standard deductions. Washington, DC: U.S. Government Printing Office. Accessed on November 9, 2015. https://www.irs.com/articles/2015-federal-tax-rates-personal-exemptions-and-standard-deductions.

Internal Revenue Service. January 4, 2016. Topic 49—Capital gains and losses. Washington, DC: U.S. Government Printing Office. Accessed on January 31, 2016. https://www.irs.gov/taxtopics/tc409.html.

Internal Revenue Service. January 21, 2016. Form 1120 U.S. Corporate Income Tax Return—Instructions. Washington, DC: U.S. Government Printing Office. Accessed on January 31, 2016. https://www.irs.gov/pub/irs-pdf/i1120.pdf.

Appendix A: Basic Engineering Economic Equations and Cash Flow Diagrams

Factor	Equation	Cash Flow Diagram	As $n = \infty$
$F = P\,(F/P), i, n)$ Future worth (F) of a present value (P) n = number of time periods	$F = P_0(1+i)^n$ $F = Pe^{i_n \times n}$ (continuous compounding)	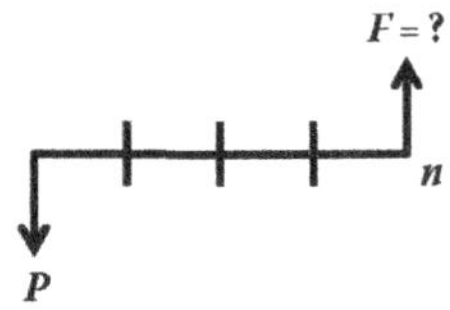	$F/P \Rightarrow \infty$
$P = F(P/F, i, n)$ Present worth (P) of a future value (F) n = number of time periods	$P = F\left[\frac{1}{(1+i)^n}\right]$	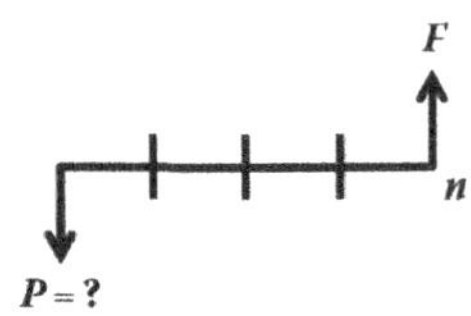	$P/F \Rightarrow \infty$
$F = A(F/A, i, n)$ Future worth (F) of a periodic uniform series (A) n = number of time periods	$F = A\left[\frac{(1+i)^n - 1}{i}\right]$	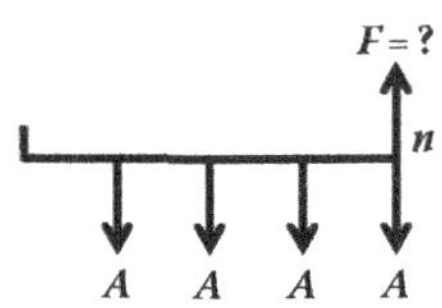	$F/A \Rightarrow \infty$
$P = A(P/A, i, n)$ Present worth (P) of a periodic uniform series (A) n = number of time periods	$P = A\left[\frac{(1+i)^n - 1}{i(1+i)^n}\right]$	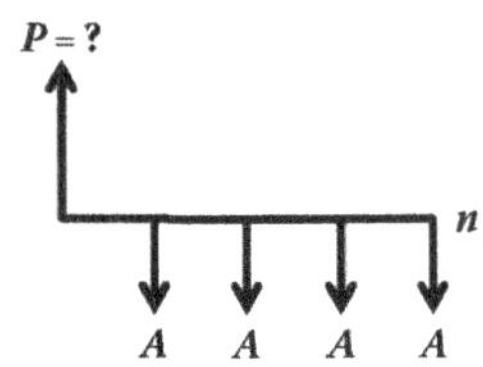	$P/A \Rightarrow \frac{1}{i}$
$A = F(A/F, i, n)$ Periodic uniform series (A) of a future worth (F) n = number of time periods	$A = F\left[\frac{i}{(1+i)^n - 1}\right]$	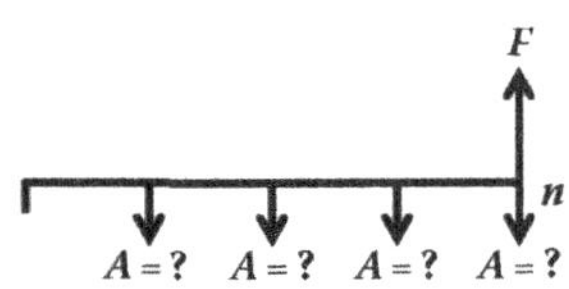	
$A = P(A/P, i, n)$ Periodic uniform series (A) of a present worth (P) n = number of time periods	$A = P\left[\frac{i(1+i)^n}{(1+i)^n - 1}\right]$	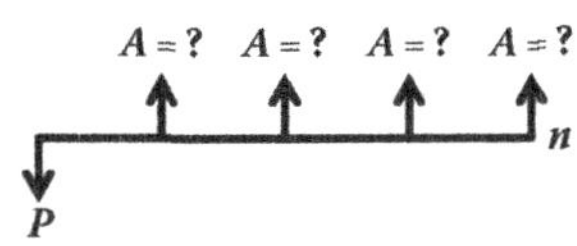	

(*Continued*)

Factor	Equation	Cash Flow Diagram	As $n = \infty$
$F = G(F/G, i, n)$ Future worth (F) of an arithmetic gradient series (G) n = number of payments plus one	$F = G\left[\left(\frac{1}{i}\right)\frac{(1+i)^n - 1}{i} - n\right]$	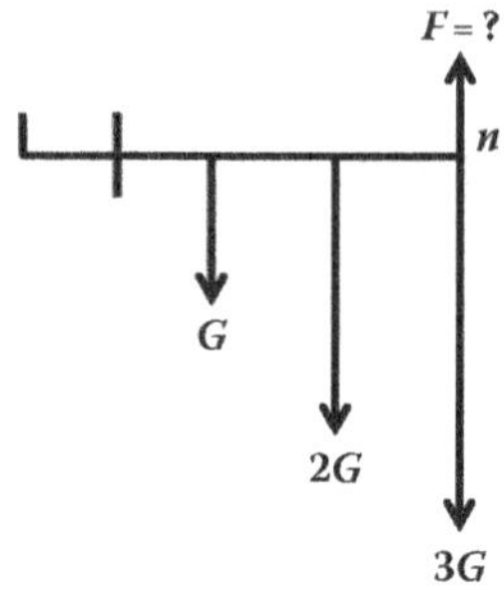	$F/G \Rightarrow \infty$
$P = G(P/G, i, n)$ Present worth (P) of an arithmetic gradient series (G) n = number of payments plus one	$P = \frac{G}{i}\left[\frac{(1+i)^n - 1}{i(1+i)^n} - \frac{n}{(1+i)^n}\right]$	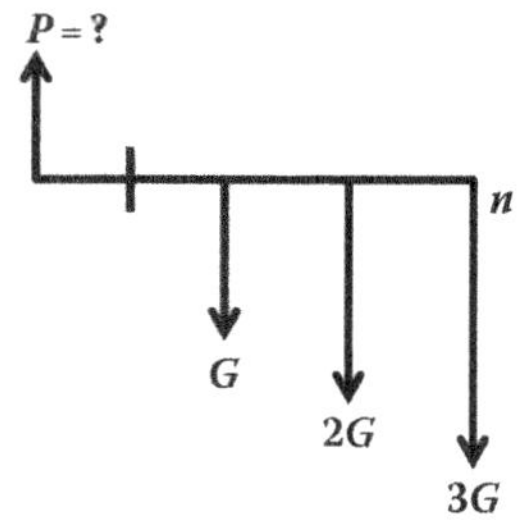	$P/G \Rightarrow \frac{1}{i^2}$
$A = G(A/G, i, n)$ Periodic uniform series (A) equivalent of arithmetic gradient series (G)	$A = G\left[\frac{1}{i} - \frac{n}{(1+i)^n - 1}\right]$	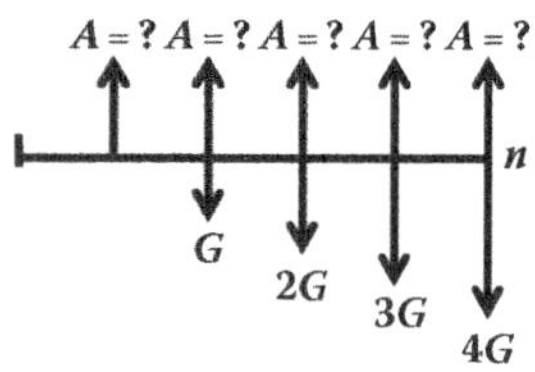	

Present worth (P) of a geometric gradient series
n = number of end of period payments

1. When $r > i$, then $w = \frac{1+r}{1+i} - 1$, $P = \frac{C}{1+i}(F/A, w, n) = \frac{C}{1+i}\left[\frac{(1+w)^n - 1}{w}\right]$ $\qquad P/C \Rightarrow \infty$

2. When $r < i$, then $w = \frac{1+i}{1+r} - 1$ and $P = \frac{C}{1+r}(P/A, w, n) = \frac{C}{1+r}\left[\frac{(1+w)^n - 1}{w(1+w)^n}\right]$ $\qquad P/C \Rightarrow \frac{1}{(1+r)w}$

3. When $r = i$, then $P = \frac{C \times n}{(1+r)} = \frac{C \times n}{(1+i)}$ $\qquad P/C \Rightarrow \infty$

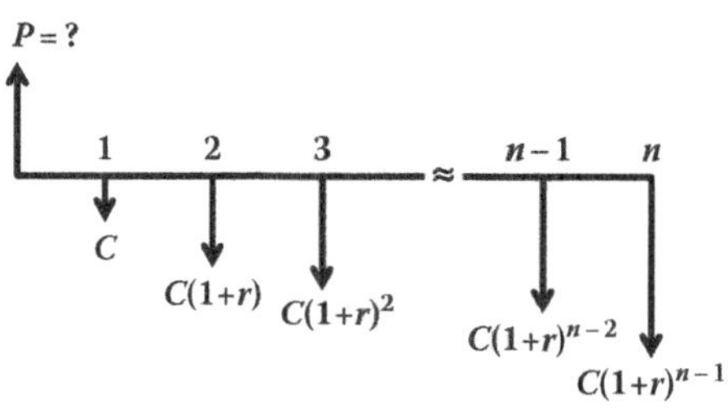

Appendix B: Interest Factor Tables

	i = 0.5%			i = 0.5%			i = 0.5%			
	Present Sum (P)			Uniform Series (A)			Future Sum (F)			
n	P/F	P/A	P/G	A/F	A/P	A/G	F/P	F/A	F/G	n
1	0.99502	0.9950	0.0000	1.00000	1.00500	0.0000	1.0050	1.0000	0.0000	1
2	0.99007	1.9851	0.9900	0.49875	0.50375	0.4987	1.0100	2.0050	1.0000	2
3	0.98515	2.9702	2.9603	0.33167	0.33667	0.9966	1.0150	3.0150	3.0050	3
4	0.98025	3.9505	5.9011	0.24813	0.25313	1.4937	1.0201	4.0301	6.0200	4
5	0.97537	4.0258	9.8026	0.19801	0.20301	1.9900	1.0252	5.0502	10.050	5
6	0.97052	5.8963	14.655	0.16460	0.16960	2.4854	1.0303	6.0755	15.100	6
7	0.96569	6.8620	20.449	0.14073	0.14573	2.9800	1.0355	7.1058	21.175	7
8	0.96089	7.8229	27.175	0.12283	0.12783	3.4738	1.0407	8.1414	28.281	8
9	0.95610	8.7790	34.824	0.10891	0.11391	3.9667	1.0459	9.1821	36.423	9
10	0.95135	9.7304	43.386	0.09777	0.10277	4.4588	1.0511	10.228	45.605	10
11	0.94661	10.677	52.852	0.08866	0.09366	4.9501	1.0564	11.279	55.833	11
12	0.94191	11.618	63.213	0.08107	0.08607	5.4405	1.0616	12.335	67.112	12
13	0.93722	12.556	74.460	0.07464	0.07964	5.9301	1.0669	13.397	79.448	13
14	0.93256	13.488	86.583	0.06914	0.07414	6.4189	1.0723	14.464	92.845	14
15	0.92792	14.416	99.574	0.06436	0.06936	6.9069	1.0776	15.536	107.31	15
16	0.92330	15.339	113.42	0.06019	0.06519	7.3940	1.0830	16.614	133.84	16
17	0.91871	16.258	128.12	0.05651	0.06151	7.8803	1.0884	17.697	139.46	17
18	0.91414	17.172	143.66	0.05323	0.05823	8.3657	1.0939	18.785	157.15	18
19	0.90959	18.082	150.03	0.05030	0.05530	8.8504	1.0994	19.879	175.94	19
20	0.90506	18.987	177.23	0.04767	0.05267	9.3341	1.1049	20.979	195.82	20
21	0.90056	19.888	195.24	0.04528	0.05028	9.8171	1.1104	22.084	216.80	21
22	0.89608	20.784	214.06	0.04311	0.04811	10.299	1.1159	23.194	238.88	22
23	0.89162	21.675	233.67	0.04113	0.04613	10.780	1.1215	23.310	262.08	23
24	0.88719	22.562	254.08	0.03932	0.04432	11.261	1.1271	25.432	286.39	24
25	0.88277	23.445	275.26	0.03765	0.04265	11.740	1.1328	26.559	311.82	25
26	0.87838	24.324	297.22	0.03611	0.04111	12.219	1.1384	27.691	338.38	26
27	0.87401	25.198	319.95	0.03469	0.03969	12.697	1.1441	28.830	366.07	27
28	0.86966	26.067	343.43	0.03336	0.03836	13.174	1.1498	29.974	394.90	28
29	0.96533	26.933	367.66	0.03213	0.03713	13.651	1.1556	31.124	424.87	29
30	0.86103	27.794	392.63	0.03098	0.03598	14.126	1.1616	32.280	456.00	30
32	0.85248	29.503	444.76	0.02889	0.03389	15.075	1.1730	34.608	521.72	32
34	0.84402	31.195	499.75	0.02706	0.03206	16.020	1.1848	36.960	592.11	34
36	0.83564	32.871	557.56	0.02542	0.03042	16.962	1.1966	39.336	667.22	36
48	0.78710	42.580	959.91	0.01849	0.02349	22.543	1.2704	54.097	1219.5	48
60	0.74137	51.725	1448.6	0.01433	0.01933	28.006	1.3488	69.770	1954.0	60
120	0.54963	90.073	4823.5	0.00610	0.01110	53.550	1.8194	163.87	8775.8	120
180	0.40748	118.50	9031.3	0.00344	0.00844	76.211	2.4540	290.81	22163.	180
240	0.30210	139.58	13415.	0.00216	0.00716	96.113	3.3102	462.04	44408.	240
300	0.22397	155.20	17603.	0.00144	0.00644	113.41	4.4649	602.99	78598.	300
360	0.16604	166.79	21403.	0.00100	0.00600	128.32	6.0225	1004.5	128903	360
INF	0.00000	200.00	40000	0.00000	0.00500	200.00	INF	INF	INF	INF

	$i = 0.75\%$			$i = 0.75\%$			$i = 0.75\%$			
	Present Sum (P)			Uniform Series (A)			Future Sum (F)			
n	P/F	P/A	P/G	A/F	A/P	A/G	F/P	F/A	F/G	n
1	0.99256	0.9925	0.0000	1.00000	1.00750	0.0000	1.0075	1.0000	0.0000	1
2	0.98517	1.9777	0.9851	0.49813	0.50563	0.4981	1.0150	2.0075	1.0000	2
3	0.97783	2.9555	2.9408	0.33085	0.33835	0.9950	1.0226	3.0225	3.0075	3
4	0.97055	3.9261	5.8525	0.24721	0.25471	1.4906	1.0303	4.0452	6.0300	4
5	0.96333	4.8894	9.7058	0.19702	0.20452	2.9850	1.0380	5.0755	10.075	5
6	0.95616	5.8456	14.486	0.16357	0.17107	2.4782	1.0458	6.1136	15.150	6
7	0.94904	6.7946	20.180	0.13967	0.14717	2.9701	1.0537	7.1594	21.264	7
8	0.94198	7.7366	26.774	0.12176	0.12926	3.4607	1.0616	8.2131	28.424	8
9	0.93496	8.6715	34.254	0.10782	0.11532	3.9501	1.0695	9.2747	36.637	9
10	0.92800	9.5995	42.606	0.09667	0.10417	4.4383	1.0775	10.344	45.911	10
11	0.92109	10.520	51.817	0.08755	0.09505	4.9252	1.0856	11.421	56.256	11
12	0.91424	11.434	61.874	0.07995	0.08745	5.4109	1.0938	12.507	67.678	12
13	0.90743	12.342	72.763	0.07352	0.08102	5.8954	1.1020	13.601	80.185	13
14	0.90068	13.243	84.472	0.06801	0.07551	6.3786	1.1102	14.703	93.787	14
15	0.89397	14.137	96.987	0.06324	0.07074	6.8605	1.1186	15.813	108.49	15
16	0.88732	15.024	110.29	0.05906	0.06656	7.3412	1.1269	16.932	124.30	16
17	0.88071	15.905	124.38	0.05537	0.06287	7.8207	1.1354	18.059	141.23	17
18	0.87416	16.779	139.24	0.05210	0.05960	8.2989	1.1439	19.194	159.29	18
19	0.86765	17.646	154.86	0.04917	0.05667	8.7759	1.1525	20.338	178.49	19
20	0.86119	18.508	171.23	0.04653	0.05403	9.2516	1.1611	21.491	198.82	20
21	0.85478	19.362	188.32	0.04415	0.05165	9.7621	1.1698	22.654	220.32	21
22	0.84842	20.211	206.14	0.04198	0.04938	10.199	1.1786	23.822	242.97	22
23	0.84210	21.053	224.66	0.04000	0.04750	10.671	1.1875	25.001	266.79	23
24	0.83583	21.889	243.89	0.03818	0.04568	11.142	1.1964	26.188	291.79	24
25	0.82961	22.718	263.80	0.03652	0.04402	11.611	1.2053	27.384	317.98	25
26	0.82343	23.542	284.38	0.03498	0.04248	12.080	1.2144	28.590	345.36	26
27	0.81730	24.359	305.63	0.03355	0.04105	12.547	1.2235	29.804	373.96	27
28	0.81122	25.170	327.54	0.03223	0.03973	13.012	1.2327	31.028	403.76	28
29	0.80518	25.975	350.08	0.03100	0.03850	13.447	1.2419	32.260	434.79	29
30	0.79919	26.755	373.26	0.02985	0.03735	13.940	1.2512	33.502	267.05	30
32	0.78733	28.355	421.46	0.02777	0.03527	14.863	1.2701	36.014	535.31	32
34	0.77565	29.912	472.07	0.02593	0.03343	15.781	1.2892	38.564	608.61	34
36	0.76415	31.446	524.99	0.02430	0.03180	16.694	1.3086	41.152	687.02	36
48	0.69861	40.184	886.84	0.01739	0.02489	22.069	1.4314	57.520	1269.4	48
60	0.63870	48.173	1313.5	0.01326	0.02076	27.266	1.5656	75.424	2056.5	60
120	0.40794	78.941	3995.5	0.00517	0.01267	50.652	2.4513	193.51	9801.9	120
180	0.26055	98.593	6892.6	0.00264	0.01014	69.909	3.8380	378.40	26454.	180
240	0.16641	111.14	9494.1	0.00150	0.00900	85.421	6.0091	667.88	57051.	240
300	0.10629	119.16	11636.	0.00089	0.00839	97.654	9.4084	1121.1	109483	300
360	0.06789	124.28	13312.	0.00055	0.00805	107.11	14.730	1830.7	196099	360
INF	0.00000	133.33	17777.	0.00000	0.00750	133.33	INF	INF	INF	INF

	i = 1%			*i* = 1%			*i* = 1%			
	Present Sum (*P*)			Uniform Series (*A*)			Future Sum (*F*)			
n	*P/F*	*P/A*	*P/G*	*A/F*	*A/P*	*A/G*	*F/P*	*F/A*	*F/G*	*n*
1	0.99010	0.9901	0.0000	1.00000	1.01000	0.0000	1.0100	1.0000	0.0000	1
2	0.98030	1.9704	0.9803	0.49751	0.50751	0.4975	1.0201	2.0100	1.0000	2
3	0.97059	2.9409	2.9214	0.33002	0.34002	0.9933	1.0303	3.0301	3.0100	3
4	0.96098	3.9019	5.8044	0.24628	0.25628	1.4875	1.0406	4.0604	6.0401	4
5	0.95147	4.8534	9.6102	0.19604	0.20604	1.9801	1.0510	5.1010	10.100	5
6	0.94205	5.7954	14.320	0.16225	0.17255	2.4709	1.0615	6.1520	15.201	6
7	0.93272	6.7821	19.916	0.13863	0.14863	2.9602	1.0721	7.2135	21.353	7
8	0.92348	7.6516	25.381	0.12069	0.13069	3.4477	1.0828	8.2856	28.567	8
9	0.91434	8.5660	33.695	0.10674	0.11674	3.9336	1.0936	9.3685	36.852	9
10	0.90529	9.4713	41.843	0.09558	0.10558	4.4179	1.1046	10.462	46.221	10
11	0.89632	10.367	50.806	0.08645	0.09645	4.9005	1.1156	11.566	56.683	11
12	0.88745	11.255	60.568	0.07885	0.08885	5.3814	1.1268	12.682	68.250	12
13	0.87866	12.133	71.112	0.07241	0.08241	5.8607	1.1380	13.809	80.932	13
14	0.86996	13.003	82.422	0.06690	0.07690	6.3383	1.1494	14.947	94.742	14
15	0.86135	13.865	94.481	0.06212	0.07212	6.8143	1.1609	16.096	109.69	15
16	0.85282	14.717	107.27	0.05794	0.06794	7.2886	1.1725	17.257	125.78	16
17	0.84438	15.562	120.78	0.05426	0.06426	7.7613	1.1843	18.430	143.04	17
28	0.83602	16.398	134.99	0.05098	0.06098	8.2323	1.1961	19.614	161.47	18
19	0.82774	17.226	149.89	0.04805	0.05805	8.7016	1.2081	20.810	181.09	19
20	0.81954	18.045	154.46	0.04542	0.05542	9.1693	1.2201	22.019	201.90	20
21	0.81143	18.857	181.69	0.04303	0.05303	9.6354	1.2323	23.239	223.91	21
22	0.80340	19.660	198.56	0.04086	0.05086	10.099	1.2447	24.471	247.15	22
23	0.79544	20.455	216.06	0.03889	0.04889	10.562	1.2571	25.716	271.63	23
24	0.87857	21.243	234.18	0.03737	0.04704	11.023	1.2697	26.973	297.34	24
25	0.77977	22.023	252.89	0.03541	0.04541	11.483	1.2824	28.243	324.32	25
26	0.77205	22.795	272.19	0.03387	0.04387	11.940	1.2952	28.525	352.56	26
27	0.76440	23.559	292.07	0.03245	0.04245	12.397	1.3082	30.820	382.08	27
28	0.75684	24.316	312.50	0.03112	0.04112	12.851	1.3212	32.129	412.91	28
29	0.74934	25.065	333.48	0.02990	0.03990	13.304	1.3345	33.450	445.03	29
30	0.74192	25.807	355.00	0.02875	0.03875	13.755	1.3478	34.784	478.48	30
32	0.72730	27.269	399.58	0.02667	0.03667	14.663	1.3749	37.493	549.40	32
34	0.71297	38.702	446.15	0.02484	0.03484	15.554	1.4025	40.257	625.77	34
36	0.69892	30.107	494.62	0.02321	0.03321	16.428	1.4307	43.076	707.68	36
48	0.62026	37.974	820.14	0.01633	0.02633	21.597	1.6122	61.222	1233.2	48
60	0.55045	44.955	1192.8	0.01224	0.02224	26.533	1.8167	81.669	2166.9	60
120	0.30299	69.700	3334.1	0.00435	0.01435	47.834	3.3003	230.03	11003.	120
180	0.16678	83.321	5330.0	0.00200	0.01200	63.969	5.9958	499.58	31958.	180
240	0.09181	90.819	6878.6	0.00101	0.01101	75.739	10.892	989.25	74925.	240
300	0.05053	94.946	7978.6	0.00053	0.01053	84.032	19.788	1878.8	157885	300
360	0.02782	97.218	8720.4	0.00029	0.01029	89.699	35.949	3494.9	313496	360
INF	0.00000	100.00	10000.	0.00000	0.01000	100.00	INF	INF	INF	INF

	i = 1.5%			i = 1.5%			i = 1.5%			
	Present Sum (*P*)			Uniform Series (*A*)			Future Sum (*F*)			
n	*P/F*	*P/A*	*P/G*	*A/F*	*A/P*	*A/G*	*F/P*	*F/A*	*F/G*	*n*
1	0.98522	0.9852	0.0000	1.00000	1.00500	0.0000	1.0150	1.0000	0.0000	1
2	0.97066	1.9558	0.9706	0.49628	0.51128	0.4962	1.0302	2.0150	1.0000	2
3	0.95632	2.9122	2.8833	0.32838	0.34338	0.9900	1.0456	3.0452	3.0150	3
4	0.94218	3.8543	5.7098	0.24444	0.25944	1.4813	1.0613	4.0909	6.0602	4
5	0.92826	4.7826	9.4228	0.19409	0.20909	1.9702	1.0772	5.1522	10.151	5
6	0.91454	5.6791	13.995	0.16053	0.17553	2.4565	1.0934	6.2295	15.303	6
7	0.90103	6.5982	19.401	0.13656	0.15156	2.9404	1.1098	7.3229	21.532	7
8	0.88771	7.4859	25.615	0.11858	0.13358	3.4218	1.1264	8.4328	28.855	8
9	0.97459	8.3605	32.612	0.10461	0.11961	3.9007	1.1433	9.5593	37.288	9
10	0.86167	9.2221	40.367	0.09343	0.10843	4.3772	1.1605	10.702	46.848	10
11	0.84893	10.071	48.856	0.08429	0.09929	4.8511	1.1779	11.863	57.550	11
12	0.83639	10.907	58.057	0.07668	0.09168	5.3226	1.1956	13.041	69.414	12
13	0.82403	11.731	67.945	0.07024	0.08524	5.7916	1.2135	14.236	82.455	13
14	0.81185	12.543	78.499	0.06472	0.07972	6.2582	1.2317	15.450	96.692	14
15	0.79985	13.343	89.697	0.05994	0.07494	6.7223	1.2502	16.682	112.14	15
16	0.78803	14.131	101.51	0.05577	0.07077	7.1839	1.2689	17.932	128.82	16
17	0.77639	14.907	113.94	0.05208	0.06708	7.6430	1.2880	19.201	146.75	17
18	0.76491	15.672	126.94	0.04881	0.06381	8.0997	1.3073	20.489	165.95	18
19	0.75361	16.426	140.50	0.04588	0.06088	8.5539	1.3268	21.796	186.44	19
20	0.74347	17.168	154.61	0.04325	0.05825	9.0056	1.3458	23.123	208.24	20
21	0.73150	17.900	169.24	0.04087	0.05587	9.4549	1.3670	24.470	231.36	21
22	0.72069	18.620	184.38	0.03870	0.05370	9.9018	1.3875	25.837	255.83	22
23	0.71004	19.330	200.00	0.03673	0.05173	10.346	1.4083	27.225	281.67	23
24	0.69954	20.030	216.09	0.03492	0.04992	10.788	1.4295	28.633	308.90	24
25	0.68921	20.719	232.63	0.03326	0.04826	11.227	1.4509	30.063	337.53	25
26	0.67902	21.398	249.60	0.03173	0.04673	11.664	1.4727	31.514	367.59	26
27	0.66899	22.067	267.00	0.03032	0.04532	12.099	1.4948	32.986	399.11	27
28	0.65910	22.726	284.79	0.02900	0.04400	12.531	1.5172	34.481	432.09	28
29	0.64936	23.376	302.97	0.02778	0.04278	12.961	1.5399	35.998	466.58	29
30	0.63976	24.015	321.53	0.02664	0.04164	13.388	1.5630	37.538	502.57	30
31	0.62099	25.267	359.69	0.02458	0.03958	14.235	1.6103	40.688	579.21	31
34	0.60277	26.481	399.16	0.02276	0.03776	15.073	1.6590	43.933	662.20	34
36	0.58509	27.660	429.83	0.02115	0.03615	15.900	1.7091	47.276	751.73	36
48	0.48936	34.042	703.54	0.01437	0.02937	20.666	2.0434	69.565	1437.6	48
60	0.40930	39.380	988.16	0.01039	0.02539	25.093	2.4432	96.214	2414.3	60
120	0.16752	55.498	2359.7	0.00302	0.01802	42.518	5.9693	331.29	14083.	120
180	0.06857	62.095	3316.9	0.00110	0.01610	53.416	14.584	905.62	48375.	180
240	0.02806	64.795	3870.6	0.00043	0.01543	59.736	35.632	2308.8	137924	240
300	0.01149	659.00	4163.6	0.00017	0.01517	63.180	87.058	5737.2	362474	300
360	0.00470	66.353	4310.7	0.00007	0.01507	64.966	212.70	14113.	916906	360
INF	0.00000	66.666	4444.4	0.00000	0.01500	66.666	INF	INF	INF	INF

	i = 2%			i = 2%			i = 2%			
	Present Sum (P)			Uniform Series (A)			Future Sum (F)			
n	P/F	P/A	P/G	A/F	A/P	A/G	F/P	F/A	F/G	n
1	0.98039	0.9803	0.0000	1.00000	1.02000	0.0000	1.0200	1.0000	0.0000	1
2	0.96117	1.9415	0.9611	0.49505	0.51505	0.4950	1.0404	2.0200	1.0000	2
3	0.94232	2.8838	2.8458	0.32675	0.34675	0.9868	1.0612	3.0604	3.0200	3
4	0.92385	3.8077	5.6173	0.24262	0.26262	1.4752	1.0824	4.1216	6.0804	4
5	0.90573	4.7134	9.2402	0.19216	0.21216	1.9604	1.1040	5.2040	10.202	5
6	0.88797	5.6014	13.680	0.15853	0.17853	2.4422	1.1261	6.3018	15.406	6
7	0.87056	6.4719	18.903	0.13451	0.15451	2.9208	1.1486	7.4342	21.714	7
8	0.85349	7.3254	24.877	0.11651	0.13651	3.3960	1.1716	8.5829	29.148	8
9	0.83676	8.1622	31.572	0.10252	0.12252	3.8680	1.1950	8.6546	37.731	9
10	0.82035	8.9825	38.955	0.09133	0.11133	4.3367	1.2189	10.949	47.486	10
11	0.80426	9.7868	46.997	0.08218	0.10218	4.8021	1.2433	12.168	58.435	11
12	0.78849	10.575	55.671	0.07456	0.09456	5.2642	1.2682	13.412	70.604	12
13	0.77303	11.348	64.947	0.06812	0.08812	5.7230	1.2936	14.680	84.016	13
14	0.75788	12.106	75.799	0.06260	0.08260	6.1786	1.3194	15.973	98.696	14
15	0.74301	12.849	85.202	0.05783	0.07783	6.6309	1.3458	17.293	114.67	15
16	0.72845	13.577	96.128	0.05365	0.07365	7.0799	1.3727	18.639	131.96	16
17	0.71416	14.291	107.55	0.04997	0.06997	7.5256	1.4004	20.012	150.60	17
18	0.70016	14.992	119.45	0.04670	0.06670	7.9681	1.4282	21.412	170.81	18
19	0.68643	15.678	131.81	0.04378	0.06378	8.4073	1.4568	22.840	192.01	19
20	0.67297	16.351	144.60	0.04116	0.06116	8.8432	1.4859	24.297	214.86	20
21	0.65978	17.011	157.79	0.03878	0.05878	9.2759	1.5156	25.783	239.16	21
22	0.65684	17.658	171.37	0.03663	0.05663	9.7054	1.5459	27.299	264.94	22
23	0.63416	18.292	185.33	0.03467	0.05467	10.131	1.5769	28.845	292.24	23
24	0.62172	18.913	199.63	0.03287	0.05287	10.554	1.6084	30.421	321.09	24
25	0.60953	19.523	214.25	0.03122	0.05122	10.974	1.6406	32.030	351.51	25
26	0.59758	20.131	229.19	0.02970	0.04970	11.391	1.6734	33.670	383.54	26
27	0.58586	20.706	244.43	0.02829	0.04829	11.804	1.7068	35.344	417.21	27
28	0.57437	21.281	259.93	0.02699	0.04699	12.214	1.7410	37.051	452.56	28
29	0.56311	21.844	275.70	0.02578	0.04578	12.621	1.7758	38.792	489.61	29
30	0.55207	22.396	291.71	0.02455	0.04465	13.025	1.8113	40.568	528.50	30
32	0.53063	23.468	324.40	0.02261	0.04261	13.823	1.8845	44.227	611.35	32
34	0.51003	24.498	357.88	0.02082	0.04082	14.608	1.9606	48.033	701.69	34
36	0.49002	25.488	392.04	0.01923	0.03923	15.380	2.0398	51.994	799.71	36
48	0.38654	30.673	605.96	0.01260	0.03260	19.755	2.5870	79.353	1567.6	48
60	0.30478	34.760	823.69	0.00877	0.02877	23.696	3.2810	114.05	2702.5	60
120	0.09289	45.355	1710.4	0.00205	0.02205	37.711	10.765	488.25	18412.	120
180	0.02831	48.584	2174.4	0.00058	0.02058	44.755	35.320	1716.0	76802.	180
240	0.00863	49.568	2374.8	0.00017	0.02017	47.911	115.88	5744.4	275222	240
300	0.00263	49.868	2453.9	0.00005	0.02005	49.208	380.23	18961.	930086	300
INF	0.00000	50.000	2500.0	0.00000	0.02000	50.000	INF	INF	INF	INF

n	i = 2.5% Present Sum (P)			i = 2.5% Uniform Series (A)			i = 2.5% Future Sum (F)			n
n	P/F	P/A	P/G	A/F	A/P	A/G	F/P	F/A	F/G	n
1	0.97561	0.9856	0.0000	1.00000	1.02500	0.0000	1.0250	1.0000	0.0000	1
2	0.95181	1.9272	0.9518	0.49383	0.51883	0.4938	1.0506	2.0250	1.0000	2
3	0.92860	2.8560	2.8090	0.32514	0.35014	0.9835	1.0768	3.0756	3.0250	3
4	0.90595	3.7619	5.5268	0.24082	0.26582	1.4691	1.1039	4.1525	6.1006	4
5	0.88385	4.6458	9.0622	0.19025	0.21525	1.9506	1.1314	5.2563	10.253	5
6	0.86230	5.5081	13.373	0.15655	0.18155	2.4280	1.1596	6.3877	15.509	6
7	0.84127	6.3493	18.421	0.13250	0.15750	2.9012	1.1886	7.5474	21.897	7
8	0.82075	7.1701	23.166	0.11447	0.13947	3.3704	1.2184	8.7361	29.444	8
9	0.80073	7.9708	30.572	0.10046	0.12546	3.8355	1.2488	9.4545	38.180	9
10	0.78120	8.7520	38.603	0.08926	0.11426	4.2964	1.2800	11.203	48.135	10
11	0.76214	9.5142	45.224	0.08011	0.10511	4.7533	1.3120	12.483	59.338	11
12	0.74356	10.257	53.403	0.07249	0.09749	5.2061	1.3448	13.795	71.822	12
13	0.72542	10.983	62.108	0.06605	0.09105	5.6549	1.3785	15.140	85.617	13
14	0.70773	11.690	71.309	0.06054	0.08554	6.0995	1.4129	16.519	100.75	14
15	0.69047	12.381	80.975	0.05577	0.08077	6.5401	1.4483	17.931	117.27	15
16	0.67362	13.055	91.080	0.05160	0.07660	6.9766	1.4845	19.380	135.20	16
17	0.65720	13.712	101.59	0.04793	0.07293	7.4091	1.5216	20.864	154.58	17
18	0.64117	14.353	112.49	0.04467	0.06967	7.8375	1.5596	22.396	175.45	18
19	0.62553	14.979	123.75	0.04176	0.06676	8.2619	1.5986	23.946	197.84	19
20	0.61027	15.589	135.35	0.03915	0.06415	8.6823	1.6386	25.544	221.78	20
21	0.59539	16.184	147.25	0.03679	0.06179	9.0986	1.6795	27.183	247.33	21
22	0.58086	16.765	159.45	0.03465	0.05965	9.5109	1.7215	28.862	274.51	22
23	0.56670	17.332	171.92	0.03270	0.05770	9.9193	1.7646	30.584	303.37	23
24	0.55288	17.885	184.63	0.03091	0.05591	10.323	1.8087	32.349	333.96	24
25	0.53939	18.424	197.58	0.02928	0.05428	10.724	1.8539	34.157	366.31	25
30	0.47674	20.930	265.12	0.02278	0.04778	12.666	2.0975	43.902	556.10	30
35	0.42137	23.145	335.88	0.01821	0.04321	14.512	2.3732	54.928	797.12	35
36	0.41109	23.556	350.27	0.01745	0.04245	14.869	2.4325	57.301	852.05	36
40	0.37243	25.102	408.22	0.01484	0.03984	16.262	2.6850	67.402	1096.1	40
48	0.30567	27.773	524.03	0.01101	0.03601	18.868	3.2714	90.859	1714.3	48
50	0.29094	28.362	552.60	0.01026	0.03526	19.483	3.4371	97.484	1899.3	50
60	0.22728	30.908	690.86	0.00735	0.03235	22.351	4.3997	135.99	3039.6	60
100	0.08465	36.614	1125.9	0.00231	0.02731	30.752	11.813	432.54	13301.	100
120	0.05166	37.933	1269.3	0.00136	0.02636	33.463	19.358	734.32	24573.	120
180	0.01174	39.530	1496.6	0.00030	0.02530	37.861	85.171	3366.8	127475	180
240	0.00267	39.893	1570.1	0.00007	0.02507	39.357	374.73	14949.	588381	240
INF	0.00000	40.000	1600.0	0.00000	0.02500	40.000	INF	INF	INF	INF

	i = 3%			i = 3%			i = 3%			
	Present Sum (P)			Uniform Series (A)			Future Sum (F)			
n	P/F	P/A	P/G	A/F	A/P	A/G	F/P	F/A	F/G	n
1	0.97087	0.9708	0.0000	1.00000	1.03000	0.0000	1.0030	1.0000	0.0000	1
2	0.94260	1.9134	0.9426	0.49261	0.52261	0.4926	1.0609	2.0300	1.0000	2
3	0.91514	2.8286	2.7728	0.32353	0.35353	0.9803	1.0927	3.0909	3.0300	3
4	0.88849	3.7171	5.4283	0.23903	0.26903	1.4630	1.1255	4.1836	6.1209	4
5	0.86261	4.5797	8.8887	0.18835	0.21835	1.9409	1.1592	5.3091	10.304	5
6	0.83748	5.4171	13.076	0.15460	0.18460	2.4138	1.1940	6.4684	15.613	6
7	0.81309	6.2302	17.954	0.13051	0.16051	2.8818	1.2298	7.6624	22.082	7
8	0.78941	7.0196	23.480	0.11246	0.14246	3.3449	1.2667	8.8923	29.744	8
9	0.76642	7.7861	29.611	0.09843	0.12843	3.8031	1.3047	10.159	38.636	9
10	0.74409	8.5302	36.308	0.08723	0.11723	4.2565	1.3439	11.463	48.796	10
11	0.72242	9.2526	43.533	0.07808	0.10808	4.7049	1.3842	12.807	60.259	11
12	0.70138	9.9540	51.248	0.07046	0.10046	5.1485	1.4257	14.192	73.067	12
13	0.68095	10.635	59.419	0.06403	0.09403	5.5872	1.4685	15.617	87.259	13
14	0.66112	11.296	68.014	0.05853	0.08853	6.0210	1.5125	17.086	102.87	14
15	0.64186	11.937	77.000	0.05377	0.08377	6.4500	1.5579	18.598	119.96	15
16	0.62317	12.561	86.347	0.04961	0.07961	6.8742	1.6047	20.156	138.56	16
17	0.60502	13.166	96.028	0.04595	0.07595	7.2935	1.6528	21.761	158.72	17
18	0.58739	13.753	106.01	0.04271	0.07271	7.7081	1.7024	23.414	180.48	18
19	0.57029	14.323	116.27	0.03981	0.06981	8.1178	1.7535	25.116	203.89	19
20	0.55368	14.877	126.79	0.03722	0.06722	8.5228	1.8061	26.870	229.01	20
21	0.53755	15.415	137.55	0.03487	0.06487	8.9230	1.8602	28.676	255.88	21
22	0.52189	15.936	148.50	0.03275	0.06275	9.3185	1.9161	30.536	284.55	22
23	0.50669	16.443	159.65	0.03081	0.06081	9.7093	1.9735	32.452	315.09	23
24	0.49193	16.935	170.97	0.02905	0.05905	10.095	2.0327	34.426	347.54	24
25	0.47761	17.413	182.43	0.02743	0.05743	10.476	2.0937	36.459	381.97	25
26	0.46369	17.876	194.02	0.02594	0.05594	10.853	2.1565	38.553	418.43	26
28	0.43708	18.764	217.53	0.02329	0.05329	11.593	2.2879	42.930	497.69	28
30	0.41199	19.600	241.36	0.02102	0.05102	12.314	2.4272	47.575	585.84	30
35	0.35538	21.487	301.62	0.01654	0.04654	14.037	2.8138	60.462	848.73	35
36	0.34503	21.832	313.70	0.01580	0.04580	14.368	2.8982	63.275	909.19	36
40	0.30665	23.114	351.75	0.01326	0.04326	15.650	3.2620	75.401	1180.0	40
45	0.26444	24.518	420.63	0.01079	0.04079	17.155	3.7816	92.719	1590.6	45
48	0.24200	25.266	455.02	0.00958	0.03958	18.008	4.1322	104.40	1880.2	48
50	0.22811	25.729	477.48	0.00887	0.03887	18.557	4.3839	112.79	2093.2	50
60	0.16973	27.675	583.05	0.00613	0.03613	21.067	5.8916	163.05	3435.1	60
70	0.12630	29.123	676.08	0.00434	0.03434	23.214	7.9178	230.59	5353.1	70
80	0.09398	30.200	756.08	0.00311	0.03311	23.035	10.640	321.36	8045.4	80
90	0.06693	31.002	823.63	0.00226	0.03226	26.566	14.300	443.34	11778.	90
100	0.05203	31.598	879.85	0.00165	0.03165	27.844	19.218	607.28	16909.	100
120	0.02881	32.373	963.86	0.00089	0.03089	29.773	34.711	1123.7	33456.	120
180	0.00489	33.170	1076.3	0.00015	0.03015	32.448	204.50	6783.4	220115	180
INF	0.00000	33.333	1111.1	0.00000	0.03000	33.333	INF	INF	INF	INF

	i = 4%			*i* = 4%			*i* = 4%			
	Present Sum (*P*)			Uniform Series (*A*)			Future Sum (*F*)			
n	*P/F*	*P/A*	*P/G*	*A/F*	*A/P*	*A/G*	*F/P*	*F/A*	*F/G*	*n*
1	0.96154	0.9615	0.0000	1.00000	1.04000	0.0000	1.0400	1.0000	0.0000	1
2	0.92456	1.8860	0.9245	0.49020	0.53020	0.4902	1.0816	2.0400	1.0000	2
3	0.88900	2.7750	2.7025	0.32035	0.36035	0.9738	1.1248	3.1216	3.0400	3
4	0.85480	3.6299	5.2669	0.23549	0.27549	1.4510	1.1669	4.2464	6.1616	4
5	0.82193	4.4518	8.5546	0.18463	0.22463	1.9216	1.2166	5.4163	10.408	5
6	0.79031	5.2421	12.506	0.15076	0.19076	2.3857	1.2653	6.6329	15.824	6
7	0.75992	6.0020	17.065	0.12661	0.16661	2.8433	1.3159	7.8982	22.457	7
8	0.73069	6.7327	22.180	0.10853	0.14853	3.2944	1.3685	9.2142	30.355	8
9	0.70259	7.4353	27.801	0.09499	0.13499	3.7390	1.4233	10.582	39.569	9
10	0.67556	8.1109	33.881	0.08329	0.12329	4.1772	1.4802	12.006	50.152	10
11	0.65958	8.7604	40.377	0.07415	0.11415	4.6090	1.5394	13.486	62.158	11
12	0.62460	9.3850	47.247	0.06655	0.10655	5.0343	1.6010	15.025	75.645	12
13	0.60057	9.9856	54.454	0.06014	0.10014	5.4532	1.6650	16.626	90.670	13
14	0.57748	10.563	61.981	0.05467	0.09467	5.8658	1.7316	18.291	107.29	14
15	0.55526	11.118	69.735	0.04994	0.08994	6.2720	1.8009	20.023	125.59	15
16	0.53391	11.652	77.744	0.04582	0.08582	6.6720	1.8729	21.824	145.61	16
17	0.51337	12.165	85.958	0.04220	0.08220	7.0656	1.9479	23.697	167.43	17
18	0.49363	12.659	94.349	0.03899	0.07899	7.4530	2.0258	25.645	191.13	18
19	0.47464	13.133	102.89	0.03614	0.07614	7.8341	2.1068	27.671	216.78	19
20	0.45639	13.590	111.56	0.03358	0.07358	8.2091	2.1911	29.788	244.45	20
21	0.43883	14.029	120.34	0.03128	0.07128	8.5779	2.2787	31.969	274.23	21
22	0.42196	14.451	129.20	0.02920	0.06920	8.9406	2.3699	34.248	306.19	22
23	0.40573	14.856	138.12	0.02731	0.06731	9.2972	2.4647	36.617	340.44	23
24	0.39012	15.247	147.10	0.02559	0.06559	9.6479	2.5633	39.082	377.06	24
25	0.37512	15.622	156.10	0.02401	0.06401	9.9925	2.6658	41.645	416.14	25
26	0.36069	15.982	165.12	0.02257	0.06257	10.331	2.7724	44.311	457.79	26
28	0.33348	16.663	183.14	0.02001	0.06001	10.990	2.9987	49.967	549.19	28
30	0.30832	17.292	201.06	0.01783	0.05783	11.627	3.2434	56.084	652.12	30
35	0.25342	18.664	224.87	0.01358	0.05358	13.119	3.9460	73.652	966.30	35
36	0.24367	18.908	253.40	0.01289	0.05289	13.401	4.1039	77.598	1039.0	36
40	0.20829	19.792	286.51	0.01052	0.05052	14.476	4.8010	95.025	1375.6	40
45	0.17120	20.720	325.40	0.00826	0.04826	15.704	5.8411	121.02	1900.7	45
48	0.15219	21.195	347.24	0.00718	0.04718	16.383	6.5705	139.26	2281.5	48
50	0.14071	21.482	361.16	0.00655	0.04655	16.812	7.1066	152.66	2566.6	50
55	0.11300	22.108	383.08	0.00523	0.04523	17.807	8.6463	191.15	3403.9	55
60	0.09506	22.623	422.99	0.00420	0.04420	18.697	10.519	237.99	4449.7	60
70	0.06422	23.394	472.47	0.00275	0.04275	20.196	15.571	364.29	7357.2	70
80	0.04338	23.915	511.11	0.00181	0.04181	21.371	23.049	551.24	11781.	80
90	0.02931	24.267	540.73	0.00121	0.04121	22.282	34.119	827.98	18449.	90
100	0.01980	24.505	463.12	0.00081	0.04081	22.980	50.504	1237.6	28440.	100
120	0.00904	24.774	592.24	0.00036	0.04036	23.905	110.66	2741.5	65539.	120
180	0.00086	24.978	620.59	0.00003	0.04003	24.845	1164.1	29078.	724456	180
INF	0.00000	25.000	625.00	0.00000	0.04000	25.000	INF	INF	INF	INF

	i = 5%			i = 5%			i = 5%			
	Present Sum (P)			Uniform Series (A)			Future Sum (F)			
n	P/F	P/A	P/G	A/F	A/P	A/G	F/P	F/A	F/G	n
1	0.95238	0.9523	0.0000	1.00000	1.05000	0.0000	1.0500	1.0000	0.0000	1
2	0.90703	1.8594	0.9070	0.48780	0.53780	0.4878	1.1025	2.0500	1.0000	2
3	0.86384	2.7232	2.6347	0.31721	0.36721	0.9674	1.1576	3.1525	3.0500	3
4	0.82270	3.5459	5.1028	0.23201	0.28201	1.4309	1.2155	4.3101	6.2025	4
5	0.78353	4.3294	8.2369	0.18097	0.23097	1.9025	1.2762	5.5256	10.512	5
6	0.74622	5.0756	11.968	0.14702	0.19702	2.3579	1.3401	6.8019	16.038	6
7	0.71068	5.7863	16.232	0.12282	0.17282	2.8052	1.4071	8.1420	22.840	7
8	0.67684	6.4632	20.970	0.10472	0.15472	3.2445	1.4774	9.5491	30.982	8
9	0.64461	7.1078	26.126	0.09069	0.14069	3.6757	1.5513	11.026	40.531	9
10	0.61391	7.7217	31.652	0.07950	0.12950	4.0990	1.6288	12.577	51.557	10
11	0.58468	8.3054	37.498	0.07039	0.12039	4.5144	1.7103	14.206	64.135	11
12	0.55684	8.8632	43.624	0.06283	0.11283	4.9219	1.7958	15.917	78.342	12
13	0.53032	9.3935	49.987	0.05646	0.10646	5.3215	1.8856	17.713	94.259	13
14	0.50507	9.8986	56.553	0.05102	0.10102	5.7132	1.9799	19.598	111.97	14
15	0.48102	10.379	63.288	0.04634	0.09634	6.0973	2.0789	21.578	131.57	15
16	0.45811	10.837	70.159	0.04227	0.09227	6.4736	2.1828	23.657	153.15	16
17	0.43630	11.274	77.140	0.03870	0.08870	6.8422	2.2920	25.840	176.80	17
18	0.41552	11.689	84.204	0.03555	0.08555	7.2033	2.4066	28.132	202.64	18
19	0.39573	12.085	91.327	0.03275	0.08275	7.5569	2.5269	30.539	230.78	19
20	0.37689	12.462	98.488	0.03024	0.08024	7.9029	2.6533	33.066	261.31	20
21	0.35894	12.821	105.66	0.02800	0.07800	8.2416	2.7859	35.719	294.38	21
22	0.34185	13.163	112.84	0.02597	0.07597	8.5729	2.9252	38.505	330.10	22
23	0.32557	13.488	120.00	0.02414	0.07414	8.8970	3.0715	41.430	368.61	23
24	0.31007	13.798	127.14	0.02247	0.07247	9.2139	3.2251	44.502	410.04	24
25	0.29530	14.093	134.22	0.02095	0.07095	9.5237	3.3863	47.727	454.54	25
25	0.28124	14.375	141.25	0.01956	0.06956	9.8265	3.3556	51.113	502.26	26
28	0.25509	14.898	155.11	0.01712	0.06712	10.411	3.9201	58.402	608.05	28
30	0.23138	15.372	168.62	0.01505	0.06505	10.969	4.3219	66.438	728.77	30
35	0.18129	16.374	200.58	0.01107	0.06107	12.249	5.5160	90.320	1106.4	35
36	0.17266	16.546	206.62	0.01043	0.06043	12.487	5.7918	95.836	1196.7	36
40	0.14205	17.159	229.54	0.00828	0.05828	13.377	7.0399	120.80	1616.0	40
45	0.11130	17.774	255.31	0.00626	0.05626	14.364	8.9850	159.70	2294.0	45
48	0.09614	18.077	269.24	0.00532	0.05532	14.894	10.401	188.02	2800.5	48
50	0.08720	18.255	277.91	0.00478	0.05478	15.223	11.467	209.34	3186.9	50
55	0.06833	18.633	297.51	0.00367	0.05367	15.966	14.635	272.71	4354.2	55
60	0.05354	18.929	314.34	0.00283	0.05283	16.606	18.679	353.58	5871.6	60
70	0.03287	19.432	340.84	0.00170	0.05170	17.621	30.426	588.52	10370.	70
80	0.02018	19.596	359.64	0.00103	0.05203	18.352	49.561	971.22	17824.	80
90	0.01239	19.752	372.74	0.00063	0.05063	18.781	80.730	1594.6	30092.	90
100	0.00760	19.847	381.74	0.00038	0.05038	19.233	131.50	2610.0	50200.	100
120	0.00287	19.942	391.97	0.00014	0.05014	19.655	348.91	6958.2	136765.	120
INF	0.00000	20.000	40.000	0.00000	0.05000	20.000	INF	INF	INF	INF

	i = 6%			*i* = 6%			*i* = 6%			
	Present Sum (*P*)			Uniform Series (*A*)			Future Sum (*F*)			
n	*P/F*	*P/A*	*P/G*	*A/F*	*A/P*	*A/G*	*F/P*	*F/A*	*F/G*	*n*
1	0.94340	0.9434	0.0000	1.00000	1.05000	0.0000	1.0600	1.0000	0.0000	1
2	0.89000	1.8333	0.8900	0.48544	0.54544	0.4854	1.1236	2.0600	1.0000	2
3	0.83962	2.6730	2.5692	0.31411	0.37411	0.9611	1.1910	3.1836	3.0600	3
4	0.79209	3.4651	4.9455	0.22859	0.28859	1.4272	1.2624	4.3746	6.2436	4
5	0.74726	4.2123	7.9345	0.17740	0.23740	1.8836	1.3382	5.6370	10.618	5
6	0.70496	4.9173	11.459	0.14336	0.20336	2.3304	1.4185	6.9753	16.255	6
7	0.66506	5.5823	15.449	0.11914	0.17914	2.7675	1.5036	8.3938	23.230	7
8	0.62741	6.2097	19.841	0.10104	0.16104	3.1952	1.5938	9.8974	31.524	8
9	0.59190	6.8016	24.576	0.08702	0.14702	3.6133	1.6894	11.491	41.521	9
10	0.55839	7.3600	29.602	0.07587	0.13587	4.0220	1.7908	13.180	53.013	10
11	0.52679	7.8868	34.870	0.06679	0.12679	4.4212	1.8983	14.971	66.194	11
12	0.49697	8.3838	40.336	0.05928	0.11928	4.8112	2.0122	16.869	81.165	12
13	0.46884	8.8526	45.962	0.05296	0.11296	5.1919	2.1329	18.882	98.035	13
14	0.44230	9.2949	51.712	0.04758	0.10758	5.5635	2.2609	21.015	116.91	14
15	0.41727	9.7122	57.554	0.04296	0.10296	5.9259	2.3965	23.276	137.93	15
16	0.39365	10.105	63.459	0.03895	0.09895	6.2794	2.5403	25.672	161.20	16
17	0.37136	10.477	69.401	0.03544	0.09544	6.6239	2.6927	28.212	186.88	17
18	0.35034	10.827	75.356	0.03236	0.09236	6.9597	2.8543	30.905	215.09	18
19	0.33051	11.158	81.306	0.02962	0.08962	7.2867	3.0256	33.760	246.00	19
20	0.31180	11.469	97.230	0.02718	0.08718	7.6051	3.2071	36.785	279.76	20
21	0.29416	11.764	93.113	0.02500	0.08500	7.9150	3.3995	39.992	315.54	21
22	0.22751	12.041	98.941	0.02305	0.08305	8.2166	3.6035	42.392	356.53	22
23	0.26180	12.303	104.70	0.02128	0.08128	8.5099	3.8197	46.995	399.92	23
24	0.24698	12.550	110.38	0.01968	0.07968	8.7950	4.0489	50.815	446.92	24
25	0.23300	12.783	115.97	0.01823	0.07823	9.0722	4.2918	54.864	497.74	25
26	0.21981	13.003	121.46	0.01690	0.07690	8.3414	4.5493	59.156	552.60	26
27	0.20737	13.210	126.86	0.01570	0.07570	9.6029	4.8223	63.705	611.76	27
28	0.19563	13.406	132.14	0.01459	0.07459	9.8568	5.1116	68.528	675.46	28
29	0.18456	13.590	137.31	0.01358	0.07358	10.103	5.4183	73.639	743.99	29
30	0.17411	13.764	142.35	0.01265	0.07265	10.342	5.7434	79.058	817.63	30
35	0.13011	14.498	155.74	0.00897	0.06897	11.431	7.6860	111.43	1273.9	35
40	0.09722	15.046	185.95	0.00646	0.06646	12.359	10.285	154.76	1912.7	40
45	0.07265	15.455	203.11	0.00470	0.06470	13.141	13.764	212.74	2795.7	45
50	0.05429	15.761	217.45	0.00344	0.06344	13.786	18.420	290.33	4005.6	50
55	0.04057	15.990	229.32	0.00254	0.06254	14.341	24.650	394.17	5652.8	55
60	0.03031	16.161	239.04	0.00188	0.06188	14.790	32.987	533.12	7885.4	60
65	0.02265	16.289	246.94	0.00139	0.06139	15.160	44.145	719.08	10901.	65
70	0.01693	16.384	253.32	0.00103	0.06103	15.461	59.075	967.93	14965.	70
80	0.00945	16.509	262.54	0.00057	0.06057	15.903	105.79	1746.6	27776.	80
90	0.00528	16.578	268.39	0.00032	0.06032	16.189	189.46	3141.0	50851.	90
100	0.00295	16.617	272.04	0.00018	0.06018	16.371	339.30	5638.3	92306.	100
120	0.00092	16.651	275.68	0.00006	0.06006	16.556	1088.1	18119.	299997.	120
INF	0.00000	16.666	277.77	0.00000	0.06000	16.666	INF	INF	INF	INF

	i = 7%			i = 7%			i = 7%			
	Present Sum (P)			Uniform Series (A)			Future Sum (F)			
n	*P/F*	*P/A*	*P/G*	*A/F*	*A/P*	*A/G*	*F/P*	*F/A*	*F/G*	*n*
1	0.93458	0.9345	0.0000	1.00000	1.07000	0.0000	1.0700	1.0000	0.0000	1
2	0.87344	1.8080	0.8734	0.48309	0.55309	0.4830	1.1449	2.0700	1.0000	2
3	0.81630	2.6243	2.5060	0.31105	0.38105	0.9549	1.2250	3.2149	3.0700	3
4	0.76290	3.3872	4.7947	0.22523	0.29523	1.4155	1.3108	4.4399	6.2849	4
5	0.71299	4.1002	7.6466	0.17389	0.24389	1.8649	1.4025	5.7507	10.724	5
6	0.66634	4.7665	10.978	0.13980	0.20980	2.3032	1.5007	7.1532	16.475	6
7	0.62275	5.3893	14.714	0.11555	0.18555	2.7303	1.6057	8.6540	23.628	7
8	0.58201	5.9713	18.788	0.09747	0.16747	3.1465	1.7181	10.259	32.282	8
9	0.54393	6.5152	23.140	0.08349	0.15349	3.5517	1.8384	11.978	42.542	9
10	0.50835	7.0235	27.715	0.07238	0.14238	3.9460	1.9671	13.816	54.520	10
11	0.47509	7.4986	32.466	0.06336	0.13336	4.3296	2.1048	15.783	68.337	11
12	0.44401	7.9426	37.350	0.05590	0.12590	4.7025	2.2521	17.888	84.120	12
13	0.41496	8.3576	42.330	0.04965	0.11965	5.0648	2.4098	20.140	102.00	13
14	0.38782	8.7454	47.371	0.04434	0.11434	5.4167	2.5785	22.550	122.15	14
15	0.36245	0.1079	52.446	0.03979	0.10979	5.7582	2.7590	24.129	144.70	15
16	0.33873	9.4466	57.527	0.03586	0.10586	6.0896	2.9521	27.888	169.83	16
17	0.31657	9.7632	62.592	0.03243	0.10243	6.4110	3.1588	30.840	197.71	17
18	0.29586	10.059	67.621	0.02941	0.09941	6.7224	3.3799	33.999	228.55	18
19	0.27651	10.335	72.599	0.02675	0.09675	7.0241	3.6165	37.379	262.55	19
20	0.25842	10.594	77.509	0.02439	0.09439	7.3163	3.8696	40.995	299.93	20
21	0.24151	10.835	82.339	0.02229	0.09229	7.5990	4.1405	44.865	349.93	21
22	0.22571	11.061	87.079	0.02041	0.09041	7.8424	4.4304	49.005	385.79	22
23	0.21095	11.272	91.720	0.01871	0.08871	8.1368	4.7405	53.436	434.80	23
24	0.19715	11.469	06.254	0.01719	0.08719	8.3923	5.0723	58.176	477.23	24
25	0.18425	11.653	100.67	0.01581	0.08581	8.6391	5.4274	63.249	546.41	25
26	0.17220	11.825	104.98	0.01456	0.08456	8.8773	5.8073	68.676	609.66	26
27	0.16093	11.986	109.16	0.01343	0.08343	9.1072	6.2138	74.483	678.34	27
28	0.15040	12.137	113.22	0.01239	0.08239	9.3289	6.6488	80.697	752.82	28
29	0.14056	12.277	117.16	0.01145	0.08145	9.5427	7.1142	87.346	833.52	29
30	0.13137	12.409	120.97	0.01059	0.08059	9.7486	7.6122	94.460	920.86	30
35	0.09366	12.947	138.13	0.00723	0.07723	10.668	10.676	138.23	1474.8	35
40	0.06678	13.331	152.29	0.00501	0.07501	11.423	14.974	199.63	2280.5	40
45	0.04761	13.605	163.75	0.00350	0.07350	12.036	21.002	285.74	3439.2	45
50	0.03395	13.800	172.90	0.00246	0.07246	12.528	29.457	406.52	5093.2	50
55	0.02420	13.949	180.12	0.00174	0.07174	12.921	41.315	575.92	7441.8	55
60	0.01726	14.039	185.76	0.00123	0.07123	13.232	57.946	813.52	10764.	60
65	0.01230	14.109	190.14	0.00087	0.07087	13.476	81.272	1146.7	15453.	65
70	0.00877	15.160	193.51	0.00062	0.07062	13.666	113.93	1614.1	22059	70
80	0.00446	14.222	198.07	0.00031	0.07031	13.927	224.23	3189.0	44415.	80
90	0.00227	14.253	200.70	0.00016	0.07016	14.081	441.10	6287.1	88531.	90
100	0.00115	14.269	202.20	0.00008	0.07008	14.170	867.71	12381.	175452	100
120	0.00030	14.281	203.51	0.00002	0.07002	14.250	3357.7	47954.	683345	120
INF	0.00000	14.285	204.08	0.00000	0.07000	14.285	INF	INF	INF	INF

	i = 8%			*i* = 8%			*i* = 8%			
	Present Sum (*P*)			Uniform Series (*A*)			Future Sum (*F*)			
n	*P/F*	*P/A*	*P/G*	*A/F*	*A/P*	*A/G*	*F/P*	*F/A*	*F/G*	*n*
1	0.92593	0.9259	0.0000	1.00000	1.08000	0.0000	1.0800	1.0000	0.0000	1
2	0.85734	1.7832	0.8573	0.48077	0.56077	0.4807	1.1664	2.0800	1.0000	2
3	0.79383	2.5771	2.4450	0.30803	0.38803	0.9487	1.2597	3.2464	3.0800	3
4	0.73503	3.3121	4.6500	0.22192	0.30192	1.4039	1.3604	3.5061	6.3264	4
5	0.68058	3.9927	7.3724	0.17046	0.25046	1.8464	1.4693	5.8666	10.832	5
6	0.63107	4.6228	10.523	0.13632	0.21632	2.2763	1.5868	7.3359	16.699	6
7	0.58349	5.2063	14.024	0.11207	0.19207	2.6936	1.7138	8.9228	24.035	7
8	0.54027	5.7466	17.806	0.09401	0.17401	3.0985	1.8509	10.636	32.957	8
9	0.50025	6.2468	21.808	0.08008	0.16008	3.4910	1.9990	12.487	43.594	9
10	0.46319	6.7100	25.976	0.06903	0.14903	3.8713	2.1589	14.486	56.082	10
11	0.42888	5.1389	30.265	0.06008	0.14003	4.2395	2.3316	16.645	70.568	11
12	0.39711	7.5360	34.633	0.05270	0.13270	4.5957	2.5181	18.977	87.214	12
13	0.36770	4.9037	39.046	0.04652	0.12652	4.9402	2.7196	21.495	106.19	13
14	0.34046	8.2442	43.472	0.04130	0.12130	5.2730	2.9371	24.214	127.68	14
15	0.31524	8.5594	47.885	0.03683	0.11683	5.5944	3.1721	27.152	151.90	15
16	0.29189	8.8513	52.264	0.03298	0.11298	5.9046	3.4259	30.324	179.05	16
17	0.27027	9.1216	56.588	0.02963	0.10963	6.2037	3.7000	33.750	209.37	17
18	0.25025	9.3718	60.842	0.02670	0.10670	6.4920	3.9960	37.450	243.12	18
19	0.23171	9.6036	65.013	0.02413	0.10413	6.7696	4.3157	41.446	280.57	19
20	0.21455	9.8181	69.089	0.02185	0.10185	7.0369	4.6609	45.762	322.02	20
21	0.19866	10.016	73.062	0.01981	0.09983	7.2940	5.0338	50.422	367.78	21
22	0.18394	10.200	76.925	0.01803	0.09803	7.5411	5.4365	55.456	418.20	22
23	0.17032	10.371	80.672	0.01642	0.09642	7.7786	5.8714	60.893	473.66	23
24	0.15770	10.528	84.299	0.01498	0.09498	8.0066	6.3411	66.764	534.55	24
25	0.14602	10.674	87.804	0.01368	0.09368	8.2253	6.8484	7.3105	601.32	25
26	0.13520	10.810	91.184	0.01251	0.09251	8.4351	7.3963	79.954	674.43	26
27	0.12519	10.935	94.439	0.01145	0.09145	8.6362	7.9880	87.350	754.38	27
28	0.11591	11.051	97.568	0.01049	0.09049	8.8288	8.6271	95.338	841.73	28
29	0.10733	11.158	100.57	0.00962	0.08962	9.0132	9.3172	103.96	937.07	29
30	0.09938	11.257	103.45	0.00883	0.08883	9.1897	10.062	113.28	1041.0	30
35	0.06763	11.654	116.09	0.00580	0.08580	9.9610	14.785	172.31	1716.4	35
40	0.04603	11.924	126.04	0.00386	0.08386	10.569	21.724	259.05	2738.2	40
45	0.03133	12.108	133.73	0.00259	0.08259	11.044	31.920	386.50	4268.8	45
50	0.02132	12.233	139.59	0.00174	0.08174	11.410	46.901	573.77	6547.1	50
55	0.01451	12.318	144.00	0.00118	0.08118	11.690	68.913	848.92	9924.0	55
60	0.00988	12.376	147.30	0.00080	0.08080	11.901	101.25	1253.2	14915.	60
65	0.00672	12.416	149.73	0.00054	0.08054	12.060	148.78	1847.2	22278.	65
70	0.00457	12.442	151.53	0.00037	0.08037	12.178	218.60	2720.0	33126	70
80	0.00212	12.473	153.80	0.00017	0.08017	12.330	471.95	5866.9	72586.	80
90	0.00098	12.487	154.99	0.00008	0.08008	12.411	1018.9	12723.	157924.	90
100	0.00045	12.494	155.61	0.00004	0.08004	12.454	2199.7	27484.	342306.	100
INF	0.00000	12.500	156.25	0.00000	0.08000	12.500	INF	INF	INF	INF

	i = 9%			*i* = 9%			*i* = 9%			
	Present Sum (*P*)			Uniform Series (*A*)			Future Sum (*F*)			
n	*P/F*	*P/A*	*P/G*	*A/F*	*A/P*	*A/G*	*F/P*	*F/A*	*F/G*	*n*
1	0.91743	0.9174	0.0000	1.00000	1.09000	0.0000	1.0900	1.0000	0.0000	1
2	0.84168	1.7591	0.8416	0.47847	0.56847	0.4783	1.1881	2.0900	1.0000	2
3	0.77218	2.5312	2.3860	0.30505	0.39505	0.9426	1.2950	3.2781	3.0900	3
4	0.70843	3.2397	4.5113	0.21867	0.30867	1.3925	1.4115	4.5731	6.3681	4
5	0.64993	3.8896	7.1110	0.16709	0.25709	1.8282	1.5486	5.9847	10.941	5
6	0.59627	4.4859	10.092	0.13292	0.22292	2.2497	1.6771	7.5233	16.925	6
7	0.54703	5.0329	13.374	0.10869	0.19869	2.6574	1.8280	9.2004	24.449	7
8	0.50187	5.5348	16.887	0.09067	0.18067	3.0511	1.9925	11.028	33.649	8
9	0.46043	5.9952	20.571	0.07680	0.16680	3.4312	2.1718	13.021	44.678	9
10	0.42241	6.4176	24.372	0.06582	0.15582	3.7977	2.3673	15.192	57.699	10
11	0.38753	6.8051	28.248	0.05695	0.14695	4.1509	2.5804	17.560	72.892	11
12	0.35553	7.1607	32.159	0.04965	0.13965	4.4910	2.8126	20.140	90.452	12
13	0.32618	7.4869	36.073	0.04357	0.13357	4.8181	3.0658	22.953	110.59	13
14	0.29925	7.7861	39.963	0.03843	0.12843	5.1326	3.3417	26.019	133.54	14
15	0.27454	8.0606	43.806	0.03406	0.12406	5.4346	3.6424	29.360	159.56	15
16	0.25187	8.3125	47.584	0.03030	0.12030	5.7244	3.9703	33.003	188.92	16
17	0.23107	8.5436	51.282	0.02705	0.11705	6.0023	4.3276	36.973	221.93	17
18	0.21199	8.7556	54.886	0.02421	0.11421	6.2686	4.7171	41.301	258.90	18
19	0.19449	8.9501	58.386	0.02173	0.11173	6.5235	5.1416	46.018	300.20	19
20	0.17843	9.1285	61.777	0.01955	0.10955	6.7674	5.6044	51.160	346.22	20
21	0.16370	9.2922	65.050	0.01762	0.10762	7.0005	6.1088	56.764	397.38	21
22	0.15018	9.4424	68.204	0.01590	0.10590	7.2232	6.6586	62.873	454.14	22
23	0.13778	9.5802	71.235	0.01438	0.10438	7.4357	7.2578	69.531	517.02	23
24	0.12640	9.7066	74.143	0.01302	0.10302	7.6384	7.9110	76.789	586.55	24
25	0.15597	9.8225	76.926	0.01181	0.10181	7.8316	8.6230	84.700	663.34	25
26	0.10639	9.9289	79.586	0.01072	0.10072	8.0155	9.3991	93.324	748.04	26
27	0.09761	10.026	82.124	0.00973	0.09973	8.1906	10.245	102.72	841.36	27
28	0.08955	10.116	84.541	0.00885	0.09885	8.3571	11.167	112.96	944.09	28
29	0.08215	10.198	86.842	0.00806	0.09806	8.5153	12.172	124.13	1057.0	29
30	0.07537	10.273	89.028	0.00734	0.09734	8.6656	13.267	136.30	1181.1	30
35	0.04899	10.566	98.359	0.00464	0.09464	9.3082	20.414	215.71	2007.9	35
40	0.03184	10.757	105.37	0.00296	0.09296	9.7957	31.409	337.88	3309.8	40
45	0.02069	10.881	110.55	0.00190	0.09190	10.160	48.327	525.85	5342.8	45
50	0.01345	10.961	114.32	0.00123	0.09123	10.429	74.357	815.08	8500.9	50
55	0.00874	11.014	117.03	0.00079	0.09079	10.626	114.40	1260.0	13389.	55
60	0.00568	11.048	118.96	0.00051	0.09051	10.768	176.03	1944.7	20942.	60
65	0.00369	11.070	120.33	0.00033	0.09033	10.870	270.84	2998.2	32592.	65
70	0.00240	11.084	121.29	0.00022	0.09022	10.942	416.73	4619.2	50546.	70
80	0.00101	11.099	122.43	0.00009	0.09009	11.029	986.55	10950.	120784.	80
90	0.00043	11.106	122.97	0.00004	0.09004	11.072	2335.5	25939	287213.	90
100	0.00018	11.109	123.23	0.00002	0.09002	11.093	5529.0	61422.	681363.	100
INF	0.00000	11.111	123.45	0.00000	0.09000	11.111	INF	INF	INF	INF

	i = 10%			i = 10%			i = 10%			
	Present Sum (P)			Uniform Series (A)			Future Sum (F)			
n	P/F	P/A	P/G	A/F	A/P	A/G	F/P	F/A	F/G	n
1	0.90909	0.9090	0.0000	1.00000	1.10000	0.0000	1.1000	1.0000	0.0000	1
2	0.82645	1.7355	0.8264	0.47619	0.57619	0.4761	1.2100	2.1000	1.0000	2
3	0.75131	2.4868	2.3290	0.30211	0.40211	0.9365	1.3310	3.3100	3.1000	3
4	0.68301	3.1698	4.3781	0.21547	0.31547	1.3811	1.4641	4.6410	6.4100	4
5	0.62092	3.7907	6.8618	0.16380	0.26380	1.8101	1.6105	6.1051	11.051	5
6	0.56447	4.3552	9.6842	0.12961	0.22961	2.2235	1.7715	7.1756	17.156	6
7	0.51316	4.8684	12.763	0.10541	0.20541	2.6216	1.9487	9.4871	24.871	7
8	0.46651	5.3349	16.028	0.08744	0.18744	3.0044	2.1435	11.435	34.358	8
9	0.42410	5.7590	19.421	0.07364	0.17364	3.3723	2.3579	13.579	45.794	9
10	0.38554	6.1445	22.891	0.06275	0.16275	3.7254	2.5937	15.937	59.374	10
11	0.35049	6.4950	26.396	0.05396	0.15396	4.0640	2.8531	18.531	75.311	11
12	0.31863	6.8136	29.901	0.04676	0.14676	4.3884	3.1384	21.384	93.842	12
13	0.28966	7.1033	33.377	0.04708	0.14078	4.6987	3.4522	24.522	115.22	13
14	0.26333	7.3666	36.800	0.03575	0.13575	4.9955	3.7975	27.975	139.75	14
15	0.23939	7.6060	40.152	0.03147	0.13147	5.2789	4.1772	31.772	167.72	15
16	0.21763	7.8237	43.416	0.02782	0.12782	5.5493	4.5949	35.949	199.49	16
17	0.19784	8.0215	46.581	0.02466	0.12466	5.8071	5.0544	40.544	235.44	17
18	0.17986	8.2014	40.639	0.02193	0.12193	6.0525	5.5599	45.599	275.99	18
19	0.16351	8.3649	52.582	0.01955	0.11955	6.2681	6.1159	51.159	321.59	19
20	0.14864	8.5136	55.406	0.01746	0.11746	6.5080	6.7275	57.275	372.75	20
21	0.13513	8.6486	58.109	0.01562	0.11562	6.7188	7.4002	64.002	430.02	21
22	0.12285	8.7715	60.689	0.01401	0.11401	6.9188	8.1402	71.402	494.02	22
23	0.11168	8.8832	63.146	0.01257	0.11257	7.1084	8.9543	59.543	565.43	23
24	0.10153	8.9847	65.481	0.01130	0.11130	7.2880	9.8497	88.497	644.97	24
25	0.09230	9.0770	67.696	0.01017	0.11017	7.4579	10.834	98.347	733.47	25
26	0.08391	9.1609	69.794	0.00916	0.10916	6.4186	11.918	109.18	831.81	26
27	0.07628	9.3272	71.777	0.00826	0.10826	7.7704	13.110	121.10	940.99	27
28	0.06934	9.3065	73.659	0.00745	0.10745	7.9137	14.421	134.21	1062.1	28
29	0.06304	9.3696	75.414	0.00673	0.10673	8.0488	15.863	148.63	1196.3	29
30	0.05731	9.4269	77.076	0.00608	0.10608	8.1762	17.449	164.49	1344.9	30
35	0.03558	9.6441	83.987	0.00369	0.10369	8.7086	28.102	271.01	2360.2	35
40	0.02209	9.7790	88.952	0.00226	0.10226	9.0962	45.259	442.59	4025.9	40
45	0.01372	9.8628	92.454	0.00139	0.10139	9.3740	72.890	718.90	6739.0	45
50	0.00852	9.9148	94.888	0.00086	0.10086	9.5704	117.39	1163.0	11139.	50
55	0.00529	9.9471	96.561	0.00053	0.10053	9.7075	189.05	1880.5	18255.	55
60	0.00328	9.9671	97.701	0.00033	0.10033	9.8022	304.48	3034.8	29748.	60
65	0.00204	9.9796	98.470	0.00020	0.10020	9.8671	490.37	4893.7	48287.	65
70	0.00127	9.9873	98.987	0.00013	0.10013	9.9112	789.74	7887.4	78174.	70
80	0.00049	9.9951	99.560	0.00005	0.10005	9.9609	2048.4	20474.	203940.	80
90	0.00019	9.9981	99.811	0.00002	0.10002	9.9830	5313.0	53120.	530302.	90
INF	0.00000	10.000	100.00	0.00000	0.10000	10.000	INF	INF	INF	INF

	i = 11%			i = 11%			i = 11%			
	Present Sum (P)			Uniform Series (A)			Future Sum (F)			
n	P/F	P/A	P/G	A/F	A/P	A/G	F/P	F/A	F/G	n
1	0.90090	0.9009	0.0000	1.00000	1.11000	0.0000	1.1100	1.0000	0.0000	1
2	0.81162	1.7125	0.8116	0.47393	0.58393	0.4739	1.2321	2.1100	1.0000	2
3	0.73119	2.4437	2.2740	0.29921	0.40921	0.9305	1.3676	3.3421	3.1100	3
4	0.65873	3.1024	4.2502	0.21233	0.32233	1.3699	1.5180	4.7097	6.4521	4
5	0.59345	3.6959	6.6240	0.16057	0.27057	1.7922	1.6850	6.2278	11.161	5
6	0.53464	4.2305	9.2972	0.12638	0.23638	2.1976	1.8704	7.9128	17.389	6
7	0.48166	4.7122	12.187	0.10222	0.21222	2.5863	2.0761	9.7832	25.302	7
8	0.43393	5.1461	15.224	0.08432	0.19432	2.9584	2.3045	11.859	35.085	8
9	0.39092	5.5370	18.352	0.07060	0.18060	3.3144	2.5580	14.164	46.945	9
10	0.35218	5.8892	21.521	0.05980	0.16980	3.6544	2.8394	16.722	61.109	10
11	0.31728	6.2065	24.694	0.05112	0.16112	3.9788	3.1517	19.561	77.831	11
12	0.28584	6.4923	27.838	0.04403	0.15403	4.2879	3.4984	22.713	97.392	12
13	0.25751	6.7498	30.929	0.03815	0.14815	4.5821	3.8832	26.211	120.10	13
14	0.23199	6.9818	33.944	0.03323	0.14323	4.8618	4.3104	30.094	146.31	14
15	0.20900	7.1908	36.870	0.02907	0.13907	5.1274	4.7845	34.405	176.41	15
16	0.18829	7.3791	39.695	0.02552	0.13552	5.3793	5.3108	39.189	210.81	16
17	0.16963	7.5487	42.409	0.02247	0.13247	5.6180	5.8950	44.500	250.00	17
18	0.15282	7.7016	45.007	0.01984	0.12984	5.8438	6.5435	50.395	294.50	18
19	0.13768	7.8392	47.485	0.01756	0.12756	6.0573	7.2633	56.939	344.90	19
20	0.12403	7.9633	49.842	0.01558	0.12558	6.2589	8.0623	64.202	401.84	20
21	0.11174	8.0750	52.077	0.01384	0.12384	6.4491	8.9491	72.265	466.04	21
22	0.10067	8.1757	54.191	0.01231	0.12231	6.6282	9.9335	81.214	538.31	22
23	0.09069	8.2664	56.186	0.01097	0.12097	6.7969	11.026	91.147	619.52	23
24	0.08170	8.3481	58.065	0.00979	0.11979	6.9555	12.239	102.17	710.67	24
25	0.07361	8.4217	59.832	0.00874	0.11874	7.1044	13.585	114.41	812.84	25
26	0.06631	8.4880	61.490	0.00781	0.11781	7.2443	15.079	127.99	927.26	26
27	0.05974	8.5478	63.043	0.00699	0.11699	7.3753	16.738	143.07	1055.2	27
28	0.05382	8.6016	64.496	0.00626	0.11626	7.4981	18.579	159.81	1198.3	28
29	0.04849	8.6501	65.854	0.00561	0.11561	7.6131	20.623	178.39	1358.1	29
30	0.04368	8.6937	67.121	0.00502	0.11502	7.7205	22.892	199.02	1536.5	30
35	0.02592	8.8552	72.253	0.00293	0.11293	8.1594	38.574	341.59	2787.1	35
40	0.01538	8.9510	75.778	0.00172	0.11172	8.4659	65.000	581.82	4925.6	40
45	0.00913	9.0079	78.155	0.00101	0.11101	8.6762	109.53	986.63	8560.3	45
50	0.00542	9.0416	79.734	0.00060	0.11060	8.8185	184.56	1668.7	14715.	50
55	0.00322	9.0616	70.771	0.00035	0.11035	8.9134	311.00	2818.2	25120.	55
60	0.00191	9.0735	81.446	0.00021	0.11021	8.9762	524.05	4755.0	42682.	60
65	0.00113	9.0806	81.881	0.00012	0.11012	9.0172	883.06	8018.7	72307.	65
70	0.00067	9.0848	82.161	0.00007	0.11007	9.0438	1488.0	13518.	122258.	70
INF	0.00000	9.0909	82.644	0.00000	0.11000	9.0909	INF	INF	INF	INF

	i = 12%			*i* = 12%			*i* = 12%			
	Present Sum (*P*)			**Uniform Series (*A*)**			**Future Sum (*F*)**			
n	***P/F***	***P/A***	***P/G***	***A/F***	***A/P***	***A/G***	***F/P***	***F/A***	***F/G***	***n***
1	0.89286	0.8928	0.0000	1.00000	1.12000	0.0000	1.1200	1.0000	0.0000	1
2	0.79719	1.6900	0.7971	0.47170	0.59170	0.4717	1.2544	2.1200	1.1000	2
3	0.71178	2.4018	2.2207	0.29635	0.41635	0.9246	1.4049	3.3744	3.1200	3
4	0.63552	3.0373	4.1273	0.20923	0.32923	1.3588	1.5735	4.7793	6.4944	4
5	0.56743	3.6047	6.3970	0.15741	0.27741	1.7745	1.7623	6.3528	11.273	5
6	0.50663	4.1114	8.9301	0.12323	0.24323	2.1720	1.9738	8.1151	17.626	6
7	0.45235	4.5637	11.644	0.09912	0.21912	2.5514	2.2106	10.089	25.741	7
8	0.40388	4.9676	14.471	0.08130	0.20130	2.9131	2.4759	12.299	35.830	8
9	0.36061	5.3282	17.356	0.06768	0.18768	3.2574	2.7730	14.775	48.130	9
10	0.32197	5.6502	20.254	0.05698	0.17698	3.5846	3.1058	17.548	62.906	10
11	0.28748	5.9377	23.128	0.04842	0.16842	3.8952	3.4785	20.654	80.454	11
12	0.25668	6.1943	25.952	0.04144	0.16144	4.1896	3.8959	24.133	101.10	12
13	0.22917	6.4235	28.702	0.03568	0.15568	4.4683	4.3634	28.029	125.24	13
14	0.20462	6.6281	31.362	0.03087	0.15087	4.7316	4.8871	32.392	153.27	14
15	0.18270	6.8108	33.920	0.02682	0.14682	4.9803	5.4735	37.279	185.66	15
16	0.16312	6.9739	36.367	0.02339	0.14339	5.2146	6.1303	42.753	222.94	16
17	0.14564	7.1196	38.697	0.02046	0.14046	5.4353	6.8660	48.883	265.69	17
18	0.13004	7.2496	40.908	0.01794	0.13794	5.6427	7.6899	55.749	314.58	18
19	0.11611	7.3657	42.997	0.01576	0.13576	5.8375	8.6127	63.439	370.33	19
20	0.10367	7.4694	44.967	0.01388	0.13388	6.0202	9.6462	72.052	433.77	20
21	0.09256	7.5620	46.818	0.01224	0.13224	6.1913	10.803	81.698	505.82	21
22	0.08264	7.6446	48.554	0.01081	0.13081	6.3514	12.100	92.502	587.52	22
23	0.07379	7.7184	50.177	0.00956	0.12956	6.5010	13.552	104.60	680.01	23
24	0.06588	7.7843	51.692	0.00846	0.12846	6.6406	15.178	118.15	784.62	24
25	0.05882	7.8431	53.104	0.00750	0.12750	6.7708	17.000	133.33	902.78	25
26	0.05252	7.8956	54.417	0.00665	0.12665	6.8921	19.040	150.33	1036.1	26
27	0.04689	7.9425	55.636	0.00590	0.12590	7.0049	21.324	169.37	1186.4	27
38	0.04187	7.9844	56.767	0.00524	0.12524	7.1097	23.883	190.69	1355.8	28
29	0.03738	8.0218	57.814	0.00466	0.12466	7.2071	26.749	214.58	1546.5	29
30	0.03338	8.0551	58.782	0.00414	0.12414	7.2974	29.959	241.33	1761.1	30
35	0.01894	8.1755	62.605	0.00232	0.12232	7.6576	52.799	431.66	3305.5	35
40	0.01075	8.2437	65.113	0.00130	0.12130	7.8987	93.051	767.09	6059.1	40
45	0.00610	8.2525	66.734	0.00074	0.12074	8.0572	163.98	1358.2	10943.	45
50	0.00346	8.3045	67.762	0.00042	0.12042	8.1597	289.00	2400.0	19583.	50
55	0.00196	8.3169	68.408	0.00024	0.12024	8.2251	509.32	4236.0	34841.	55
60	0.00111	8.3240	68.810	0.00013	0.12013	8.2664	897.59	7471.6	61763.	60
65	0.00063	8.3280	69.058	0.00008	0.12008	8.2922	1581.8	13173.	109241.	65
70	0.00036	8.3303	69.210	0.00004	0.12004	8.3082	2787.8	23223.	192944.	70
INF	0.00000	8.3333	69.4444	0.00000	0.12000	8.3333	INF	INF	INF	INF

	i = 13%			*i* = 13%			*i* = 13%			
	Present Sum (*P*)			Uniform Series (*A*)			Future Sum (*F*)			
n	*P/F*	*P/A*	*P/G*	*A/F*	*A/P*	*A/G*	*F/P*	*F/A*	*F/G*	*n*
1	0.88496	0.8849	0.0000	1.00000	1.13000	0.0000	1.1300	1.0000	0.0000	1
2	0.78315	1.6681	0.7831	0.46948	0.59948	0.4694	1.2769	2.1300	1.0000	2
3	0.69305	2.3611	2.1692	0.29352	0.42352	0.9187	1.4429	3.4069	3.1300	3
4	0.61332	2.9744	4.0092	0.20619	0.33619	1.3478	1.6304	4.8498	6.5369	4
5	0.54276	3.5172	6.1802	0.15431	0.28431	1.7571	1.8424	6.4802	11.386	5
6	0.48032	3.9975	8.5818	0.12015	0.25015	2.1467	2.0819	8.3227	17.857	6
7	0.42506	4.4226	11.132	0.09611	0.22611	2.5171	2.3526	10.404	26.189	7
8	0.37616	4.7987	13.765	0.07839	0.20839	2.8685	2.6584	12.757	36.594	8
9	0.33288	5.1316	16.428	0.06487	0.19487	3.2013	3.0040	15.415	49.351	9
10	0.29459	5.4262	19.079	0.05429	0.18429	3.5161	3.3945	18.419	64.767	10
11	0.26070	5.6869	21.686	0.04584	0.17584	3.8134	3.8358	21.814	83.187	11
12	0.23071	5.9176	24.224	0.03899	0.16899	4.0935	4.3345	25.650	105.00	12
13	0.20416	6.1218	26.674	0.03335	0.16335	4.3572	4.8980	29.984	130.65	13
14	0.18068	6.3024	29.023	0.02867	0.15867	4.6050	5.5347	34.882	160.63	14
15	0.15989	6.4623	31.261	0.02474	0.15474	4.8374	6.2542	40.417	195.51	15
16	0.14150	6.6038	33.384	0.02143	0.15143	5.0552	5.0673	46.671	235.93	16
17	0.12522	6.7290	35.387	0.01861	0.14861	5.2589	7.9860	53.739	282.60	17
18	0.11081	6.8399	37.271	0.01620	0.14620	5.4491	9.0242	61.725	336.34	18
19	0.09806	6.9379	39.036	0.01413	0.14413	5.6265	10.197	70.749	398.07	19
20	0.08678	7.0247	40.685	0.01235	0.14235	5.7917	11.523	80.946	468.82	20
21	0.07680	7.1015	42.221	0.01081	0.14081	5.9453	13.021	92.469	549.76	21
22	0.06796	7.1695	43.648	0.00948	0.13948	6.0880	14.713	105.49	642.23	22
23	0.06014	7.2296	44.971	0.00832	0.13832	6.2204	16.626	120.20	747.73	23
24	0.05323	7.2828	45.196	0.00731	0.13731	6.3430	18.788	136.83	867.93	24
25	0.04710	7.3299	47.326	0.00643	0.13643	6.4565	21.230	155.62	1004.7	25
26	0.04168	7.3716	48.368	0.00565	0.13565	6.5614	23.990	176.85	1160.3	26
27	0.03689	7.4085	49.327	0.00498	0.13498	6.6581	27.109	200.84	1337.2	27
28	0.03264	7.4412	50.209	0.00439	0.13439	6.7474	30.633	227.95	1538.0	28
29	0.02880	7.4700	51.017	0.00387	0.13387	6.8296	34.615	358.58	1766.0	29
30	0.02557	7.4956	51.759	0.00341	0.13341	6.9052	39.115	293.19	2024.6	30
35	0.01388	7.5855	54.614	0.00183	0.13183	7.1998	72.068	546.68	3936.0	35
40	0.00753	7.6343	56.408	0.00099	0.13099	7.3887	132.78	1013.7	7490.0	40
45	0.00409	7.6608	57.514	0.00053	0.13053	7.5076	244.64	1874.1	14070.	45
50	0.00222	7.6752	58.187	0.00029	0.13029	7.5811	450.73	3459.5	26227.	50
55	0.00120	7.6830	58.590	0.00016	0.13016	7.6260	830.45	6380.4	48656.	55
60	0.00065	7.6872	58.831	0.00009	0.13009	7.6530	1530.0	11761.	90015.	60
65	0.00035	7.6895	58.973	0.00005	0.13005	7.6692	2819.0	21677.	166247	65
70	0.00019	7.6908	59.056	0.00003	0.13003	7.6788	5193.8	39945.	306732.	70
INF	0.00000	7.6923	59.171	0.00000	0.13000	7.6923	INF	INF	INF	INF

	i = 14%			i = 14%			i = 14%			
	Present Sum (P)			Uniform Series (A)			Future Sum (F)			
n	P/F	P/A	P/G	A/F	A/P	A/G	F/P	F/A	F/G	n
1	0.87719	0.8771	0.0000	1.00000	1.14000	0.0000	1.1400	1.0000	0.0000	1
2	0.76947	1.6466	0.7694	0.46729	0.60729	0.4672	1.2996	2.1400	1.0000	2
3	0.67497	2.3216	2.1194	0.29073	0.43073	0.9129	1.4815	3.4396	3.1400	3
4	0.59208	2.9137	3.8956	0.20320	0.34320	1.3370	1.6889	4.9211	6.5796	4
5	0.51937	3.4330	5.9731	0.15128	0.29128	1.7398	1.9254	6.6101	11.500	5
6	0.45559	3.8886	8.2510	0.11716	0.25716	2.1218	2.1949	8.5355	18.110	6
7	0.39964	4.2883	10.648	0.09319	0.23319	2.4832	2.5022	10.730	26.646	7
8	0.35056	4.6388	13.102	0.07557	0.21557	2.8245	2.8525	13.232	37.276	8
9	0.30751	4.9463	15.562	0.06217	0.20217	3.1463	3.2519	16.085	50.609	9
10	0.26974	5.2161	17.990	0.05171	0.19171	3.4490	3.7072	19.337	66.695	10
11	0.23662	5.4527	20.356	0.04339	0.18339	3.7333	4.2262	23.044	86.032	11
12	0.20756	5.6601	22.639	0.03667	0.17667	3.9997	4.8179	27.270	109.07	12
13	0.18207	5.8423	24.824	0.03116	0.17116	4.2490	5.4924	32.088	136.24	13
14	0.15971	6.0020	26.900	0.02661	0.16661	4.4819	6.2613	37.581	168.43	14
15	0.14010	6.1421	28.862	0.02281	0.16281	4.6990	7.1379	43.842	206.01	15
16	0.12289	6.2650	30.705	0.01962	0.15692	4.9011	8.1372	50.980	249.86	16
17	0.10780	6.3728	32.430	0.01692	0.15692	5.0888	8.2764	59.117	300.84	17
18	0.09456	6.4674	34.038	0.01462	0.15462	5.2629	10.575	68.394	359.95	18
19	0.08295	6.5503	35.531	0.01266	0.15266	5.4242	12.055	78.969	428.35	19
20	0.07276	6.6231	36.913	0.01099	0.15099	5.5734	13.743	91.024	507.32	20
21	0.06383	6.8969	38.190	0.00954	0.14954	5.7111	15.667	104.76	598.34	21
22	0.05599	6.7429	39.365	0.00830	0.14830	5.8380	17.861	120.42	703.11	22
23	0.04911	6.7920	40.446	0.00723	0.14723	5.9549	20.361	138.29	823.55	23
24	0.04308	6.8351	41.437	0.00630	0.14630	6.0623	23.212	158.65	961.84	24
25	0.03779	6.8729	42.344	0.00550	0.14550	6.1610	26.461	181.87	1120.5	25
26	0.03315	6.9060	43.172	0.00480	0.14480	6.2514	30.166	208.33	1302.3	26
27	0.02908	6.9351	43.928	0.00419	0.14419	6.3342	34.389	238.49	1510.7	27
28	0.02551	6.9606	44.617	0.00366	0.14366	6.4099	39.204	272.88	1749.2	28
29	0.02237	6.9830	45.244	0.00320	0.14320	6.4791	44.693	312.09	2022.1	29
30	0.01963	7.0026	45.813	0.00280	0.14280	6.5422	50.950	356.78	2334.1	30
35	0.01019	7.0700	47.951	0.00144	0.14144	6.7824	98.100	693.57	4704.0	35
40	0.00529	7.1050	49.237	0.00075	0.14075	6.9299	188.88	1342.0	9300.1	40
45	0.00275	7.1232	49.996	0.00039	0.14039	7.0187	363.67	2590.5	18182.	45
50	0.00143	7.1326	50.437	0.00020	0.14020	7.0713	700.23	4994.5	35318.	50
55	0.00074	7.1375	50.691	0.00010	0.14010	7.1020	1348.2	9623.1	68343.	55
60	0.00039	4.1401	50.835	0.00005	0.14005	7.1197	2595.9	18535.	131965	60
65	0.00020	7.1414	50.917	0.00003	0.14003	7.1298	4998.2	35693.	254496.	65
70	0.00010	7.1421	50.963	0.00001	0.14001	7.1355	9623.6	68733.	490351.	70
INF	0.00000	7.1428	51.020	0.00000	0.14000	7.1428	INF	INF	INF	INF

	$i = 15\%$			$i = 15\%$			$i = 15\%$			
	Present Sum (*P*)			Uniform Series (*A*)			Future Sum (*F*)			
n	*P/F*	*P/A*	*P/G*	*A/F*	*A/P*	*A/G*	*F/P*	*F/A*	*F/G*	*n*
1	0.86957	0.8695	0.0000	1.00000	1.15000	0.0000	1.1500	1.0000	0.0000	1
2	0.75614	1.6257	0.7561	0.46512	0.61512	0.4651	1.3225	2.1500	1.0000	2
3	0.65752	2.2832	2.0711	0.28798	0.43798	0.9071	1.5208	3.4725	3.1500	3
4	0.57175	2.8549	3.7864	0.20027	0.35027	1.3262	1.7490	4.9933	6.6225	4
5	0.49718	3.3521	5.7751	0.14832	0.29832	1.7228	2.0113	6.7423	11.615	5
6	0.43233	3.7844	7.9367	0.14424	0.26424	2.0971	2.3130	8.7537	18.358	6
7	0.37594	4.1604	10.192	0.09036	0.24036	2.4498	2.6600	11.066	27.122	7
8	0.32690	4.4873	12.480	0.07285	0.22285	2.7813	3.0590	13.726	38.178	8
9	0.28426	4.7715	14.754	0.05957	0.20957	3.0922	3.5178	16.785	51.905	9
10	0.24718	5.0187	16.979	0.04925	0.19925	3.3832	4.0455	20.303	68.691	10
11	0.21491	5.2337	19.128	0.04107	0.19107	3.6549	4.6523	24.349	88.995	11
12	0.18691	5.4206	21.184	0.03448	0.18448	3.9082	5.3502	29.001	113.34	12
13	0.16253	5.5831	23.135	0.02911	0.17911	4.1437	6.1527	34.351	142.34	13
14	0.14133	5.7244	24.972	0.02469	0.17469	4.3624	7.0757	40.504	176.69	14
15	0.12289	5.8473	26.693	0.02102	0.17102	4.5649	8.1370	47.580	217.20	15
16	0.10686	5.9542	28.296	0.01795	0.16795	4.7522	9.3576	55.717	264.78	16
17	0.09293	6.0471	29.782	0.01537	0.16537	4.9250	10.761	65.075	320.50	17
18	0.08081	6.1279	31.156	0.01319	0.16319	5.0843	12.375	75.836	385.57	18
19	0.07027	6.1982	32.421	0.01134	0.16134	5.2307	14.231	88.211	461.41	19
20	0.06110	6.2593	33.582	0.00976	0.15976	5.3651	16.366	102.44	549.62	20
21	0.05313	6.3124	34.644	0.00842	0.15842	5.4883	18.821	118.81	652.06	21
22	0.04620	6.3586	35.615	0.00727	0.15727	5.6010	21.644	137.63	770.87	22
23	0.04017	6.3988	36.498	0.00628	0.15628	5.7039	24.891	159.27	908.50	23
24	0.03493	6.4337	37.302	0.00543	0.15543	5.7978	28.625	184.16	1067.7	24
25	0.03038	6.4641	38.031	0.00470	0.15470	5.8834	32.919	212.79	1251.9	25
26	0.02642	6.4905	38.691	0.00407	0.15407	5.9612	37.856	245.71	1464.7	26
27	0.02297	6.5135	39.289	0.00353	0.15353	6.0319	43.535	283.56	1710.4	27
28	0.01997	6.5335	39.828	0.00306	0.15306	6.9060	50.065	327.10	1994.0	28
29	0.01737	6.5508	40.314	0.00265	0.15265	6.1540	57.575	377.17	2321.1	29
30	0.01510	6.5659	40.752	0.00230	0.15230	6.2066	66.211	434.74	2698.3	30
35	0.00751	6.6166	42.358	0.00113	0.15113	6.4018	133.17	881.17	5641.1	35
40	0.00373	6.6417	43.283	0.00056	0.15056	6.5167	267.86	1779.0	11593.	40
45	0.00186	6.6542	43.805	0.00028	0.15028	6.5829	538.76	3585.1	23600.	45
50	0.00092	6.6605	44.095	0.00014	0.15014	6.6204	1083.6	7217.7	47784.	50
55	0.00046	6.6636	44.255	0.00007	0.15007	6.6414	2179.6	14524.	95461.	55
60	0.00023	6.6651	44.343	0.00003	0.15003	6.6529	4384.0	29220.	184400.	60
65	0.00011	6.6659	44.390	0.00002	0.15002	6.6592	8817.7	5877.8	391424.	65
INF	0.00000	6.6666	44.444	0.00000	0.15000	6.6666	INF	INF	INF	INF

	i = 20%			i = 20%			i = 20%			
	Present Sum (P)			Uniform Series (A)			Future Sum (F)			
n	P/F	P/A	P/G	A/F	A/P	A/G	F/P	F/A	F/G	n
1	0.83333	0.8333	0.0000	1.00000	1.20000	0.0000	1.2000	1.0000	0.0000	1
2	0.69444	1.5277	0.6944	0.45455	0.65455	0.4545	1.4400	2.2000	1.0000	2
3	0.57870	2.1064	1.8518	0.27473	0.47473	0.8791	1.7280	3.6400	3.2000	3
4	0.48225	2.5887	3.2986	0.18629	0.38629	1.2742	2.0736	5.3680	6.8400	4
5	0.40188	2.9906	4.9061	0.13438	0.33438	1.6405	2.4883	7.4416	12.208	5
6	0.33490	3.3255	6.5806	0.10071	0.30071	1.9788	2.9859	9.9299	19.649	6
7	0.27908	3.6045	8.2551	0.07742	0.27742	2.2901	3.5831	12.915	29.579	7
8	0.23257	3.8371	9.8830	0.06061	0.26061	2.5756	4.2998	16.499	42.495	8
9	0.19381	4.0309	11.433	0.04808	0.24808	2.8364	5.1597	20.798	58.994	9
10	0.16151	4.1924	12.887	0.03852	0.23852	3.0738	6.1917	25.958	79.793	10
11	0.13459	4.3270	14.233	0.03110	0.23110	3.2892	7.4300	32.150	105.75	11
12	0.11216	4.4392	15.466	0.02526	0.22526	3.4841	8.9161	39.580	137.90	12
13	0.09346	4.5326	16.588	0.02062	0.22062	3.6597	10.699	48.496	177.48	13
14	0.07789	4.6105	17.600	0.01689	0.21689	3.8174	12.839	59.195	225.98	14
15	0.06491	4.6754	18.509	0.01388	0.21388	3.9588	15.407	72.035	285.17	15
16	0.05409	4.7295	19.320	0.01144	0.21144	4.0861	18.488	87.442	357.21	16
17	0.04507	4.7746	20.041	0.00944	0.20944	4.1975	22.186	105.93	444.65	17
18	0.03756	4.8121	20.680	0.00781	0.20781	4.2975	26.623	128.22	550.58	18
19	0.03130	4.8435	21.243	0.00646	0.20646	4.3860	31.948	154.74	678.70	19
20	0.02608	4.8695	21.739	0.00536	0.20536	4.4643	38.337	186.68	833.44	20
21	0.02174	4.8913	22.174	0.00444	0.20444	4.5333	46.005	225.02	1020.1	21
22	0.01811	4.9094	22.554	0.00369	0.20369	4.5941	55.206	271.03	1245.1	22
23	0.01509	4.9245	22.886	0.00307	0.20307	4.6475	66.247	326.23	1516.1	23
24	0.01258	4.9371	23.176	0.00255	0.20255	4.6942	79.496	392.48	1842.4	24
25	0.01048	4.9475	23.427	0.00212	0.20212	4.7351	95.396	471.98	2334.9	25
26	0.00874	4.9563	23.646	0.00176	0.20176	4.7708	114.47	567.37	2706.8	26
27	0.00728	4.9636	23.835	0.00147	0.20147	4.8020	137.37	681.85	3274.2	27
28	0.00607	4.9696	23.999	0.00122	0.20122	4.8291	164.84	819.22	3956.1	28
29	0.00506	4.9747	24.140	0.00102	0.20102	4.8526	197.81	984.06	4775.3	29
30	0.00421	4.9789	24.262	0.00085	0.20085	4.8730	237.37	1181.8	5759.4	30
35	0.00169	4.9915	24.661	0.00034	0.20034	4.9407	590.66	2948.3	14566.	35
40	0.00068	4.9966	24.846	0.00014	0.20014	4.9727	1469.7	7343.8	36519.	40
45	0.00027	4.9986	24.931	0.00005	0.20005	4.9876	3657.2	18281.	91181.	45
50	0.00011	4.9994	24.969	0.00002	0.20002	4.9945	9100.4	45497.	227236.	50
55	0.00004	4.9997	24.986	0.00001	0.20001	4.9975	22644.	113219.	565820.	55
INF	0.00000	5.0000	25.000	0.00000	0.20000	5.0000	INF	INF	INF	INF

	i = 25%			i = 25%			i = 25%			
	Present Sum (P)			Uniform Series (A)			Future Sum (F)			
n	P/F	P/A	P/G	A/F	A/P	A/G	F/P	F/A	F/G	n
1	0.80000	0.8000	0.0000	1.00000	1.25000	0.0000	1.2500	1.0000	0.0000	1
2	0.64000	1.4400	0.6400	0.44444	0.69444	0.4444	1.5625	2.2500	1.0000	2
3	0.51200	1.9520	1.6640	0.26230	0.51230	0.8524	1.9531	3.8125	3.2500	3
4	0.40960	2.3616	2.8928	0.17344	0.42344	1.2249	2.4414	5.7656	7.0625	4
5	0.32768	2.6892	4.2035	0.12185	0.37185	1.5630	3.0517	8.2070	12.828	5
6	0.26214	2.9514	5.5142	0.08882	0.33882	1.8683	3.8147	11.258	21.035	6
7	0.20972	3.1611	6.7725	0.06634	0.31634	2.1424	4.7683	15.073	32.293	7
8	0.16777	3.3289	7.9469	0.05040	0.30040	2.3872	5.9604	19.841	47.367	8
9	0.13422	3.4631	9.0206	0.03876	0.28876	2.6047	7.4505	25.802	67.209	9
10	0.10737	3.5705	9.9870	0.03007	0.28007	2.7971	9.3132	33.252	93.011	10
11	0.08590	3.6564	10.846	0.02349	0.27349	2.9663	11.641	42.566	126.26	11
12	0.06872	3.7251	11.602	0.01845	0.26845	3.1145	14.551	54.207	168.83	12
13	0.05498	3.7801	12.261	0.01454	0.26454	3.2437	18.189	68.759	223.03	13
14	0.04398	3.8240	12.833	0.01150	0.26150	3.3559	22.737	86.949	291.79	14
15	0.03518	3.8592	13.326	0.00912	0.25912	3.4529	28.421	109.68	378.74	15
16	0.02815	3.8874	13.748	0.00724	0.25724	3.5366	35.527	138.10	488.43	16
17	0.02252	3.9099	14.108	0.00576	0.25576	3.6083	44.408	173.63	626.54	17
18	0.01801	3.9279	14.414	0.00459	0.25459	3.6697	55.511	218.04	800.17	18
19	0.01441	3.9423	15.674	0.00366	0.25366	3.7221	69.388	273.55	1018.2	19
20	0.01553	3.9538	14.893	0.00292	0.25292	3.7667	86.736	342.94	1291.7	20
21	0.00922	3.9631	15.077	0.00233	0.25233	3.8045	108.42	429.68	1634.7	21
22	0.00738	3.9704	15.232	0.00186	0.25186	3.8364	135.52	538.10	2064.4	22
23	0.00590	3.9763	15.362	0.00148	0.25148	3.8634	169.40	673.62	2602.5	23
24	0.00472	3.9811	15.471	0.00119	0.25119	3.8861	211.75	843.03	3276.1	24
25	0.00378	3.9848	15.561	0.00095	0.25095	3.9051	264.69	1054.7	4119.1	25
26	0.00302	3.9879	15.637	0.00076	0.25076	3.9211	330.87	1319.4	5173.9	26
27	0.00242	3.9903	15.700	0.00061	0.25061	3.9345	413.59	1650.3	6493.4	27
28	0.00193	3.9922	15.752	0.00048	0.25048	3.9457	516.98	2063.9	8143.8	28
29	0.00155	3.9938	15.795	0.00039	0.25039	3.9550	646.23	2580.9	10207.	29
30	0.00124	3.9950	15.831	0.00031	0.25031	3.9628	807.79	3227.1	12788.	30
35	0.00041	3.9983	15.936	0.00010	0.25010	3.9858	2465.1	9856.7	39287.	35
40	0.00013	3.9994	15.976	0.00003	0.25003	3.9946	7523.1	30088.	120195.	40
45	0.00004	3.9998	15.991	0.00001	0.25001	3.9980	22958.	91831.	357146.	45
INF	0.00000	4.0000	16.000	0.00000	0.25000	4.0000	INF	INF	INF	INF

	i = 30%			i = 30%			i = 30%			
	Present Sum (P)			Uniform Series (A)			Future Sum (F)			
n	P/F	P/A	P/G	A/F	A/P	A/G	F/P	F/A	F/G	n
1	0.76923	0.7692	0.0000	1.00000	1.30000	0.0000	1.3000	1.0000	0.0000	1
2	0.59172	1.3609	0.5917	0.43478	0.73478	0.4347	1.6900	2.3000	1.0000	2
3	0.45517	1.8161	1.5020	0.25063	0.55063	0.8270	2.1970	3.9900	3.3000	3
4	0.35013	2.1662	2.5524	0.16163	0.46163	1.1782	2.8561	6.1870	7.2900	4
5	0.26933	2.4355	3.6297	0.11058	0.41058	1.4903	3.7129	9.0431	13.477	5
6	0.20718	2.6427	4.6656	0.07839	0.37839	1.7654	4.8268	12.756	22.520	6
7	0.15937	2.8021	5.6218	0.05687	0.35687	2.0062	6.2748	17.582	35.276	7
8	0.12259	2.9247	6.4799	0.04192	0.34192	2.2155	8.1573	23.857	52.850	8
9	0.09430	3.0190	7.2343	0.03124	0.33124	2.3962	10.604	32.015	76.716	9
10	0.07254	3.0915	7.8871	0.02346	0.32346	2.5512	13.785	42.619	108.73	10
11	0.05580	3.1473	8.4452	0.01773	0.31773	2.6822	17.921	56.405	151.35	11
12	0.04392	3.1902	8.9173	0.01345	0.31345	2.7951	23.298	74.327	207.75	12
13	0.03302	3.2232	9.3135	0.01024	0.32024	2.8894	30.287	97.625	282.08	13
14	0.02540	3.2486	9.6436	0.00782	0.30782	2.9685	39.373	127.91	379.70	14
15	0.01954	3.2682	9.9172	0.00598	0.30598	3.0344	51.185	167.28	507.62	15
16	0.01503	3.2832	10.142	0.00458	0.30458	3.0892	66.541	218.47	674.90	16
17	0.01156	3.2948	10.327	0.00351	0.30351	3.1345	86.504	285.01	893.38	17
18	0.00889	3.3036	10.478	0.00269	0.30269	3.1718	112.45	371.51	1178.3	18
19	0.00684	3.3105	10.601	0.00207	0.30207	3.2024	146.19	483.97	1549.9	19
20	0.00526	3.3157	10.701	0.00159	0.30159	3.2275	190.05	630.16	2033.8	20
21	0.00405	3.3198	10.782	0.00122	0.30122	3.2479	247.06	820.21	2664.0	21
22	0.00311	3.3229	10.848	0.00094	0.30094	3.2646	321.18	1067.2	3484.2	22
23	0.00239	3.3253	10.900	0.00072	0.30072	3.2781	417.53	1388.4	4551.5	23
24	0.00184	3.3271	10.943	0.00055	0.30055	3.2890	542.80	1806.0	5940.0	24
25	0.00142	3.3286	10.977	0.00043	0.30043	3.2978	705.64	2348.8	7746.0	25
26	0.00109	3.3297	11.004	0.00033	0.39933	3.3049	917.33	3054.4	10094.	26
27	0.00084	3.3305	11.026	0.00025	0.30025	3.3106	1192.5	3971.7	13149.	27
28	0.00065	3.3311	11.043	0.00019	0.30019	3.3152	1550.2	5164.3	17121.	28
29	0.00050	3.3316	11.057	0.00015	0.30015	3.3189	2015.3	6714.6	22285.	29
30	0.00038	3.3320	11.068	0.00011	0.30011	3.3218	2620.0	8729.9	29000.	30
35	0.00010	3.3329	11.098	0.00003	0.30003	3.3297	9727.8	32422.	107960.	35
INF	0.00000	3.3333	11.111	0.00000	0.30000	3.3333	INF	INF	INF	INF

	i = 40%			i = 40%			i = 40%			
	Present Sum (P)			Uniform Series (A)			Future Sum (F)			
n	P/F	P/A	P/G	A/F	A/P	A/G	F/P	F/A	F/G	n
1	0.71429	0.7142	0.0000	1.00000	1.40000	0.0000	1.4000	1.0000	0.0000	1
2	0.51020	1.2244	0.5102	0.41667	0.81667	0.4166	1.9600	2.4000	1.0000	2
3	0.36443	1.5889	1.2390	0.22936	0.62936	0.7798	2.7440	4.3600	3.4000	3
4	0.26031	1.8492	2.0199	0.14077	0.54077	1.0923	3.8416	7.1040	7.7600	4
5	0.18593	2.0351	2.7637	0.09136	0.49136	1.3579	5.3782	10.945	14.864	5
6	0.13281	2.1679	3.4277	0.06126	0.46126	1.5811	7.5295	16.323	25.809	6
7	0.09586	2.2628	3.9969	0.04192	0.44192	1.7663	10.541	23.853	42.133	7
8	0.06776	2.3306	4.4712	0.02907	0.42907	1.9185	14.757	34.394	65.986	8
9	0.04840	2.3790	4.8584	0.02034	0.42034	2.0422	20.661	48.152	100.38	9
10	0.03457	2.4135	5.1696	0.01432	0.41432	2.1419	28.925	69.813	149.53	10
11	0.02469	2.4382	5.4165	0.01013	0.41013	2.2214	40.495	98.739	219.34	11
12	0.01764	2.4559	5.6106	0.00718	0.40718	2.2845	56.693	139.23	318.08	12
13	0.01260	2.4685	5.7617	0.00510	0.40510	2.3341	79.371	195.92	457.32	13
14	0.00900	2.4775	5.8787	0.00363	0.40363	2.3728	111.12	275.30	653.25	14
15	0.00643	2.4839	5.9687	0.00259	0.40259	2.4029	156.56	386.42	928.55	15
16	0.00459	2.4885	6.0376	0.00185	0.40185	2.4262	217.79	541.98	1314.9	16
17	0.00328	2.4918	6.0901	0.00132	0.40132	2.4440	304.91	759.78	1856.9	17
18	0.00234	2.4941	6.1299	0.00094	0.40094	2.4577	426.87	1064.7	2616.7	18
29	00.0167	2.4958	6.1600	0.00067	0.40067	2.4681	597.63	1491.5	3681.4	19
20	0.00120	2.4970	6.1827	0.00048	0.40048	2.4760	836.68	2089.2	5173.0	20
21	0.00085	2.4978	6.1998	0.00034	0.40034	2.4820	1171.3	2925.8	7262.2	21
22	0.00061	2.4984	6.2126	0.00024	0.40024	2.4865	1639.9	4097.2	10188.	22
23	0.00044	2.4989	6.2222	0.00017	0.40017	2.4899	2295.8	5737.1	14285.	23
24	0.00031	2.4992	6.2293	0.00012	0.40012	2.4935	3214.2	8033.0	20022.	24
25	0.00022	2.4994	6.2347	0.00009	0.40009	2.4944	4499.8	11247.	28055.	25
INF	0.00000	2.5000	6.2500	0.00000	0.40000	2.5000	INF	INF	INF	INF

	i = 50%			i = 50%			i = 50%			
	Present Sum (P)			Uniform Series (A)			Future Sum (F)			
n	P/F	P/A	P/G	A/F	A/P	A/G	F/P	F/A	F/G	n
1	0.66667	0.6666	0.0000	1.00000	1.50000	0.0000	1.5000	1.0000	0.0000	1
2	0.44444	1.1111	0.4444	0.40000	0.90000	0.4000	2.2500	2.5000	1.0000	2
3	0.29630	1.4074	1.0370	0.21053	0.71053	0.7368	3.3750	4.7500	3.5000	3
4	0.19753	1.6049	1.6296	0.12308	0.62308	1.1053	5.0625	8.1250	8.2500	4
5	0.13169	1.7366	2.1563	0.07583	0.57583	1.2417	7.5937	13.187	16.375	5
6	0.08779	1.8244	2.5953	0.04812	0.43812	1.4225	11.390	20.781	29.562	6
7	0.05853	1.8829	2.9465	0.03108	0.53108	1.5648	17.085	32.171	50.343	7
8	0.03902	1.9219	3.1296	0.02030	0.52030	1.6751	25.628	49.257	82.515	8
9	0.02601	1.9479	3.4277	0.01335	0.51335	1.7596	38.443	74.886	131.77	9
10	0.01734	1.9653	3.5838	0.00882	0.50882	1.8235	57.665	113.33	206.66	10
11	0.01156	1.9768	3.6994	0.00585	0.50585	1.8713	86.497	170.99	319.99	11
12	0.00771	1.9845	3.7841	0.00388	0.50388	1.9067	129.74	257.49	490.98	12
13	0.00514	1.9897	3.8458	0.00258	0.50258	1.9328	194.62	387.23	748.47	13
14	0.00343	1.9931	3.8903	0.00172	0.50172	1.9518	291.92	581.85	1135.7	14
15	0.00228	1.9954	3.9223	0.00114	0.50114	1.9656	437.89	873.78	1717.5	15
16	0.00152	1.9969	3.9451	0.00076	0.50076	1.9756	656.83	1311.6	2591.3	16
17	0.00101	1.9979	3.9614	0.00051	0.50051	1.9827	985.26	1968.5	3903.0	17
18	0.00068	1.9986	3.9729	0.00034	0.50034	1.9878	1477.8	2953.7	5871.5	18
19	0.00045	1.9991	3.9810	0.00023	0.50023	1.9914	2216.8	4431.6	8825.3	19
20	0.00030	1.9994	3.9867	0.00015	0.50015	1.9939	3325.2	6648.5	13257.	20
21	0.00020	1.9996	3.9907	0.00010	0.50010	1.9957	4987.8	9973.7	19905.	21
22	0.00013	1.9997	3.9935	0.00007	0.50007	1.9970	7481.8	14961.	29879.	22
23	0.00009	1.9998	3.9955	0.00004	0.50004	1.9979	11222.	22443.	44841.	23
24	0.00006	1.9998	3.9969	0.00003	0.50003	1.9985	16834.	33666.	67284.	24
25	0.00004	1.9999	3.9978	0.00002	0.50002	1.9990	25251.	50500.	100951.	25
INF	0.00000	2.0000	4.0000	0.00000	0.50000	2.0000	INF	INF	INF	INF

Appendix C: Using Spreadsheets to Solve Engineering Economic Problems

The formulas used in spreadsheets to calculate the engineering economic solutions for the factors covered in this book are the following:

Present worth = PV
Future value = FV
Annuity = PMT
Net present worth = NPV
Internal rate of return = IRR
Effective interest rate= EFFECT
Straight line depreciation = SLN
Sum-of-the-years digits depreciation = SYD
Double declining balance depreciation = DDB

To calculate the different factors, the appropriate factor is listed as (=factor) and it is entered along with all of the different variables in parenthesis according to their location in the spreadsheet.

For example, to calculate present worth, first, enter all of the relevant variables in the following order:

A1) Interest rate
A2) Periods
A3) Uniform series
A4) Future value
A5) Present value

If a formula does not use one of the variables, enter a zero instead of a value.

For the present worth, enter the following after all of the variables are entered into the spreadsheet:

=PV(A1,A2,A3,A4)

where A1 through A4 are the cells where the values are listed for the variables. The present value will be calculated by the spreadsheet and displayed on the spreadsheet rather than the formula.

The following are examples that demonstrate using spreadsheets to calculate the answers to engineering economic problems. The first set of numbers demonstrates entering the data into the spreadsheet and the second set of numbers is the resulting spreadsheet with the answers.

Note: Values in parenthesis represent negative numbers.

Converting a present value into a future worth	
Interest rate	B5) 0.1
Periods	B6) 5
Uniform series	B7) 0
Present value	B8) $1,000.00
Future value	=FV(B5,B6,B7,B8)
Interest rate	10%
Periods	5
Uniform series	0
Present value	$1,000.00
Future value	($1,610.51)

Converting a future value into a present worth	
Interest rate	B13) 0.10
Periods	B14) 5
Uniform series	B15) 0
Future value	B16) $1,000.00
	=PV(B13,B14,B15,B16)
Interest rate	10%
Periods	5
Uniform series	0
Future value	$1,000.00
Present value	($620.92)

Converting a uniform series into a future worth	
Interest rate	B21) 0.09
Periods	B22) 10
Uniform series	B23) $500.00
Present value	B24) 0
Future value	=FV(B21,B22,B23,B24)
Interest rate	9%
Periods	10
Uniform series	$500.00
Present value	0
Future value	($7,596.46)

Converting a future value into a uniform series	
Interest rate	B29) 0.1
Periods	B30) 20
Present value	B32) 0
Future value	B32) $100,000.00
Uniform series	=PMT(B29,B30,B31,B32)

(*Continued*)

Converting a future value into a uniform series (Continued)	
Interest rate	10%
Periods	20
Present value	0
Future value	$100,000.00
Uniform series	($1,745.96)

Converting a uniform series into a present worth	
Interest	B37) 0.06
Periods	B38) 5
Uniform series	B39) $500.00
Future value	B40) 0
Present value	=PV(B37,B38,B39, B40)
Interest rate	6%
Periods	5
Uniform series	$500.00
Present value	($2,106.18)

Converting a nonuniform series into a present worth		
Interest rate	B44) 0.12	
Period	Amount	PV
A46) 0	B46) 0	C46) =PV(B44,A46,0,B46)
A47) 1	B47) $1,000.00	C47) =PV(B44,A47,0,B47)
A48) 2	B48) $1,000.00	C48) =PV(B44,A48,0,B48)
A49) 3	B49) $2,000.00	C49) =PV(B44,A49,0,B49)
A50) 4	B50) $1,000.00	C50) =PV(B44,A50,0,B50)
A51) 5	B51) $1,000.00	C51) =PV(B44,A51,0,B51)
	Total	=SUM(C46:C51)
Interest rate	12%	
Period	Amount	PV
0	0	$0.00
1	$1,000.00	($892.86)
2	$1,000.00	($797.19)
3	$2,000.00	($1,423.56)
4	$1,000.00	($635.52)
5	$1,000.00	($567.43)
	Total	($4,316.56)

Calculating net present worth

Year	Cash Flow	MARR
0	B57) ($120,000.00)	C57) 0.2
1	B58) $48,000.00	
2	B59) $48,000.00	
3	B60) $48,000.00	
4	B61) $48,000.00	
5	B62) $60,000.00	
	=NPV(C57,B58,B59,B60,B61,B62)+B57)	

Year	Cash Flow	MARR
0	($120,000.00)	20%
1	$48,000.00	
2	$48,000.00	
3	$48,000.00	
4	$48,000.00	
5	$60,000.00	
	$28,371.91	

Calculating net future worth

MARR	B68) 0.2	
Year	**Cash Flow**	
0	B70) ($120,000.00)	=FV(B68,5,0,B70)
1	B71) $48,000.00	=FV(B68,4,0,B71)
2	B72) $48,000.00	=FV(B68,3,0,B72)
3	B73) $48,000.00	=FV(B68,2,0,B73)
4	B74) $48,000.00	=FV(B68,1,0,B74)
5	B75) $60,000.00	=B75
	Future worth	=SUM(B70:B75)

MARR	20%	
Year	**Cash Flow**	
0	(120,000)	$298,598.40
1	$48,000.00	$99,532.80
2	$48,000.00	$82,944.00
3	$48,000.00	$69,120.00
4	$48,000.00	$57,600.00
5	$60,000.00	$60,000.00
	Future worth	$70,598.00

Calculating annual equivalent worth of present and future values	
MARR	B80) 0.18
Purchase price	B81) ($65,000.00)
Salvage value	B82) $15,000.00
Life years	B83) 7
Annual profit (or cost)	B84) $17,100.00
Annual equivalent	=PMT(B80,B81,B82,B83,B84)
MARR	18%
Purchase price	($65,000.00)
Salvage value	$15,000.00
Life years	7
Annual profit (or cost)	$17,100.00
Annual equivalent	$1,181.90

Calculating rate of return	
Year	**Cash Flow**
0	B91) ($120,000.00)
1	B92) $48,000.00
2	B93) $48,000.00
3	B94) $48,000.00
4	B95) $48,000.00
5	B96) $60,000.00
Rate of return	=IRR(B91:B96,0.2)
Year	**Cash Flow**
0	($120,000.00)
1	$48,000.00
2	$48,000.00
3	$48,000.00
4	$48,000.00
5	$60,000.00
Rate of return	30%

Determining monthly payments for a loan	
Annual interest rate	B103) 0.09
Monthly interest rate	B104) 0.0075
Years	B105) 30
Months	B106) 360

(Continued)

Determining monthly payments for a loan (Continued)	
Loan principal	B107) $150,000.00
Balloon payment	B108) 0
Payment	=PMT(B104,B106,B107,0)
Annual interest rate	9.00%
Monthly interest rate	0.75%
Years	30
Months	360
Loan principal	$150,000.00
Balloon payment	0
Payment	($1,206.93)

Converting annual percentage rate to annual effective rate	
Nominal interest rate	B113) 0.12
Compounding periods	B114) 4
Effective interest rate	=EFFECT(B113,B114)
Nominal interest rate	12%
Compounding periods	4
Effective interest rate	0.12550881 or 12.55%

Calculating the interest rate when replacing a loan	
Months	B119) 60
Existing loan amount	B120) $55,000.00
Closing costs	B121) $1,590.00
New loan principal	B122) =B120+B121
Monthly payment	=PMT(B122)
Monthly interest rate	0.0052
Nominal interest rate	0.0623
Months	60
Existing loan amount	$55,000.00
Closing costs	$1,590.00
New loan principal	$56,590.00
Monthly payment	($1,100.00)
Monthly interest rate	0.0052
Nominal interest rate	6.23%

Calculating monthly interest, principal reduction, and balance for a loan	
Annual interest rate	B130) 0.09
Monthly interest rate	B131) =B130/12
Monthly payment	B132) $1,206.93
Balance from last month	B133) $150,000.00
Interest for this month	B134) =(B131 * B133)
Principal reduction	B135) =(B132 – B134)
Balance from this month	B136 =B133 – B135
Annual interest rate	9%
Monthly interest rate	0.0075
Monthly payment	$1,206.93
Balance from last month	$150,000.0
Interest for this month	$1,125.00
Principal reduction	$81.93
Balance from this month	$149,918.07

Calculating straight line depreciation		
Purchase price	B140) $110,000.00	
Salvage value	B141) $10,000.00	
Recovery period in years	B142) 5	
n	Depreciation	BVm($)
0		C144) =B140
1	B145) =SLN(B140,B141,B142)	C145) =(C144 – B145)
2	B146) =SLN(B140,B141,B142)	C146) =(C145 – B146)
3	B147) =SLN(B140,B141,B142)	C147) =(C146 – B147)
4	B148) =SLN(B140,B141,B142)	C148) =(C147 – B148)
5	B149) =SLN(B140,B141,B142)	C149) =(C148 – B149)
Purchase price	$110,000.00	
Salvage value	$10,000.00	
Recovery period in years	5	
n	Depreciation	BVm($)
0		$110,000.00
1	$20,000.00	$90,000.00
2	$20,000.00	$70,000.00
3	$20,000.00	$50,000.00
4	$20,000.00	$30,000.00
5	$20,000.00	$10,000.00

Calculating sum-of-the-years digits depreciation		
Purchase price	B153) $110,000.00	
Salvage value	B154) $10,000.00	
Recovery period in years	B156) 5	
n	Depreciation	BVm
0		C157) =B153
1	B158) =SYD(B153,B154,B155,A158)	C158) =C157 − B158
2	B159) =SYD(B153,B154,B155,A159)	C159) =C158 − B159
3	B160) =SYD(B153,B154,B155,A160)	C160) =C159 − B160
4	B161) =SYD(B153,B154,B155,A161)	C161) =C160 − B161
5	B162)=SYD(B153,B154,B155,A162)	C162) =C161 − B162
Purchase price	$110,000.00	
Salvage value	$10,000.00	
Recovery period in years	5	
n	Depreciation	BVm
0		$110,000.00
1	$33,333.33	$76,666.67
2	$26,666.67	$50,000.00
3	$20,000.00	$30,000.00
4	$13,333.33	$16,666.67
5	$6,666.67	$10,000.00

Calculating declining balance depreciation		
Purchase price	B166) $110,000.00	
Salvage value	B167) $10,000.00	
Recovery period in years	B168) 5	
Rate (%)	B169) 200	
n	Depreciation	BVm
0		C171) =B166
1	B172) =DDB(B166,B167,B168,A171,A172)	C172) =C171 − B172
2	B173) =DDB(B166,B167,B168,A172,A173)	C173) =C172 − B173
3	B174) =DDB(B166,B167,B168,A173,A174)	C174) =C173 − B174
4	B175) =DDB(B166,B167,B168,A174,A175)	C175) =C174 − B175
5	B176) =DDB(B166,B167,B168,A175,A176)	C176) =C175 − B176
Purchase price	$110,000.00	
Salvage value	$10,000.00	
Recovery period in years	5	
Rate (%)	200	
n	Depreciation	BVm
0		$110,000.00
1	$44,000.00	$66,000.00
2	$26,400.00	$39,600.00
3	$15,840.00	$23,760.00
4	$9,504.00	$14,256.00
5	$4,256.00	$10,000.00

Appendix D: Derivations of Engineering Economic Equations

D.1 SINGLE PAYMENT COMPOUND AMOUNT FACTOR (*F/P*)

$$F_n = P_0(1+i)^n$$

Derivation:

Year 1

$$F_1 = P_0(1+i)$$

Year 2

$$F_2 = P_0(1+i) + P(1+i)$$

Year 3

$$F_3 = \left[P_0(1+i) + P(1+i)i\right] + \left[P(1+i) + P(1+i)i\right]i$$

$$= P_0(1+i) + 2P(1+i)i + P(1+i)i^2$$

By factoring out $P(1 + i)$, the result would be the following:

$$F_3 = P_0(1+i)(1+2i+i^2)$$

$$= P_0(1+i)(1+i)^2$$

$$= P_0(1+i)^2$$

$$\therefore F_n = P_0(1+i)^n$$

D.2 SINGLE PAYMENT PRESENT WORTH FACTOR (*P/F*)

$$P_0 = F_n\left[\frac{1}{(1+i)^n}\right]$$

Derivation:

This factor is the inverse of the single payment compound amount factor.

D.3 UNIFORM SERIES PRESENT WORTH FACTOR (*P/A*)

$$P_0 = A\left[\frac{(1+i)^n - 1}{i(1+i)^n}\right]$$

Derivation:

This formula is based on the single payment present worth factor. If each year the value is considered to be a future value, then the present worth would be solved for using the following formula:

$$P_0 = F_1\left(\frac{1}{(1+i)^1}\right) + F_2\left(\frac{1}{(1+i)^2}\right) + F_3\left(\frac{1}{(1+i)^3}\right) + \cdots F_n\left(\frac{1}{(1+i)^n}\right)$$

Substituting (*A*) for (*F*) in this formula results in the following equation:

$$P_0 = A_1\left(\frac{1}{(1+i)^1}\right) + A_2\left(\frac{1}{(1+i)^2}\right) + A_3\left(\frac{1}{(1+i)^3}\right) + \cdots A_n\left(\frac{1}{(1+i)^n}\right)$$

If (*A*) is factored out of this equation, it yields:

$$P_0 = A\left[\frac{1}{(1+i)^1} + \frac{1}{(1+i)^2} + \frac{1}{(1+i)^3} + \cdots \frac{1}{(1+i)^n}\right]$$

Next, each side of the equation is multiplied by $\left(\frac{1}{1+i}\right)$ to obtain:

$$\frac{P_0}{(1+i)} = A\left[\frac{1}{(1+i)^2} + \frac{1}{(1+i)^3} + \frac{1}{(1+i)^4} + \cdots \frac{1}{(1+i)^n} + \frac{1}{(1+i)^{n+1}}\right]$$

Subtracting the last equation from the previous equation yields:

$$\frac{P_0}{(1+i)} - P_0 = A\left[-\frac{1}{(1+i)^i} + \frac{1}{(1+i)^{n+1}}\right]$$

Next, factor out (*P*) from this equation and it becomes:

$$P_0\left(\frac{1}{(1+i)}\right) - 1 = A\left[\frac{1}{(1+i)^{n+1}} - \frac{1}{1+i}\right]$$

If this equation is simplified, it becomes:

$$P_0\left(\frac{-i}{(1+i)}\right) = A\left(\frac{1}{1+i}\right)\left[\frac{1}{(1+i)^n} - 1\right]$$

Next, divide the equation by $\frac{-1}{(1+i)}$:

$$P_0 = A\left(\frac{1}{1+i}\right)\left[\frac{\left(\frac{1}{(1+i)^n}\right)-1}{\frac{-i}{(1+i)}}\right]$$

$$= A\left(\frac{1}{-i}\right)\left[\frac{1-(1+i)^n}{(1+i)^n}\right]$$

$$\therefore P_0 = A\left[\frac{(1+i)^n-1}{i(1+i)^n}\right]$$

D.4 CAPITAL RECOVERY FACTOR (*A/P*)

$$A = P_0\left[\frac{i(1+i)^n}{(1+i)^n-1}\right]$$

Derivation:

The capital recovery factor is the inverse of the uniform series present worth factor.

D.5 UNIFORM SERIES SINKING FUND FACTOR (*A/F*)

$$A = F_n\left[\frac{i}{(1+i)^n-1}\right]$$

Derivation:

Combine the single payment present worth factor and the uniform series present worth factor:

$$\text{Combine } P_0 = F_n\left[\frac{1}{(1+i)^n}\right] \quad \text{and} \quad A = P_0\left[\frac{i(1+i)^n}{(1+i)^n-1}\right]$$

$$A = F_n\left[\frac{1}{(1+i)^n}\right]\left[\frac{i(1+i)^n}{(1+i)^n-1}\right]$$

$$\therefore A = F_n\left[\frac{i}{(1+i)^n-1}\right]$$

D.6 UNIFORM SERIES COMPOUND AMOUNT FACTOR (*F/A*)

$$F_n = A\left[\frac{(1+i)^n - 1}{i}\right]$$

Derivation:

This factor is the inverse of the sinking fund factor.

D.7 FUTURE WORTH GRADIENT FACTOR (*F/G*)

$$F_n = G\left[\left(\frac{1}{i}\right)\frac{(1+i)^n - 1}{i} - n\right]$$

Derivation:

The future worth could be viewed as the sum of uniform annual series but for decreasing time periods beginning with $(n-1)$. If the uniform series compound amount factor (*F/A*) is repeated for each yearly value, it would be a series of (*F/A*) equations summed up. The uniform series compound amount factor is the following and below it is the series of (*F/A*) equations with (*G*) substituted for (*A*).

$$F_n = A\left[\frac{(1+i)^n - 1}{i}\right]$$

$$F_n = G\left[\frac{(1+i)^{n-1} - 1}{i}\right] + \frac{(1+i)^{n-2} - 1}{i} + \cdots + \frac{(1+i) - 1}{i}$$

Simplifying this equation yields:

$$\frac{F_n}{i} = \frac{G}{i}\left[(1+i)^{n-1} + (1+i)^{n-2} + \cdots + (1+i) - (n-1)\right]$$

The generalized formula for the uniform series compound amount factor for (*n*) periods is the following:

$$F_n = A\left[(1+i)^{n-1} + (1+i)^{n-2} + \cdots + (1+i)^2 + (1+i)\right] + 1$$

Subtracting this formula from the previous formula results in the following:

$$F_n i/G - F/A = -n$$

or

$$F_n = \frac{G}{i}\left(F/A - n\right)$$

Since

$$F/A = \left[\frac{(1+i)^n - 1}{i}\right]$$

$$\therefore F_n = G\left[\left(\frac{1}{i}\right)\frac{(1+i)^n - 1}{i} - n\right]$$

D.8 PRESENT WORTH GRADIENT FACTOR (*P/G*)

$$P_0 = \frac{G}{i}\left[\frac{(1+i)^n - 1}{i(1+i)^n} - \frac{n}{(1+i)^n}\right]$$

Derivation:

Start with the single payment present worth factor and substitute (*G*) for (*F*) in the derivation formula:

$$P_0 = G\left(\frac{1}{(1+i)^2}\right) + 2G\left(\frac{2}{(1+i)^3}\right) + 3G\left(\frac{3}{(1+i)^4}\right) + \cdots \left[(n-1)G\right]\left(\frac{n-1}{(1+i)^n}\right)$$

Factoring out (*G*) results in the following:

$$P_0 = G\left[\left(\frac{1}{(1+i)^2}\right) + \left(\frac{2}{(1+i)^3}\right) + \left(\frac{3}{(1+i)^4}\right) + \cdots \left(\frac{n-1}{(1+i)^n}\right)\right]$$

Multiply both sides by $\frac{1}{(1+i)^{-1}}$:

$$\frac{P_0}{(1+i)^{-1}} = G\left[\left(\frac{1}{(1+i)^1}\right) + \left(\frac{2}{(1+i)^2}\right) + \left(\frac{3}{(1+i)^3}\right) + \cdots \left(\frac{n-1}{(1+i)^{n-1}}\right)\right]$$

Subtract the formula with (*G*) factored out of it from the previous formula:

$$\frac{P_0}{(1+i)^{-1}} - P_0 = G\left[\left(\frac{1}{(1+i)^1}\right) + \left(\frac{(2-1)}{(1+i)^2}\right) + \left(\frac{(3-2)}{(1+i)^3}\right) + \cdots \left(\frac{(n-1)-(n-2)}{(1+i)^{n-1}} - \frac{n-1}{(1+i)^n}\right)\right]$$

Rearranging the left side of the equation and the last term on the right side of the equation results in the following:

$$P_0(1+i)^1 - P_0 = G\left[\left(\frac{1}{(1+i)^1}\right) + \left(\frac{1}{(1+i)^2}\right) + \left(\frac{1}{(1+i)^3}\right) + \cdots \left(\frac{1}{(1+i)^{n-1}} - \frac{n-1}{(1+i)^n}\right)\right]$$

If (n) is factored out of the last term and the left side of the equation is written as $P_0 + Pi - P$, it results in the following:

$$P_0 i = G\left[\left(\frac{1}{(1+i)^1}\right)+\left(\frac{1}{(1+i)^2}\right)+\left(\frac{1}{(1+i)^3}\right)+\cdots\left(\frac{1}{(1+i)^{n-1}}+\frac{1-n}{(1+i)^n}\right)\right]-\frac{G_n}{(1+i)^n}$$

Dividing both sides of the equation by (i) results in the following:

$$P_0 = \frac{G}{i}\left[\left(\frac{1}{(1+i)^1}\right)+\left(\frac{1}{(1+i)^2}\right)+\left(\frac{1}{(1+i)^3}\right)+\cdots\left(\frac{1}{(1+i)^{n-1}}+\frac{1-n}{(1+i)^n}\right)\right]-\frac{G_n}{i(1+i)^n}$$

The expression in the brackets in the previous formula is the present worth of a uniform series for (n) years. Therefore, if the expression for the (P/A) factor is substituted into the previous formula, and it results in the following:

$$P_0 = \frac{G}{i}\left[\frac{(1+i)^n-1}{i(1+i)^n}\right]-\frac{G_n}{i(1+i)^n}$$

$$= \frac{G}{i}\left[\frac{(1+i)^n-1}{i(1+i)^n}\right]-\frac{n}{i(1+i)^n}$$

$$\therefore P_0 = \frac{G}{i}\left[\frac{(1+i)^n-1}{i(1+i)^n}-\frac{n}{(1+i)^n}\right]$$

D.9 UNIFORM CAPITAL RECOVERY FACTOR (A/G)

$$A = G\left[\frac{1}{i}-\frac{n}{(1+i)^n-1}\right]$$

Derivation:

The uniform capital recovery factory is derived from the present worth gradient factor and the capital recovery factor:

$$P_0 = \frac{G}{i}\left[\frac{(1+i)^n-1}{i(1+i)^n}\right]-\frac{n}{i(1+i)^n} \text{ times } A = P_0\left[\frac{i(1+i)^n}{(1+i)^n-1}\right] \text{ results in the following:}$$

$$A = G\left[\frac{1}{i}-\frac{n}{(1+i)^n-1}\right]$$

Appendix E: Summary of Engineering Economic Equations

Section E.1 lists the engineering economics formulas. Section E.2 provides a numerical listing of all of the equations introduced in Chapters 1 through 14.

E.1 ENGINEERING ECONOMIC FORMULAS

Future worth of a present value

$$F_n = P_0\left(1+i\right)^n$$

$$F_n = P_0(F/P,i,n)$$

Present worth of a future value

$$P_0 = F_n\left(\frac{1}{\left(1+i\right)^n}\right)$$

$$P_0 = F_n(P/F,i,n)$$

Future worth of a uniform series

$$F_n = A\left[\frac{\left(1+i\right)^n - 1}{i}\right]$$

$$F_n = A\left(F/A,i,n\right)$$

Present worth of a uniform series

$$P_0 = A\left[\frac{\left(1+i\right)^n - 1}{i\left(1+i\right)^n}\right]$$

$$P_0 = A\left(P/A,i,n\right)$$

Net present worth

$$\text{NPW} = \pm P_0 \pm \sum A\left[\frac{\left(1+i\right)^n - 1}{i\left(1+i\right)^n}\right] \pm \sum F\left[\frac{1}{\left(1+i\right)^n}\right]$$

$$\text{NPW} = \pm P_0 \pm \sum A\left(P/A,i,n\right) \pm \sum F\left(P/F,i,n\right)$$

Present worth of an infinite uniform series

$$P_0 = \frac{A}{i}$$

Uniform series of an infinite present worth

$$A = P_0(i)$$

Uniform series of a future worth

$$A = F_n\left[\frac{i}{(1+i)^n - 1}\right]$$

$$A = F_n(A/F, i, n)$$

Uniform series of a present worth

$$A = P_0\left[\frac{i(1+i)^n}{(1+i)^n - 1}\right]$$

$$A = P_0(A/P, i, n)$$

Future worth of an arithmetic gradient

$$F_n = G\left[\left(\frac{1}{i}\right)\frac{(1+i)^n - 1}{i} - n\right]$$

$$F_n = G(F/G, i, n)$$

Present worth of an arithmetic gradient

$$P_0 = \frac{G}{i}\left[\left(\frac{(1+i)^n - 1}{i(1+i)^n} - \frac{n}{(1+i)^n}\right)\right]$$

$$P_0 = G(P/G, i, n)$$

Present worth of an infinite arithmetic gradient

$$P_0 = \frac{G}{i^2}$$

Arithmetic gradient of an infinite present worth

$$G = P_0(i^{-2})$$

Uniform series of an arithmetic gradient

$$A = G\left[\frac{1}{i} - \frac{n}{(1+i)^n - 1}\right]$$

$$A = G(A/G, i, n)$$

Future worth of a present value with an effective interest rate

$$F_n = P_0^{i_n \times n}$$

Annual interest rate from a nominal interest rate

$$i = \frac{i_n}{m}$$

m = number of compounding periods per year

Nominal interest rate from an annual interest rate

$$i_n = i \times m$$

m = number of compounding periods per year

Effective interest rate from nominal and annual interest rate

$$i_e = \left(1 + \frac{i_n}{m}\right)^m - 1$$

or

$$i_e = (1+i)^n - 1$$

Continuously compounded interest rate

$$\overline{i_e} = e^{i_n} - 1$$

Interpolation formulas for unknown interest rate

$$i = c + \left(\frac{a}{b}\right)d$$

Interpolation table for unknown interest rate

Table for Setting Up Interpolation Problems for Unknown Rate of Return

	i	*F/P*		
d	*c*	*A*	*a*	*b*
	Unknown *i*	*B*		
	e	*C*		

Note: $a = A - B$, $b = A - C$, $d = c - e$.

Interpolation formula for unknown number of years

$$n = c + \left(\frac{a}{b}\right)d$$

Interpolation table for unknown number of years

Table for Setting Up Interpolation Problems for Unknown Number of Years

	n	F/P	
	c	A	a
d	Unknown n	B	b
	e	C	

Note: $a = A - B$, $b = A - C$, $d = c - e$.

Geometric gradients:

Geometric gradient when $r > i$, then $w = \frac{1+r}{1+i} - 1$, and $P_0 = \frac{C}{1+i}\left(F/A, w, n\right)$

$$P_0 = \frac{C}{1+i}\left[\frac{(1+w)^n - 1}{w}\right]$$

Geometric gradient when $r < i$, then $w = \frac{1+i}{1+r} - 1$, and $P_0 = \frac{C}{1+r}\left(P/A, w, n\right)$

$$P_0 = \frac{C}{1+r}\left[\frac{(1+w)^n - 1}{w(1+w)^n}\right]$$

Geometric gradient when $r = i$, then

$$P_0 = \frac{C \times n}{(1+r)} = \frac{C \times n}{(1+i)}$$

E.2 EQUATIONS FROM CHAPTERS 1 THROUGH 14

Chapter 1

$$i_n = r + e \tag{1.1}$$

Where
- i_n is the nominal interest rate
- r is the real interest rate
- e is the inflation rate

$$\text{Net worth} = \text{Equity of the firm (summary of the value of the company and its assets)} - \text{The liabilities of the firm} \tag{1.2}$$

Chapter 2

$$\text{Interest owed at the end of 1 year} = \frac{\text{Interest rate}}{100} \times \text{Amount borrowed or loaned} \tag{2.1}$$

$$\text{Total amount owed} = \text{Amount borrowed} + \left(\frac{\text{Interest rate}}{100} \times \text{Amount borrowed}\right) \tag{2.2}$$

$$F_1 = \text{Principal} + \left(\frac{\text{Interest rate}}{100} \times \text{Principal}\right) \tag{2.3}$$

$$\text{Interest for year 2} = (\text{Principal} + \text{First year's interest}) \times \frac{\text{Interest rate}}{100} \tag{2.4}$$

$$F_2 = \text{Principal} + \text{First year's interest} + \left[(\text{Principal} + \text{First year's interest}) \times \frac{\text{Interest rate}}{100}\right] \tag{2.5}$$

$$i_n = i \times m \tag{2.6}$$

where
- n is the yearly interest rate
- i is the interest rate per interest period
- m is the number of interest periods per year

$$i_e = \left(1 + \frac{i_n}{m}\right)^m - 1 \quad \text{or} \quad i_e = (1+i)^m - 1 \tag{2.7}$$

$$\overline{i_e} = e^{i_n} - 1 \tag{2.8}$$

$$\text{Rate of return (ROR) (in percent)} = \frac{\text{Total amount of money received} - \text{Original investment}}{\text{Original investment}} \times 100\% \tag{2.9}$$

$$\text{Rate of return (ROR) (in percent)} = \frac{\text{Profit}}{\text{Original investment}} \times 100\% \tag{2.10}$$

Chapter 3

$$F_1 = P + P(i) = P_0(1+i) \tag{3.1}$$

$$F_2 = P_0(1+i)(1+i) \tag{3.2}$$

$$F_3 = P_0(1+i)(1+i)(1+i) \tag{3.3}$$

$$F_n = P_0(1+i)^n \tag{3.4}$$

$$F_n = P_0 e^{i_n \times n} \tag{3.5}$$

$$F_n = P_0(F/P, i, n) \tag{3.6}$$

$$P_0 = F_n\left[\frac{1}{(1+i)^n}\right] \tag{3.7}$$

$$P_0 = F_n(F/P,i,n) \tag{3.8}$$

$$\text{NPW} = \pm P_0 \pm \sum F_n\left[\frac{1}{(1+i)^n}\right] \tag{3.9}$$

$$\text{NPW} = \pm P_0 \pm \sum F_n(P/F,i,n) \tag{3.10}$$

$$i = \frac{i_n}{m} \tag{3.11}$$

where
i is the interest rate per compounding period
i_n is the nominal interest rate (interest rate per year)
m is the number of compounding periods per year

$$n = \frac{\text{Number of compounding periods}}{\text{Year}} \times \text{Total number of years} \tag{3.12}$$

$$i = \sqrt[n]{\frac{F_n}{P_0}} - 1 \tag{3.13}$$

$$i = c + \left(\frac{a}{b}\right)d \tag{3.14}$$

$$n = c + \left(\frac{a}{b}\right)d \tag{3.15}$$

Chapter 4

$$n = \text{Last payment period} - \text{Equation time zero } (ETZ) \tag{4.1}$$

$$F_n = A\left[\frac{(1+i)^n - 1}{i}\right] \tag{4.2}$$

$$F_n = A(F/A,i,n) \tag{4.3}$$

$$P_0 = F_1\left(\frac{1}{(1+i)^1}\right) + F_2\left(\frac{1}{(1+i)^2}\right) + F_3\left(\frac{1}{(1+i)^3}\right) + \cdots F_n\left(\frac{1}{(1+i)^n}\right) \tag{4.4}$$

$$P_0 = A_1\left(\frac{1}{(1+i)^1}\right) + A_2\left(\frac{1}{(1+i)^2}\right) + A_3\left(\frac{1}{(1+i)^3}\right) + \cdots A_n\left(\frac{1}{(1+i)^n}\right) \quad (4.5)$$

$$P_0 = A\left[\frac{(1+i)^n - 1}{i(1+i)^n}\right] \quad (4.6)$$

$$P_0 = A(P/A,i,n) \quad (4.7)$$

$$A = F_n\left[\frac{i}{(1+i)^n - 1}\right] \quad (4.8)$$

$$A = F_n(A/F,i,n) \quad (4.9)$$

$$A = P_0\left[\frac{i(1+i)^n}{(1+i)^n - 1}\right] \quad (4.10)$$

$$A = P_0(A/P,i,n) \quad (4.11)$$

$$\text{Interest per period} = \text{Remaining principal} \times \text{Period interest rate} \quad (4.12)$$

$$\text{Principal paid per period} = \text{Payment} - \text{Interest paid in that period} \quad (4.13)$$

$$\text{Remaining balance} = \text{Starting principal for period} - \text{Principal paid that period} \quad (4.14)$$

$$\begin{aligned}\text{Present worth of the remaining balance} &= \text{Present worth of the principal} \\ &\quad - \text{Present worth of the amount paid in } n-1 \text{ periods}\end{aligned} \quad (4.15)$$

$$P_0 = \frac{A}{i} \quad (4.16)$$

$$A = P_0(i) \quad (4.17)$$

Chapter 5

$$F_n = G\left[\left(\frac{1}{i}\right)\left(\frac{(1+i)^n - 1}{i} - n\right)\right] \quad (5.1)$$

$$F_n = G(F/G,i,n) \quad (5.2)$$

$$P_0 = \frac{G}{i}\left[\frac{(1+i)^n - 1}{i(1+i)^n} - \frac{n}{(1+i)^n}\right] \tag{5.3}$$

$$P_0 = G(P/G,i,n) \tag{5.4}$$

$$P_0 = A(P/A,i,n) + G(P/G,i,n) \tag{5.5}$$

$$A = G\left[\frac{1}{i} - \frac{n}{(1+i)^n - 1}\right] \tag{5.6}$$

$$A = G(A/G,i,n) \tag{5.7}$$

$$P_0 = \frac{G}{i^2} \tag{5.8}$$

$$G = P_0(i^2) \tag{5.9}$$

When $r > i$, then

$$w = \frac{1+r}{1+i} - 1, \quad P_0 = \frac{C}{1+i}(F/A,i,n)$$

$$= \frac{C}{1+i}\left[\frac{(1+w)^n - 1}{w}\right] \qquad P/C \Rightarrow \infty \tag{5.10}$$

When $r < i$, then

$$w = \frac{1+i}{1+r} - 1, \quad \text{and} \quad P_0 = \frac{C}{1+r}(P/A,i,n)$$

$$= \frac{C}{1+r}\left[\frac{(1+w)^n - 1}{w(1+w)^n}\right] \qquad P/C \Rightarrow \frac{1}{(1+r)w} \tag{5.11}$$

When $r = i$, then

$$P_0 = \frac{C \times n}{(1+r)} - \frac{C \times n}{(1+i)} \qquad P/C \Rightarrow \infty \tag{5.12}$$

Chapter 6

$$P_0 = F_n\left(P/F, \frac{i}{m}, n \times m\right) \tag{6.1}$$

$$F_n = P_0\left(F/P, \frac{i}{m}, n \times m\right) \tag{6.2}$$

Chapter 7

No new formulas

Chapter 8

$$EUAW = -P_0(A/P,i,n) + SV(A/F,i,n) \tag{8.1}$$

$$EUAW = \left[-P_0 + SV(P/F,i,n)\right](A/P,i,n) \tag{8.2}$$

$$EUAW = (P_0 - SV)(A/P,i,n) + SV(i) \tag{8.3}$$

Chapter 9

$$0 = \pm P_0 \pm \sum F(P/F,i,n) \pm \sum A(P/A,i,n) \pm \sum G(P/G,i,n) \tag{9.1}$$

Interpolation table for net present worth unknown interest

$$0 = \pm P_0(A/P,i,n) \pm A \pm \sum G(A/G,i,n) \pm \sum F(A/F,i,n) \tag{9.2}$$

Table for Developing Interpolation Problems for Unknown Rate of Return

	i	Net Present Worth		
d	i_1	A	a	b
	Unknown ROR	0		
	i_2	C		

Note: $a = A - B$, $b = A - C$, $d = i_1 - i_2$.

Chapter 10

No new formulas

Chapter 11

$$\text{Total annual cost} = \text{Annual equivalent capital cost} + \text{Cost per variable unit} \times \text{Number of variable units per year} \tag{11.1}$$

Chapter 12

$$B/C \geq 1.0 \text{ project is justified} \tag{12.2}$$

$$B/C \leq 1.0 \text{ project is not justified} \tag{12.3}$$

$$\text{Conventional } B/C = \frac{\text{Net savings to users}}{\text{Owner's net capital cost} + \text{Owner's net operating and maintenance costs}} \tag{12.4}$$

$$\text{Conventional } B/C = \frac{U_n}{C_n + M_n} \quad \text{or} \quad = \frac{B_n}{C_n + M_n} \tag{12.5}$$

$$\text{Modified } B/C = \frac{U_n - M_n}{C_n} \quad \text{or} \quad = \frac{B_n - M_n}{C_n} \tag{12.6}$$

Chapter 13

$$\text{Production depreciation} = \frac{\text{Cost} - \text{Salvage value}}{\text{Number of hours or units of production}} \tag{13.1}$$

$$\text{Straight line depreciation} = \frac{P - F}{n} \tag{13.2}$$

$$BV_m = P - m \times D \tag{13.3}$$

$$\text{Double declining balance depreciation} = \frac{2}{n} BV \tag{13.4}$$

$$BV_m = P - \text{Depreciation to date} \tag{13.5}$$

$$BV_m = BV_{m-1} - R\left(P - \sum D\right) \tag{13.6}$$

$$\text{SOYD depreciation} = \frac{\text{Years remaining}}{\text{Sum-of-the-years digits (SOYD)}}(P - F) \tag{13.7}$$

$$= \frac{m}{\frac{n(n+1)}{2}}(P - F) \tag{13.8}$$

$$BV_m = P - \left[\frac{m(n - \left(\frac{m}{2}\right) + 0.5)}{\text{SOYD}} \right](P - F) \tag{13.9}$$

Chapter 14

$$\text{Taxable business income} = \text{Adjusted gross income} - \text{All expenditures (except capital expenditures)} - \text{Depreciation} \tag{14.1}$$

$$\text{Capital gain} = \text{Selling price} - \text{Adjusted basis of purchase price (original price} + \text{cost of all renovations)} \tag{14.2}$$

$$\text{Capital loss} = \text{Book value (cost} - \text{depreciation to date)} - \text{Selling price} \tag{14.3}$$

$$\text{Recaptured depreciation} = \text{Selling price of asset} - \text{Book value (original cost} - \text{depreciation to date)} \tag{14.4}$$

$$\text{Federal corporate income taxes} = (\text{Gross income} - \text{Business expenses} - \text{Depreciation}) \times \text{Tax rate} \tag{14.5}$$

$$\text{Effective federal corporate income tax rate} = \frac{\text{Federal income taxes}}{\text{Adjusted gross income}} \tag{14.6}$$

$$\text{Individual taxable income} = \text{Adjusted gross income} - \text{Deductions for exemptions} - \text{Standard deduction or itemized deduction} \tag{14.7}$$

$$\text{Effective federal individual income tax rate} = \frac{\text{Individual income taxes}}{\text{Adjusted gross income}} \tag{14.8}$$

Index

A

B

C